PEST CONTROL
Strategies for the Future

AGRICULTURAL BOARD
DIVISION OF BIOLOGY AND AGRICULTURE
NATIONAL RESEARCH COUNCIL

NATIONAL ACADEMY OF SCIENCES Washington, D.C. 1972

NOTICE: The study reported herein was undertaken under the aegis of the National Research Council with the express approval of its Governing Board. Such approval indicated that the Board considered the problem to be of national significance, that elucidation of the problem required scientific or technical competence, and that the resources of NRC were particularly suitable to the conduct of the project. The institutional responsibilities of the NRC were then discharged in the following manner:

The participants in the symposium were selected for their individual scholarly competence and judgment with due consideration for the balance and breadth of disciplines. Responsibility for all aspects of this report rests with the organizing committee and participants to whom we express our sincere appreciation.

Although reports are not submitted for approval to the Academy membership or to the Council, each is reviewed by a second group of scientists according to procedures established and monitored by the Academy's Report Review Committee. Such reviews are intended to determine, *inter alia*, whether the major questions and relevant points of view have been addressed and whether the reported findings, conclusions and recommendations arose from the available data and information. The report is distributed only after satisfactory completion of the review process.

This symposium was supported by the Ford Foundation, the U.S. Department of Agriculture, the U.S. Department of the Interior, the Monsanto Fund, the National Agricultural Chemicals Association, the National Science Foundation, and the Rockefeller Foundation.

International Standard Book Number 0-309-01945-1

Library of Congress Catalog Card Number 79-188497

Available from

Printing and Publishing Office
National Academy of Sciences
2101 Constitution Avenue
Washington, D.C. 20418

Printed in the United States of America

Preface

The availability of effective pesticidal chemicals has greatly contributed to the advancement of agriculture in the United States. For more than a century pesticides have played an important role in controlling insects, mites, nematodes, fungi, weeds, and rodents, thereby contributing to the health and well-being of the nation's citizens. Agricultural chemicals have become an integral part of modern agricultural technology, and pesticides have freed man from certain communicable diseases by eliminating the intermediate host or vector. Such diseases as malaria, typhus, and yellow fever, once greatly feared, have been either curtailed or eradicated from many parts of the world.

In recent years the scientific community and the general public have become greatly concerned about pesticide residues that, although present levels are low, seem to be injurious to certain species of wildlife. There is a continuing question as to their effect on human health, and a great need to achieve a better understanding of the metabolism and natural degradation of pesticides, the movement of pesticides in the environment, and the alternatives to chemical pesticides. There is need for emphasis on integrated pest control and on development, selection, and use of pesticides that are more specific and hence less toxic to nontarget species.

In view of these considerations, the Agricultural Board appointed a task force to assess the present means of pest control and to make such recommendations as seemed desirable. This in turn led to the establishment of a committee to organize a symposium under the general title "Pest Control Strategies for the Future."

The symposium summarizes the present state of knowledge in the area, analyzes available information in the light of public concern, examines changes in or alternatives to the use of pesticides, and discusses the use of pesticides in conjunction with other pest control methods.

Organizing Committee

ROBERT L. METCALF, *Chairman*
EMERY N. CASTLE
BOYSIE E. DAY
ROY HANSBERRY
WAYLAND J. HAYES, JR.
ARTHUR KELMAN
E. F. KNIPLING
L. D. NEWSOM
F. W. SLIFE
CARROLL M. WILLIAMS

Contents

Introduction

Donald S. Farner

A remarkable and, in a sense, unique central nervous system has permitted the human species through the course of its existence on this earth to acquire, communicate, store, and retrieve information at an ever-increasing rate. The technology thus made possible has permitted the human species to exert a greater impact on the global environment than perhaps any other species during the history of life on this planet. The per capita capacity to manipulate energy and materials increases as a logarithmic function of time. One facet of this technology, modern medicine, has responded almost miraculously to the primitive drive to survive and to preserve life. However, because of the absence of a reciprocal human drive, modern medicine has not been able to respond similarly with respect to the control of production of life. This imbalance in the effects of one aspect of modern technology has been a major contributor to the development of serious overpopulation. As populations have grown, another aspect of our technology, modern agriculture, has responded miraculously in its attempts to meet their food and fiber requirements. It is nevertheless clear that there is a finite limit to population size beyond which modern agriculture and its adjunct technologies can no longer supply adequate food. It is possible that we have now reached or even surpassed that limit. Barring an early, and as yet unpredictable production of food from highly artificial culture systems or a breakthrough in production by nonliving systems, it seems to me that we are confronted with the necessity of squeezing out additional improvements in agricultural and related technologies as the only humanely acceptable route between now and the time that the world's societies accept and practice the available technological mechanisms to attain zero or negative propulation growth rates.

The production of food is accomplished primarily through managed, simplified ecosystems. In these simplified ecosystems man finds himself in competition with other species. More food may be attained by reducing or eliminating or decreasing this competition. Thus, a critically important aspect of agricultural technology has been the development of pest control measures. Although a variety of measures, and combinations thereof, have been evolved, it is the use of chemical pesticides that has been, in many ways, most dramatic and controversial.

An orchard, a cotton field, a rice paddy, or a managed fish pond can be viewed advantageously as an ecosystem with respect to its structure, flow of materials, and flow of energy. But such ecosystems are, more accurately, parts of second- and third-order ecosystems. And herein lie many of the major problems. In the microscopic and myopic view of the single small system, can lie the basis for complacency; here, also, in the examination of effects at the level of the second- or third-order ecosystem and movements of materials among its components, lie the bases for alarm and controversy.

Truly, we are confronted with a dilemma. Side-effects of chemical pesticides are not sufficiently known, and although there have been significant strides, we are still very far from a truly predictive ecology. It is clear that we can permit this lack of knowledge to serve neither as a basis for categorical abandonment of control practices (the position of the extreme environmental activist) nor as the basis for the total *laissez faire* approach that ignores the ramifying effects of subtle environmental insults.

In my view, there is a critical need for two developments:

- We must have an effective algebra of decision-making—an apparatus that can produce the best decision in light of an always deficient but always improving body of knowledge of the beneficial and deleterious effects of control practices. It must be an apparatus that works in terms of successive proximations as more knowledge and new options become available. The difficulties of fitting such a decision-making apparatus to a law-making process that, in this field, is sometimes chaotic is well known.

- We must continue to evolve, through research and development, a much broader spectrum of control strategies to provide additional and optional inputs into the decision-making apparatus.

Whether or not a zero or negative rate of population growth can be attained by the technological measures available to control birth rates or whether it will come by massive malnutrition, recurrence of high infant mortality, and epidemic diseases is a moot question. We have no choice, in my view, but to assume that the former is possible and that, in fact, agriculture and related technologies can provide the essential additional food until zero or negative population growth is attained. This focuses clearly the importance of this Symposium on Pest Control Strategies for the Future.

I
Pest Control and Human Ecology

Pest Control in Health and Agriculture: International Challenges

John J. McKelvey, Jr.

The most important set of challenges in pest control are those that the insects present to man. Because insects predominate under the generic term *pest*, this paper will concern itself almost exclusively with insects. I would dispute those who claim that the dubious honor of being foremost among the pests belongs to another group of organisms, such as pathogens, rodents, or weeds.

One question we should ask ourselves: Have we really done anything significant in controlling insects for the betterment of our own health and agriculture? As of fifty years ago we had not, according to that scholarly and immortal entomologist, S. A. Forbes, who said (as quoted by Decker, 1968):

> The struggle between man and insects began long before the dawn of civilization, has continued without cessation to the present time, and will continue, no doubt, as long as the human race endures. We commonly think of ourselves as the lords and conquerors of nature, but insects had thoroughly mastered the world and taken full possession of it long before man began the attempt. They had, consequently, all the advantage of a possession of the field when the contest began, and they have disputed every step of our invasion of their original domain so persistently and so successfully that we can even yet scarcely flatter ourselves that we have gained any very important advantage over them. . . . If they want our crops they still help themselves to them. If they wish the blood of our domestic animals, they pump it out of the veins of our cattle and our horses at their leisure and under our very eyes. If they choose to take up their abode with us we cannot wholly keep them out of the houses we live in. We cannot even protect

our very persons from their annoying and pestiferous attacks, and since the world began we have never yet exterminated—probably we never shall exterminate—so much as a single insect species.

Forbes presented the situation in harsh terms, but his comments are as true now as they were in his generation.

PESTS OF HUMAN BEINGS

We cannot dismiss Forbes' gloom as readily as we might like. However, we might take justifiable pride in the fact that, of the 147 countries and territories in this world originally malarious, 23 countries now claim to have eradicated the disease (World Health Organization, 1969a). But malaria, about which we became complacent after the spectacular progress made in its control with the use of DDT, is now on the increase, explosively so in some areas. Ceylon, which recorded only 17 cases a few years ago, registered 4 million cases in a total population of 13 million during the past two years (Carter, 1970).

Sleeping sickness in Africa and the tsetse flies that transmit it may seem obsolete terms to the average U.S. citizen, so successfully has man coped with this debilitating and killing disease. But the six to seven species of fly that transmit sleeping sickness still range over the same 4 to 4½ million square miles of tropical Africa in which they caused epidemic outbreaks years ago. In 1969, a distinguished public servant from the United States, W. T. Davis, and his wife contracted the disease while on safari in the Serengeti of Tanzania. She died of it (N.Y. Times, 1969). And many Ethiopians contracted sleeping sickness in a recent outbreak that developed in the southern part of their country.

We cannot even pretend to be successful in the control of river-blindness, found in association with black flies along all the main river systems of northern tropical Africa and also endemic at certain foci south of the equator. Blindness rates of over 10 percent (sometimes as high as 30 percent) are common today in villages subject to hyperendemic situations. Heavy infections, especially if acquired early in life and followed by repeated superinfections, may result not only in blindness but also in a skin condition called craw-craw that, apart from being disfiguring, causes itching so intense as occasionally to lead to suicide. Mass treatment for the control of the worms that cause river-blindness is not practical at present. Drugs capable of mass application are needed. Until they are available, the thrust must be to attack the larval stage of the insect vector with insecticides. DDT is ideal for this purpose in large, steadily-flowing rivers, but more suitable insecticides and formulations are needed for the small, intermittent streams. The vectors of river-blindness, *Simulium damnosum* and *S. nevei*, are intolerable pests in their own right. When they swarm, they drive whole populations from the river valleys of the savannahs (Waddy, 1969).

Chagas' disease, which threatens heart damage to 35 million people in the tropics of Latin America, has no adequate control; neither does its vector, the kissing bug (World Health Organization, 1969b). Trachoma, which accounts for so much blindness in India, still relies in part on the common housefly for its transmission.

Use of insecticides and the altering of our external environment in other ways to reduce the populations of insect vectors of human disease do afford us temporary respite, as do drugs injected into our most intimate "environment," the bloodstream, to control the diseases carried there. But the congested cities of the tropics still offer scope for the frontier pests of long ago to re-establish themselves in new ecological niches and, once there, to transmit disease again. Certain species of tsetse flies are finding attractive the shade trees, bush areas, and fence rows of villages they once spurned. Disease-bearing mosquitoes gravitate to the open drains and sewers of Bangkok and Calcutta, for example, making them focal points for the spread of elephantiasis and dengue fever.

PESTS OF OUR CROPS

If the insects "want our crops they still help themselves to them." What else can we expect? For one thing, our own population expansion demands that we continually bring under cultivation lands that may still be considered the insects' domain. For another, as we organize, concentrate, and mechanize our agriculture, we present to our insect enemies better-fed food plants and more of them per unit area. As we strive for high-yielding crop plants, we may automatically bring today's pests under control in the short run. But as we narrow the germplasm base of these same plants, we foster a buildup of tomorrow's pests. Those that may be of minor importance today respond in their genetic makeup and adjust to altered plant populations.

In the summer of 1970 a surprise attack by the pathogen *Hel-*

minthosporium maydis brought about an epiphytotic of southern corn leaf blight that reduced the U.S. corn crop (estimated at 4.11 billion bushels in July) by 710 million bushels. At the December 1970 market price of $1.50 per bushel, the loss amounted to about $1 billion. Primarily responsible was the new, highly virulent race of a fungus that attacked corn having the T-type cytoplasm so extensively used in hybrid seed production. This occurrence should remind us of the everpresent need to be alert to that age-old phenomenon, the "running out," often abruptly, of varieties that have resisted pest or pathogen (Tatum, 1971).

PESTS AND DISEASES OF OUR LIVESTOCK

The tasks ahead in dealing with the pests of our domesticated livestock are equally as awesome as those relating to pests that affect man directly or damage his crops. Is it not still true that if the insects "wish the blood of our domestic animals, they pump it out of the veins of our cattle and our horses at their leisure and under our very eyes?" The whole group of insects that come under the classification of biting flies, the Tabanids, do this. They need investigation. To my knowledge, there is no satisfactory control as yet for the human bot fly, *Dermatobia hominis*, in Central America. The brown ticks that transmit East Coast fever to the dairy and the beef cattle in the highlands of East Africa respond to a rigorous schedule of cattle dipping at weekly intervals, but this is expensive and also leads to the accumulation of pesticide in the treated animals. Such unsatisfactory aspects of controlling these ticks are encouraging leading scientists to search for drugs or to develop a practical vaccine that will directly combat the protozoan, *Theileria parva*, that causes the disease. The resistance of ticks of the genus *Boophilus* to pesticides currently in use and the pesticide residues on dairy farms where insecticides are used to control the common fly constitute other urgent problems of international dimensions.

PEST CONTROL

Pesticides

As the profession of economic entomology has grown, strategies of pest control have evolved. But, in my view, sound strategies will not permit us to discard well-established control practices in favor of new practices (I refer to pesticides) yet to be proven as safe and as effective. As Dr. Philip Handler, President of the National Academy of Sciences, has said (Handler, 1971):

> We need to know more than is known today concerning pesticides, food additives, drugs, diverse industrial practices, radiation hazards, atmospheric phenomena or the alleged fragility of ecosystems if we are to make sound judgments and establish rational public policy. I share the concern for the possible hazards of DDT—but not the hysteria of those who demand an absolute prohibition against its use before an acceptable substitute is available. The predicted death or blinding by parathion of dozens of Americans last summer must rest on the consciences of every car owner whose bumper sticker urged a total ban on DDT. No human has yet been known to be damaged in consequences of normal usage of DDT; its untoward effects on bird and fish life appear to reflect heavy overdosage rather than proper use.

In 1970, Dr. L. D. Newsom, head of the Department of Entomology of Louisiana State University, delivered a paper before the Tall Timbers Conference on Ecological Animal Control by Habitat Management wherein he discussed the use of biological control, microbial pathogens, varietal resistance, autocidal methods, sex-attractants, growth-regulating hormones, and genetic manipulation as mechanisms for combating insect pests (Newsom, 1970). In speaking to the subject of future prospects for insect control, he stated that "a serious problem now facing the entomological profession is the probability that interest and research in the use of conventional chemical insecticides for pest control may be neglected as badly during the next two decades as has been the case with alternative methods during the era now ending." This, he warned, "must not be allowed to happen because the intelligent use of chemical insecticides is an ecologically sound and necessary component of modern pest management systems."

As consumers we must be realistic about the conflict of interests we promote in our tandem demands, on the one hand, for high-quality foods and freedom from disease-bearing pests and, on the other, for zero use of pesticides. Although there is considerable over-reaction to noxious side effects and great sensitivity to the word "pesticide" today, leading entomologists at the University of Illinois, at Cornell University, and at the University of California (Berkeley and Riverside campuses), have joined in a cooperative program boldly designed to afford opportunities to young entomologists to acquire the skills necessary to construct new insecticides that will be selective and biodegradable and will have minimal deleterious effects on man's environment.

Plant Resistance

An important element in a strategy for insect control is breeding plants for resistance to insect attack. The value of this approach has often been obscured because of the long-term nature of the research involved, and because chemicals seemed to promise us a panacea to liberate us from our insect enemies. Also, the 10 to 20 percent reduction in pest populations this method affords seemed too often to be not good enough; we have ignored this achievement while striving for 80 or 90 percent reduction by other methods. Yet, when combined with other means of managing our pest populations, such miniscule reductions may be crucial. The combined efforts of entomologists, plant breeders, and biochemists at the USDA laboratories in Brownsville, Texas, and in the departments of entomology at the universities of Texas, Mississippi, and California (Davis campus), may enable the cotton-grower to eliminate several applications of chemical sprays considered essential today to control the cotton bollworm, *Heliothis zea*. These scientists may be able to combat this pest by offering it cotton plants without nectaries, with smooth instead of fuzzy leaves (to which the moths may find it difficult to attach their eggs), and plants of high gossypol content that will cause high mortality to larvae hatching from these eggs. This would be a major achievement, since the cotton crop accounts for more than 50 percent of the use of all pesticides in the United States and the bulk of this pesticide load is directed against *Heliothis*, the bollworm (Koehler, 1970). Parallel progress is taking place in Asia, where certain hybrid varieties of sorghum are now available that prevent development of the sorghum shoot fly and where there are new varieties of rice that exhibit resistance to leafhoppers.

Multidisciplinary Communication

There is good communication among specialists in insect control. Men well-grounded in biochemistry, physiology, endocrinology, and other basic disciplines are working on programs that coalesce with those that concern specialists in the applied fields of medicine and agriculture. Isolating insect sex-attractants, which, when liberated in nature, will upset the normal mating schedules of insect pests and help regulate their population increase, is receiving that sort of cooperative endeavor. To meet the objective of turning insect growth-regulating hormones into insecticides, members of the academic community from central Europe, from Africa, and from the United States communicate with each other and work with the experts in industry who are concerned with production and marketing. The goals, if clearly problem-oriented ones, will command cooperative multidisciplinary teamwork.

International Efforts

Our strategies must coincide with the range of our insect pests, which are often international. These strategies must make use of the available manpower and facilities of international organizations. The institutional network for conducting research in economic entomology and for training entomologists is comprehensive and is expanding. The international institutes for basic food crop improvement located in the Philippines, Mexico, Colombia, and Nigeria afford opportunities for devising methods of insect control limited only by the natural range of the insect pest, not by national boundaries. The laboratories of economic entomology that the young universities have spawned in the so-called developing countries provide other strong teaching and research units in such diverse areas as Ibadan, Nigeria; Ludhiana, India; Los Baños, the Philippines; and Chapingo, Mexico. The University of Nairobi in Kenya has just established an International Center for Insect Physiology and Ecology that starts from fundamental scientific bases but goes into the problems of devising practical measures of control to combat tsetse flies, ticks, mosquitoes, and armyworms.

Certain pests—armyworms and bollworms among them—that feed on a wide range of hosts and that travel internationally may not become manageable until the concept of pest population management itself undergoes alteration to include the challenge of these general feeders. The concept of pest population management seems restricted today to the management of a number of pests attacking a single crop. It may need redefinition to include managing one insect that attacks a variety of crops. Accelerating studies in Arkansas on the migratory habits of the cotton bollworm, intensifying research in North Carolina on the host range of the same insect, and expanding tests in Mexico of corn varieties that may resist its attack—all would undergird an integrated program designed to bring the bollworm under control.

Professional Corps

The size, quality, and deployment of our professional corps of entomologists must be considered as integral aspects of sound pest control strategy. We have not yet placed pesticide usage where it properly belongs—in the hands of professional, practicing, consulting entomologists—and have let it pass by default to the farmer, businessman, and housewife. This is one place where entomologists may be greatly needed. "Had we a corps of specialists," according to Dr. Handler (Handler, 1971), "trained in entomology, insect and other pest physiology, and the properties and proper usage of pesticides, their attendant disadvantages could be minimized. Perhaps some safer pesticides could still be obtained—over the counter, as it were—but society could be protected if the others could only be used by licensed, certified specialists who would know what uses were legal, adhere to maximum dosage schedules, and treat the agents they use with proper respect until such time as they can be replaced by suitable biological control measures. Creation of such a corps is an appropriate function of the university."

There are not enough entomologists to man the fundamental disciplines, either. In a personal communication with the author, Dr. L. H. Rolston indicated that the Caribbean Plant Protection Commission has reported heavy attacks of stink bugs in Montserrat that caused losses in tomato production up to 2,000 lb per hectare. The three most common stink bugs in question are undescribed and unnamed. An agricultural adviser from one of the banks in the area found them damaging rice, cotton, and sesame. At least two species of rice water weevil are pruning the roots of Nicaraguan rice. Neither of these species is named. Presumably these pests will remain unnamed and unstudied for lack of funds and personnel in taxonomy.

ULTIMATE GOAL

A realistic strategy for the management of an insect population depends on a thorough understanding of the goal to be achieved. We should, of course, be striving wherever possible for the extermination of our insect pests—an idealistic, arrogant goal that we (a biological species ourselves) know full well will be next to impossible to achieve. The brilliant success of screwworm control on the island of Curaçāo and in the states of Florida and Texas by genetic and reproductive manipulation of these flies, which lay their eggs in cattle wounds, tells us so. Banished from the first two areas, the screwworm as a species lingers on in the third.

Alternatively we could relax to a soothing analysis of the insect problem as presented in the book, *Life on a Little-Known Planet*. Here Harvard entomologist Howard Ensign Evans makes the fundamental point that we need to know a great deal more than we do about insect life on this earth. However, in one passage he quips, "It's a lousy world. The body louse becomes resistant to DDT, the plant louse thrives in suburbia's shade trees and dribbles on our shiny new automobiles, and the book louse doubtless is thriving in the midst of our current information explosion though he is too small to be noticed in our busy world" (Evans, 1968). The tenor of comments such as these lull some into believing that the public need only concern itself in the abstract that the body louse has become resistant to DDT; the entomologist will provide alternative and better insecticides for the body louse when DDT fails. It may be quite safe to think of typhus, the disease that helped decide the outcome of the Crimean War, as one for the story books and as just another disease against which one receives routine injections; nit-picking perhaps should be just a figure of speech that has lost its literal connotation.

Does the nuisance value of the plant lice, the aphids, which figure so importantly in the transmission of so many virus diseases on our economic crops, become the main concern because entomologists and plant pathologists have coped successfully with the serious, at times devastating, ravages of these pests? How lucky we are, and how lucky is the book louse too, that we can take him or leave him, as we can do now with the silverfish, the carpetbeetles, clothes moths, and perhaps even the cockroaches, with confidence that entomologists will be able to hold them to tolerable population levels. We need such confidence and understanding from the public, but we also need support.

Somewhere close to the frustrations that stem from not being able to exterminate that last insect and far, far from the complacency that comes from too close an association with insects and resignation to their inevitable presence, lie the bulk of the international challenges. We must marshal all of our skills and energies to combat our insect enemies if we are to maintain our health and agriculture.

CONCLUSIONS

Have we really done anything significant to cope with the challenges our insect enemies present? In the millions of years since man became an identifiable creature he has forced the insects to accommodate in their domain today's population of 3½ to 4 billion individuals of his species. The awesome challenge—an international one, I think you will agree— is to get the insects to share their abode graciously with the equivalent of another planet of people only 25 years from now—within the lifetime of most of us here—to make a total of 7 to 10 billion people. We can expect no relief from this challenge, barring a holocaust, from any other quarter. We can force the insects to allow us health, prosperity, and comfort under our own projected population load, but only if

1. We are prepared to analyze our past successes and failures and be guided in the future accordingly;
2. We not only keep alert to the pest problems we face today but also anticipate those we are likely to face tomorrow;
3. We build sound strategies of insect control that include
   - Using judiciously our most powerful weapons—new and old—including insecticides,
   - Weaving partially effective control measures together into a fabric that smothers our prime pests,
   - Setting important problem-solving goals and implementing these with all the multidisciplinary force available,
   - Internationalizing our efforts as our target pests may dictate, and
   - Amplifying our corps of entomologists, improving its training, and deploying it so that it effectively meets the pest challenge;
4. Finally, we continue to strain to reach that ultimate goal—cruel though it may be—of extermination of our most noxious pests, while recognizing that for some time to come we will be coexisting with them.

REFERENCES

Carter, Walter. 1970. The dynamics of entomology. Bull. Entomol. Soc. Am. 16, No. 4:181–185, Dec. 1970.

Evans, Howard Ensign. 1968. Life on a little-known planet. E. P. Dutton, New York.

Forbes, S. A. As quoted in G. C. Decker. 1968. Don't let the insects rule. J. Agric. Food Chem. 6, No. 2:98–103, Feb. 1968.

Handler, Philip. 1971. The federal government and the scientific community. Science 171, No. 3967:144–151.

Koehler, C. S. 1970. Domestic pesticide consumption, commodities treated, and the more important pests against which these treatments are directed. Conference on the Present Status of Research on Crop Resistance to Insect Pests held at The Rockefeller Foundation, June 1970.

Newsom, L. D. 1970. The end of an era and future prospects for insect control. Proceedings: Tall Timbers Conference on Ecological Animal Control by Habitat Management, No. 2, Feb. 1970, pp. 117–136.

The New York Times, Sept. 7, 1969.

Tatum, L. A. 1971. The southern corn leaf blight epidemic. Science 171, No. 3976:1113–1116.

Waddy, B. B. 1969. Prospects for the control of onchocercosis in Africa with special reference to the Volta River Basin. Bull. World Health Organ. 40:843–858.

World Health Organization. 1969a. World health: The magazine of the World Health Organization. June 1969, p. 22.

World Health Organization. 1969b. Venezuela arms against Chagas' disease. World health: The magazine of the World Health Organization, Nov. 1969. pp. 26–27.

Ecology of Agriculture in the World of Today and Tomorrow

A. H. Bunting

SUMMARY

In the wealthy industrialized countries, where human population is increasing slowly, the agricultural subsystem of the human ecosystem is increasingly efficient and constantly tends to produce more food and raw materials than are required. The number of people directly supported by primary production is relatively small and decreasing. Farming will consequently tend to obtain larger and larger yields from constant or even decreasing land areas. The main future tasks are to ensure that the needs and supplies are appropriately adjusted, rather than to attain maximum productivity over unnecessarily large areas; and to regulate the consequent problems of land use.

In the poor, largely rural and nonindustrial countries, where farm technologies are backward and yields small, human populations are increasing so much more rapidly than off-farm employment that the numbers of people who depend for survival solely on agriculture seem likely to increase for at least 50 years to come. Here it is essential to increase farm output by increasing both area farmed (where possible) and particularly yield per acre, both for food and for raw materials for local agro-based industries.

In all systems of farming, many other organisms, both endogenous and exogenous—weeds, birds, mammals, insects, arachnids, nematodes, fungi, bacteria, and viruses—compete with man at the primary and secondary production levels, both before and after harvest and during storage. The means used to lessen their effects have for many years included biocides of varying degrees of selectivity together with appropriate management and the use of resistant or tolerant plants and animals. In advanced agriculture, particularly in free enterprise systems, the desire of the farmer for maximum economic return may lead him to aim to control harmful organisms (pests) more completely than the interest of the community as a whole may require, particularly with sporadic exogenous organisms, but experience in Britain suggests that such decisions can readily be influenced by social means. In the agriculture of the poorer countries, the needs are so great that no sensibly-used mode of effective economic control can be spared. In general, as new technology evolves it continually alters both the microecology and the macroecology, so that pest problems and the means of regulating them will never be static. Hence biocides are likely to continue to be necessary, but the biocides themselves and the ways of using them will change and improve.

AGRICULTURAL ECOLOGY

I consider ecology to be the study of biological systems—the relations between individuals, populations, and communities of living organisms and their environment. In agriculture we are particularly concerned with production and distribution processes in biological systems that are purposefully controlled by man, and with the ways in which environmental factors are used in them. Much classical and academic ecology traditionally excludes man, except perhaps as a regrettable external source of nuisance, disturbance, or destruction, but we cannot exclude our own species when we consider the ecology of agriculture.

EVOLUTION, CHANGE, AND DIVERSITY IN AGRICULTURE

Origin and Spread of Farming

Agriculture today is the principal biological means by which the human species, as a social organism, uses environmental resources to produce foods, beverages, drugs, stimulants, and industrial raw materials of many kinds.

No more than about 10,000 years have passed since man began to supplement gathering, hunting, and fishing, in certain seasonally arid regions of the Old and New Worlds (Braidwood and Howe, 1960; Flannery, 1969; McNeish *et al.*, 1967), by the deliberate production of plants and animals. This invention led to a great increase in settlement, and in due course to civilization, science, and culture. With a more assured food supply, human numbers began to increase, and agriculture and agricultural people spread throughout the world. Hundreds of new, domesticated species of plants and animals were evolved. When writing and numbers were invented, it became possible to record and diffuse knowledge and to put barter on to a quantitative basis. Iron tools greatly increased man's power to manage and to transform his environment. Iron Age farm technology dominated agriculture throughout the greater part of the world for nearly 2,000 years, up to the time of the great improvers in the 17th and 18th centuries, and in nonindustrial countries most farming is still at this stage. In the industrial countries, however, science and engineering have begun to transform agriculture during the last 200 years, and in our time they have begun to affect farming in the nonindustrial countries also.

Farming and Ecological Change

Agriculture has greatly changed the natural environment, and, conversely, most domesticated species live in ecological circumstances entirely different from those of their wild ancestors.

Vegetation

Except in some extreme climates, which man has seldom used because they are too cold, too dry, or too wet, all the vegetation of the earth has been affected by man's activities. For example, the ecological air-traveler can see for himself that virtually all of the so-called primeval mixed equatorial forest of the Congo Basin, which is so often cited as a stable climax, is secondary.

Nearer home, the beautiful agricultural landscapes of the deciduous broad-leaved woodland regions of Western Europe and North America are man-made. The scenes of former ecological disasters are particularly interesting. The glorious landscapes of Lincolnshire and of much of Holland (van Hulten and van Ommeren, 1970) and parts of the Po Valley in Italy (Vicinelli, 1970) have been created on reclaimed salt marshes. The well-loved Norfolk Broads occupy the sites of late medieval peat cutting (Ellis, 1965). In the East Anglian Breckland, early farmers cleared birch forest in and near the areas of industrial dereliction around some ancient flint mines (Godwin, 1956). The soil became so severely podzolized that, after the farmers had been forced to move on, secondary succession could not re-establish the original vegetation, but was deflected instead to a scrubby heath that today is of the greatest biological interest.

During the past hundred years we have entered a second phase. Advances in farm technology and communications have enabled us to concentrate agriculture in more favorable sections, while wild and often unmanaged communities have taken over the less fertile, colder, or more elevated regions.

Climate and Soils

Though many have suggested that agriculture has major effects on climate, if there are any they cannot readily be separated from the long- and short-term fluctuations recorded in history and in pollen sequences (Godwin, 1956). Poor farming in a dry period may have advanced the southern margin of the Sahara, but any such change is minute compared with the long-term climatic changes reflected in the distribution of fixed dunes (Andrews, 1948). Local harmful effects of agriculture on soils (like accelerated erosion, and the accumulation of salt in incorrectly irrigated areas) must be heavily outweighed by the steady and sometimes even dramatic improvement of soils by science and agriculture in many parts of the world during the past 100 years.

Some General Ecological Features of Agricultural Ecosystems

By their very nature, the advances of agricultural science always tend to ecological change. The ecosystems they produce are unstable, but biologically they are highly successful; primary productivity in correctly-managed farm ecosystems is generally far greater than anything nature can do without man's aid (Westlake, 1963). To achieve this, man regulates the structure, function, and duration of agricultural ecosystems and, in particular, controls competition within them. Most domesticated species are raised in and have been adapted to environments very different from those of their wild relatives and ancestors. Our cereals, descendants of certain grasses of dry regions of southwest Asia or middle America, are cultivated from the Equator almost to the northern and southern limits of agriculture. Cocoa, coffee, rubber, and other understory species of tropical forests are grown successfully as exclusive dominants, often half a world away from their original homes. Further, many of the most successful agricultural ecosystems, particularly with annual crops and short-term pastures, are designed to exploit the extremely rapid accumulation of biomass that is characteristic of the early stages of secondary succession. Indeed, many of our crops descend from plants adapted to exploit disturbed and unstable situations—in other words, from weeds (Anderson, 1954). Methods of handling even perennial species, such as apples and coffee, as secondary pioneers, are rapidly being developed, partly in order to escape from the difficult disease and pest problems that accumulate in more permanent plantations. In fact, managed change is often easier to handle practically, as well as far more profitable, than a stable equilibrium. In most of agriculture, diversity is associated, not with stability, but with change: We ride the whirlwind.

Diversity and Change in Agriculture Today and Tomorrow

On a world scale, the most striking features of agriculture today are diversity and change, to which three main groups of factors contribute—geographical/climatic factors, the stage of economic and technical development, and human demography. These factors determine the diverse states of farm technology and the diverse objectives and problems of agricultural development. All this diversity is comprehended within the worldwide and continually changing ecosystem dominated and managed by the human species and linked together by continually-changing economic, trade, and political relationships and by ever-improving communications, along which ideas and ideals, particularly those of change and the improvement of the human condition, move even more rapidly than money, goods, and people. In Guy Hunter's striking phrase, "the transistor is hung round the donkey's neck" (Hunter, 1970).

Regions of Advanced Agriculture

Climate and Geography

The countries in which agriculture is most productive at the present time lie in the temperate and Mediterranean climate zones. The agricultural organisms and methods of the temperate zone are adapted to relatively short seasons, limited by cold at both ends, except in glasshouses and other artificial environments. Over much of the temperate zone, water shortage is not a marked obstacle to plant growth, and since evaporation exceeds precipitation during the summer, leaching is uncommon during the cropping season. In Mediterranean zones, irrigation is widely used to offset summer drought. In both zones, animals are often kept during part or all of the year in special buildings.

Economic Organization

The countries of advanced agriculture are either themselves largely industrial, or are closely associated economically with advanced industrial regions. They can and do draw on the whole world for food and agricultural raw materials. The proportion of the working population that depends directly on biological production is small and declining (in Britain now about 3 percent) (Jones, 1970); most people are engaged in industry, trade, and services, and they live and work in towns or cities. Leisure is increasingly important. Average GNP per head is $1,000 per year or more, and though income may be very unevenly distributed, famine is unknown, and most people have enough to eat.

Population

Populations are growing relatively slowly (usually at less than 1.5 percent a year) and are indeed tending to decline in some areas, even though the average expectation of life is increasing. In Britain, it has more than doubled in the past century, and now, at 68.5 years for men and 74 years for women, it has passed the allotted span of threescore years and ten.

Agriculture and Farm Technology

Virtually all farm production is for sale off the farm: Farm products are readily translocated within the ecosystem. Because farm labor is scarce and expensive, farm technology is mechanized and labor–intensive; it is supported by vigorous science–based plant and animal breeding industries and it uses fertilizers and other industrial products on an increasing scale. The inputs are costly, but large yields and the regulation of prices and output make the products relatively cheap.

The first purchasers of the produce, who determine many farming objectives, are the transport and processing industries, which are essential to feed the urban population in a seasonal climate. This is *par excellence* the agriculture of cities.

Individual enterprises are increasingly specialized, so as to use environment, capital equipment, and personal skills as economically as possible. In mechanized systems, annual crops have to be handled in single-species cultures in which all individuals are sown at the same time; few crops are raised in mixtures. By past standards, yields are large and increasing: Some national averages of cereal yield are already around 1½ tons per acre. Yet average production of agricultural biomass per year represents no more than 0.3 percent of available radiant energy against a theoretical maximum of perhaps 5 percent (Penman, 1968; also see Clark, 1970). Efficient as this agriculture seems to us, biologically it still has a long way to go, as the best farmers clearly demonstrate. Indeed, in many regions of developed agriculture, the industry has constantly to be prevented from producing more than society can consume. Land on which crop production becomes more difficult or less economical of resources is continually released for more extensive agricultural use, for forestry and water conservation, for nonagricultural residence, and for amenity and wild life, and some is even abondoned altogether. Agricultural pressure on land is decreasing.

The Future

The social objects for the future must be to adjust production more precisely to need and so to lessen the waste of overproduction and to deal wisely with the consequences of steadily improving farm technology, particularly the changes in human activity and in the use of land and natural resources. In all this, it may well be that the interests of society as a whole are not identical with those of the primary producer or the landowner.

Regions of Little-Developed Agriculture

Climate and Geography

Most of the developing countries lie in the seasonally arid or humid tropics, where the growth of plants is regulated by drought and only rarely by cold. However, since precipitation exceeds evaporation during the growing season, crops may be harmed by waterlogging and the leaching of nutrients, particularly nitrogen. As a result, many tropical agricultural systems are caught between drought and flooding, so that both irrigation and drainage may be extremely important. Further, though man's time continues to be measured out in days, biological events occur more rapidly in the warmer tropical temperatures, so that peaks of work are more pronounced, and the effects of harmful organisms are often more sudden and more devastating than in temperate climates.

Economic Organization

In the developing countries about two thirds of the total population depends directly on agriculture for survival. In addition much of the relatively weak secondary and tertiary (manufacture, trade, and services) sectors of the economy also depend on agriculture. Industry and communications are relatively little developed. Average income is around $100 per year, and in many areas many people are undernourished or malnourished, at least during part of the year.

Population and Employment

Though the average expectation of human life is considerably shorter in the developing world than in the richer countries, it is increasing rapidly as disease is controlled and food supplies improve. Birth rates in the developing world do not appear to be increasing, but they are usually considerably larger than in many of the richer countries. The resulting growth rates of population are from 2 to nearly 4 percent per year. In absolute terms, populations are growing faster than nonagricultural employment, and consequently, in spite of movement to the towns, the land will have to support far more people in the foreseeable future than it does now. Of about 1,400 million people in the developing countries in 1962, 935 million, or 67 percent, were agricultural; the Indicative World Plan (Food and Agriculture Organization, 1970) suggests that by 1985, out of a total of 2,500 million, the numbers of agricultural people will have increased to nearly 1,400 million, even though the proportion will have fallen to 55 percent. To take a particular example, in the West African region, there will be 175 million rural people in the year 2000, where there are only 105 million now. These trends seem likely to continue for at least 50 years.

The future tasks and problems of agriculture are closely related to these figures: Only through agricultural development can the growing rural population maintain itself, let alone become more prosperous. To be sure, most developing countries are doing what they can to decrease the rate of population growth and to increase nonagricultural employment, but for the next half-century the progress of the developing countries and their peoples is bound to require substantial increases in agricultural output and increased productivity per unit of land area and of time.

Agriculture and Farm Technology

Over the developing world as a whole, indigenous farming systems vary from increasingly prosperous commercial farming, for example, in Turkey or the Punjab, to the poor slash and burn systems of the Latin American or African savannahs. The first object of most of the poorer systems is subsistence for the farmer and his family, and sales off the farm and market opportunities (except for specific cash crops) tend to be small, unpredicitable, and unorganized. In ecological terms, translocation within the ecosystem is poorly developed. The technology is traditional; it uses few or no purchased inputs and little mechanical equipment, and it depends largely on humans and animals for power. There are usually pronounced seasonal peaks of demand for labor, when shortage of power limits output even though labor may be in surplus at other times (Clark and Haswell, 1968).

The complex ecological relations between labor, land, crops, and pests are strikingly illustrated by the competition between cotton and other crops in the northern states of Nigeria, where the farmer and his family have to divide precious early-season labor between the cash crop, cotton, and the staple food crops, particularly sorghum, on which they depend for survival. Not surprisingly, particularly where the growing of food is primarily the responsibility of the farmer's wife, cotton comes out the loser. Only recently have we begun to breed new cotton varieties and to devise the new agronomic and crop protection systems that will enable farmers to plant it later in the year. We are also providing new "high-yielding" varieties of corn and sorghum, which should lessen the amount of land (and of time) that the farm family has to dedicate to survival, thereby releasing more of both for other crops, including cotton. Finally, on the pest front, sorghum is an important alternative host for one of the bollworms of cotton (Lyon, 1970).

Crop yields in developing countries are small—characteristically less than one-half ton of cereal grain per hectare, representing no more than 0.06 percent of available energy (Penman, 1968). Theoretical considerations suggest that nothing inherent in the environment, even in many drier regions, opposes considerably larger yields, and this is demonstrated by the achievements of research stations, of plantation enterprises, of the best farmers, and of the Fertilizer Program of the Freedom from Hunger Campaign (Olson, 1970). Indeed, over the past 20 years, the output of food *per head* in the developing world has increased by an average of 8 percent, and the increases have been particularly striking in those areas affected by the rising groundswell of change that we have come to call the green revolution. Since the disaster year of 1966, when India harvested only about 67 million tons of food grains, the output has increased yearly, though not all the seasons have been favorable; last year (1970) it is said to have exceeded 100 million tons. In the same year, the Philippines exported rice for the first time in 70 years, and rice was in surplus on the international markets of Southeast Asia. The traditional exporters, Thailand and Burma, are consequently experiencing serious difficulties.

Though continuous monoculture (for example of rice or bananas) is often important in tropical agriculture where pressure on land is not yet too great, many of the existing systems depend on shifting cultivation to maintain the supply of plant nutrients, to limit soil erosion, and to manage harmful organisms. In much of the developing world, however, shifting cultivation has long since given way to continuous cropping, and at present levels of technology and yield this can lead to heavy pressure on the land. Mixed cropping is often used, particularly in wetter regions, to increase total output, spread risk, and lessen losses by leaching and erosion. Nevertheless, very many crops are grown in single-species stands sown at a single time, evidently because farmers find them generally easier to manage than mixtures.

The Future

In these poor and populous regions, agriculture has eight principal development goals–to feed the increasing numbers of rural people; to provide productive work or employment for them (which means that mechanization may have to be selective and delayed); to feed the population of the towns; to substitute for imports, like cotton and sugar, which can be produced economically; to provide raw materials for indigenous industries (for example, food, transport, textiles, leather, footwear, beverages, tobacco, rubber, paper and board, industrial chemical raw materials including plastics, alcohol and carbon dioxide, pharmaceuticals, and timber); to produce export crops, processed locally as far as possible, to earn foreign exchange; to provide a market for locally produced consumer goods and agricultural inputs; and to contribute to capital accumulation, in part perhaps by direct labor (Food and Agriculture Organization, 1970a). The key step in all this is the development of internal translocation—that is, sales off the farm into a reliable, honest, and remunerative market, which is mainly based on the growth of towns and of industry.

To do all this, the developing countries will require new labor-intensive farm systems that use both human and environmental resources much more effectively than the present ones. They will need to improve communications, marketing, and financial arrangements, to prepare for changing social relationships, and to resolve, through new institutions, the tensions that will inevitably be associated with technological and economic change. As in all development situations, there can be no question of maintaining an equilibrium: The need is for change. The developing world is bound to continue for many years to come in a state of dynamic change, in human affairs and relationships as well as in agricultural technology.

HARMFUL ORGANISMS IN AGRICULTURAL ECOSYSTEMS

Range of Harmful Organisms

In all farming ecosystems, many sorts of organisms compete with man at both the primary and secondary stages of production. They include insects, mites, and other arachnids; ticks and other exoparasites of animals; nematodes and other harmful and parasitic worms of both plants and animals; fungi; bacteria; viruses; higher plants (weeds and poisonous plants); birds; and mammals. From man's point of view these, and the vectors that transmit some of them, are harmful organisms, and in this Symposium we refer to them collectively as pests.

In general, we may distinguish between two broad and overlapping categories of field pests–those that are endogenous to the agricultural ecosystem and those that are exogenous. The endogenous, built-in pests include soil organisms, weeds, and others that can complete their life cycles within the crop or herd ecosystem. The exogenous pests enter the crop or herd ecosystem periodically from outside; meantime they survive on alternate hosts and in other ecological niches outside the system (many of which are beyond the control of the farmer). Many of the mathematical models of pest population dynamics refer to insect pests of this sort (Southwood and Way, 1970).

A third group of pests requires special mention because it causes such extensive, discouraging, and unnecessary losses—the storage pests of plant and animal products, principally insects, fungi, and small mammals. The control of these losses, as part of improved postharvest technology, is perhaps the most immediately rewarding aspect of pest regulation in developing countries.

The losses caused by pests and diseases cannot be measured with any certainty, but it is difficult to escape the qualitative and subjective impression that in the world as a whole field losses may represent one tenth to one fifth of the yield that might be obtained at the existing level of technology in the absence of pests, and that in the developing world storage losses may represent a similar proportion of the produce actually harvested.

Regulation of Field Pests

Few people are opposed to the regulation of field pests (except, perhaps, of birds), but it is not always easy to define or to agree on the strategy, objectives, and methods. Here, once again, the regions of advanced agriculture differ from the poorer ones, save that in both, the means of regulation must be within the farmer's financial and technical power and must offer him a reasonable prospect of a profitable return.

Objectives in Advanced Regions

In the market agriculture of advanced regions, the tendency to produce more than the system requires can theoretically be governed by competitive supply and demand relations, which should automatically determine the amount the producer spends on regulating pests, and hence the extent to which they are regulated. In practice, the competitive self-regulating mechanism is often ineffective, partly because agriculture in such regions is often subsidized, partly because farmers tend to play safe in an unpredictable weather, pest, and market situation, and partly because no good farmer likes having to compromise with harmful organisms. In such a situation, the farmer may try to regulate pests more efficiently than the national interest requires, particularly if he is pressed to do so by market demands for special quality or freedom from blemish that his competitors are prepared to meet. Mechanization makes large-scale chemical control particularly feasible and attractive (see Strickland, 1970).

In relation to endogenous pests, farmers correctly seek to hold damage down to an acceptable level of loss by routine practices; indeed much customary farm management practice, including crop rotation, is directed to this end. Regulation of exogenous pests, even if they are sporadic and unpredictable, may also require routine precautionary measures, plus emergency action when an attack breaks out. These actions can be made more effective by direct or indirect monitoring, as with potato blight in Britain, where farmers are warned of likely outbreaks. Where this cannot be done economically, it may be sensible, when at least some of the control measures have the character of technical insurance, to consider whether conventional financial insurance may not be as useful as technical protection.

Objectives in Developing Countries

In most developing regions, subsistence farming is still based on small inputs, small yields, and small but reasonably assured returns. Farmers are cautious about new practices unless they are quite certain that they will not threaten their survival. Regulation of harmful organisms is essential, but in the earlier stages of advance, the means used to increase output must be cheap and effective, even if they are less than perfect in some respects, and they must not depend on expensive mechanical equipment or on sudden and dramatic technical change. This is why plant breeding for larger yield as well as for pest regulation has been so important in the past.

Of the endogenous pests, those causing seedbed loss are particularly important, and often particularly easy to regulate, in unsophisticated arable systems. Weeds grow extremely rapidly, and since farmers have to start controlling them while they are still engaged in primary cultivation and sowing, they can very greatly accentuate the shortage of labor at this time. The nematodes of tropical soils seem still to be little known, and yet there is no doubt that they are important.

Some of the most striking successes of the past have been in the regulation of exogenous pests of crops, and this will surely remain important, since invasions and epidemics can build up so fast. Among these, grain-eating birds, which depend for most of the year on wild grasses, have become particularly troublesome in East Africa and the Sudan, especially where new irrigation gives them access to sources of food that have previously been beyond their reach for lack of drinking water. In many other regions social insects, especially grasshoppers, locusts, and termites, present large-scale problems of types unknown in the temperate regions.

Throughout the developing world, field pests and diseases of animals, particularly those that are not produced for a market, are particularly serious and particularly difficult to control because of movements associated with transhumance, distance, lack of veterinary staff, the poverty and ignorance of their owners, and the unsatisfactory standards of nutrition and husbandry. Wild animals provide reservoirs of infection for many of them, notably in the tsetse–trypanosome complex that is so intimately based on the existing wild ecosystems of the savannah regions of Africa.

BIOCIDES IN THE REGULATION OF HARMFUL ORGANISMS

Technology and Constraints

Primitive annual crop production has no effective means of dealing with field pests; the harmful organisms combine with adverse effects on the soil, and in due course the farmers are compelled to move on. Over many years, farming practices, strains, and varieties have evolved that tend to lessen these losses, as well as to minimize the effects of cropping on the soil. It is not surprising that unimproved crop varieties are often important sources of resistance to disease and pests. As technology improves, the farmer's freedom of action is broadened and the constraints imposed on him by the unimproved environment are lessened. In other words, agricultural technology constantly increases the farmer's ability to manipulate the environment; this is what farming is all about. (When people object to agricultural innovation on the ground that it alters the balance of nature, they are really telling us that they are opposed to change and prefer the old ways—a common enough human failing.) Furthermore, if advancing technology can increase yield rapidly enough, the area on which pests have to be controlled must tend to decrease.

Evolution and Use of Biocides

Man's means of regulating or eliminating harmful organisms have included biocides of many sorts for many years. In human medicine, the cruder plant and mineral products have largely given way to increasingly specific vaccines, chemotherapeutic agents, antisera, and antibiotics. Though drug resistance is well known, new medicaments are produced sufficiently frequently to keep this within bounds, and few of us will be so bold as to forecast an end to progress in this direction. Veterinary medicine benefits from these advances and, indeed, contributes significantly to them.

In the control of organisms which damage crops, we can discern a similar course of events. The biocides now in use, employed correctly, are remarkably effective, but many of them are still far from fully selective, and they are mostly applied, not by qualified practitioners, but by the farmers themselves. This can lead all too easily to enthusiastic overuse and so to the development of resistance in the target organisms (but, fortunately, in some of their predators, as well), to damage to organisms other than those at which the biocides are aimed (including man), and to the accumulation of undesirable residues in the ecosystem.

Evidently, however, such unpleasant results can be minimized or avoided, especially if the government, the manufacturers of biocides, and the farmers and their advisers are appropriately educated and brought together in mutual understanding. In the United Kingdom, agricultural chemicals are regulated by voluntary agreement—backed, it is true, by actual or potential statutory powers, but voluntary nevertheless. Comparisons cannot easily be made between the United Kingdom and the United States, because the environments, crops, and needs are so different in the two countries, but it may be useful to record that the United Kingdom, with an arable area of about 17 million acres, uses each year about 300 tons of chlorinated hydrocarbons and similar compounds (an average of about two thirds of an ounce per acre), while the United States, with 400 million acres, uses about 44,000 tons (an average of more than 3½ ounces per acre, roughly six times the figure for the United Kingdom). It is perhaps also worth pointing out that as the use of fertilizers and other agricultural chemicals has increased in the United Kingdom during the past 25 years, both the yields of crops and the average expectation of life have steadily increased also.

For the future, we may, of course, expect that the biocides themselves will continue to evolve, as they have done in human medicine. At present, about 30 new agricultural biocides are released per year. The prospects of natural attractants and repellents, of hormonal regulators, and of synthetic analogues seem likely to increase. Agricultural systems are becoming more efficient, and we are learning, through increasingly interdisciplinary research, how to combine biocides more effectively with the many other modes of regulation already in use, including inherited resistance in hosts, genetic manipulation and controlled sterilization of harmful organisms, and management practices including in appropriate cases the uses of predators and diseases of pests, which we call biological control. One can hope, too, that social direction, by legal, fiscal, and economic means, will also become more sensitive and more effective in relating the benefits of control to the real costs and to the needs of the whole community, particularly to those of the ultimate consumer and user.

Biocides and Agricultural Change

Farming ecosystems, however, are not and cannot be expected to be static, and they have to meet many objectives in addition to the regulation of harmful organisms. Improvements in technology and changes in prices, costs, or demand continually lead to changes in farm systems. For example, in Britain, the decrease in the size of our farm labor force has led to the use of pelleted monogerm sugar beet seed to avoid hand-thinning. This has at once greatly increased the importance of seedbed and seedling losses. We cannot sit back until a new round of biological studies has provided us with new modes of control; we have to get on and restrict these losses as best we can with the means at hand. Conversely, a new technological advance in some other sector may render obsolete a long-studied and carefully-planned system of specific biological control. The developing spectrum of biocides adds to the degrees of freedom at the farmer's command, while, on the contrary, the need to maintain a specific biological equilibrium must often tend to lessen them. Accordingly, I do not expect biocides to disappear from farming systems in the foreseeable future, though, like any other powerful agents, they must be used with care.

We in agriculture have a job to do everywhere, and it is a job of the greatest human importance. Though we must always seek to do it better, we cannot be deflected either by the prospect of better things to come, or by the dogmatic pessimism, however menacingly voiced, of the prophets of doom. To us, the many-headed cult of the Apocalypse is nothing new, yet somehow we survive and advance. It is, after all, the true business of science to ensure that we do.

REFERENCES

Anderson, E. 1954. Plants, man and life. Melrose, New York.

Andrews, G. 1948. The geology of the Sudan. *In* Agriculture in the Sudan. J. D. Tothill (ed.). University Press, Oxford, England.

Braidwood, R. J., and B. Howe. 1960. Prehistoric investigations in Iraqi Kurdistan (Oriental Institute of the University of Chicago: Studies in Ancient Oriental Civilisation 31). University of Chicago Press, Chicago.

Clark, C. 1970. Health, population and agricultural growth. *In* Change in agriculture. A. H. Bunting (ed.). Praeger, New York.

Clark, C., and M. R. Haswell. 1968. The economics of subsistence agriculture. Macmillan, London.

Ellis, E. A. 1965. The broads. Collins, London.

Flannery, K. V. 1969. Origins and ecological effects of early domestication in Iran and the Near East. *In* The exploitation and domestication of plants and animals. P. J. Ucko and G. W. Dimbeby (eds.). Duckworth, England.

Food and Agriculture Organization of the United Nations. 1970a. Provisional Indicative World Plan for Agricultural Development. Rome.

Food and Agriculture Organization of the United Nations. 1970b. The state of food and agriculture 1970. Rome.

Godwin, H. 1956. Origin of the British flora. University Press, Cambridge, England.

Hunter, G. 1970. Agricultural change and social development. *In* Change in agriculture. A. H. Bunting (ed.). Praeger, New York.

Jones, W. E. 1970. British agriculture since 1950. *In* Change in agriculture. A. H. Bunting (ed.). Praeger, New York.

Lyon, B. J. de V. 1970. Cotton research and its application in Nigeria. *In* Change in agriculture. A. H. Bunting (ed.). Praeger, New York.

McNeish, R. S., D. S. Byers, and others. 1967. Environment and subsistence, Vol. 1. *In* D. S. Byers (ed.) The prehistory of the Tehuacan Valley. Texas Press, Austin.

Olson, R. A. 1970. The fertiliser programme of the Freedom from Hunger Campaign. *In* Change in agriculture. A. H. Bunting (ed.). Praeger, New York.

Penman, H. L. 1968. The earth's potential. Sci. J. 4:43–7.

Southwood, H. R. E., and M. Way. 1970. Ecological background to pest control. *In* Concepts of pest management. R. L. Rabb and F. E. Guthrie (eds.). North Carolina State University, Raleigh.

Strickland, A. H. 1970. Some economic principles of pest management. *In* Concepts of pest management. R. L. Rabb and F. E. Guthrie (eds.). North Carolina State University, Raleigh.

van Hulten, M. H. M., and A. G. van Ommeren, 1970. Some aspects of the Zuyderzee reclamation scheme. *In* Change in agriculture. A. H. Bunting (ed.). Praeger, New York.

Vicinelli, P. 1970. Fruit growing in Ferrara Province, Italy. *In* Change in agriculture. A. H. Bunting (ed.). Praeger, New York.

Westlake, D. F. 1963. Comparisons of plant productivity. Biol. Rev. 38:385–425.

Role of Basic Research in Implementing Pest Control Strategies

Raymond L. Bisplinghoff and John L. Brooks

Two words in the title of this paper need to be explained. By *pest*, we will mean any plant or animal species that is unacceptably abundant. Also, we prefer to use the word *regulation* rather than *control*. It is apparent that the knowledge required to implement regulation strategies will continue to be generated by basic research examining the principles of population regulation in various natural and modified ecosystems. Research contributing relevant knowledge includes investigations of interspecific ecological dynamics within natural systems and the systematics, behavior, physiology, and biochemistry of the individual species. The maintenance of a broad basic research effort should yield suggestions for novel approaches to the challenge of regulating particular populations within particular ecosystems.

ANALYSIS AND SYSTEMS MANAGEMENT APPROACH

We believe that the role of basic research should be to examine the total ecosystem of which the pest population and its regulator are but parts, in order to ascertain the effect of this matrix. The success or failure of a tactic for the regulation of one species by another can possibly be determined by other aspects of the total system. A stable regulation of one population by another that is attainable under simplified conditions in the laboratory may not be attainable under natural conditions. In this paper we propose to examine several past attempts at biological regulation to determine how the outcome of a regulation tactic has been actually affected by the characteristics of the total system. By indicating problems encountered in past efforts, these studies suggest the kinds of problems that can be anticipated.

For examples of past attempts at biological regulation we have selected those that had as target species three kinds of exotic plants accidentally introduced into Australia, each of which flourished to the extent of becoming an egregious pest: prickly-pear cactus species of the genus *Opuntia*; St.-Johns-wort, *Hypericum perforatum*; and the cocklebur, *Xanthium strumarium*. Both *Opuntia* and *Xanthium* are New World genera, but *Hypericum* was introduced into both Australia and western United States from Europe. As is often the case when an organism is introduced into a distant area, these were not accompanied by their usual pathogens and herbivores (or predators). An increasing body of evidence, in part derived from the examination of situations such as those under discussion, indicates that the food-extracting population, be it herbivore or predator, exerts a regulating effect upon the prey population. The regulation strategy for each of these plant pests was to select from the array of herbivores feeding upon the plant in its native ecosystem those insects that restrict their feeding to that particular plant and that also seemed adapted to survive in the exotic system.

The control of the Australian prickly-pears by a phycitid moth, the caterpillars of which are gregarious tunnelers in the tissue of cacti, is the classic success story of biological regulation. The story is all the more dramatic because of the truly serious threat that these *Opuntia* posed for Australia. The some 20 species of *Opuntia* that still grow wild in Australia were probably introduced in various places and at various times in the first half of the nineteenth century. Certainly the spread of the prickly-pears seemed threatening and beyond control by 1870. By 1920, shortly before insect herbivores from America were introduced, *Opuntia* had become the dominant plant on more than 60 million acres. Although about 150 species of insects were found to feed on cacti in the Americas, only a third of these were tested for use in Australia, and of these only 12 were finally introduced. The gregarious caterpillars of the Argentinian species *Cactoblastis cactorum* proved remarkably effective. Several species of cochineal mealybugs, genus *Dactylopius*, exhibit a remarkably strict specificity for the different species of *Opuntia* and are effective regulators, which could probably have been crucial in the absence of *Cactoblastis* (Sweetman, 1958; National Academy of Sciences, 1968).

Cactoblastis cactorum became established so readily that by 1927, two years after the initial introduction, it was judged that the field populations could reproduce themselves without aid, and the cage rearing of caterpillars was deemed no longer necessary. In Australia there are two generations a year. The caterpillars feed within the cactus segments in groups of 20 to 100, emerging only to spin cocoons and pupate in debris around the cactus plants. The caterpillars eat out the interior of a joint, then burrow through to adjoining joints, and even into the underground bulbs and roots. While tissue destruction by the caterpillars themselves is extensive, the death of the cactus plant is often, if not usually, the result of the rots caused by bacteria and fungi after the caterpillars have fed.

Insect herbivores usually effect regulation by reducing the vitality of the host plants and decreasing their competitive success. To the extent that *Cactoblastis* attacks usually result in the death of the host, it resembles the regulation of insect herbivores by predators. The combined action of insect herbivore and pathogenic microorganisms undoubtedly is responsible for the dramatic reduction of the populations of the various *Opuntia* species in Australia. It is possible that research and development for future strategies of weed regulation should seek to establish whether a desirable degree of host specificity by a plant pathogen can be achieved by utilizing an association with a monophagous herbivore.

Another way in which the effectiveness of the *Cactoblastis* larvae was modified by other aspects of the ecosystem was evident in the comparative slowness of the *Cactoblastis* regulation in the southern (cooler) parts of the *Opuntia* range and in the areas where the soil was low in nitrogen. In these nitrogen-deficient soils and where the *Opuntia* grew under a tree cover, the cactus was slow-growing and chlorotic. Where the trees were cut and nitrogen fertilizer applied, the cactus grew vigorously and was then readily attacked and destroyed by *Cactoblastis* larvae (Sweetman, 1958).

The populations of all of the species of *Opuntia* in Australia are being regulated at fairly tolerable densities by *Castoblastis* and by the various species of coccineal mealybugs. These populations of mealybugs are decimated by heavy rains, so that they are most effective regulators in places and at times of low rainfall. The effectiveness of mealybug attacks on *Opuntia* is much greater on plants that have also been subject to prior attack by other herbivores (either a species of bug, or a red mite, both introduced). The plants, their vitality thus reduced, produce adventitious growth from their lower segments.

This growth is particularly susceptible to mealybug attack. The effectiveness of both *Castoblastis cactorum* and the coccineal bugs, *Dactylopius* spp., as regulators of a pest plant population is, therefore, clearly linked to the effects of other organisms of the system upon the target plant.

The second example concerns the attempts to regulate the populations of St.-Johns-wort, *Hypericum perforatum*, introduced into both Australia and western United States from the Old World. This plant grows in temperate regions with moderate cool-season rainfalls, and thrives especially on sunny, well-drained soil where the vegetation has been overgrazed by sheep or cattle. Under these circumstances *Hypericum* can become dominant, excluding desirable forage. St.-Johns-wort itself is toxic to sheep, cattle, and horses (Huffaker, 1959; Sweetman, 1958). It is estimated that *Hypericum perforatum* dominated about one third of a million acres in Victoria, Australia, alone, in 1930, just prior to the introduction of several species of insect herbivores. When *Hypericum*-eating insects were introduced into western United States, more than 5 million acres from California to Washington and Montana were dominated by *Hypericum*.

The toxin is a di-anthrone derivative called hypericin. Plants containing hypericin are avoided by both insect and vertebrate herbivores. Although some 600 species of Old World insects have been reported feeding on St.-Johns-wort throughout its range, leaf-eating chrysomelid beetles of the genus *Chrysolina* were finally chosen for use as regulators. These species of *Chrysolina* have obviously evolved a system for detoxifying hypericin. The third insect selected for use in Australia was a buprestid beetle that feeds on the roots of *Hypericum*. Possibly the hypericin content of the roots is quite low. Hypericin is known to be secreted by glands on the leafy parts of the plant (Whittaker and Feeny, 1971). The two chrysomelid beetles released in both Australia and western United States are *Chrysolina hyperici* and *Chrysolina quadrigemina* (=*C. gemellata*) (Huffaker, 1959).

These regulation effects were a complete success on a regional basis only in California, where St.-Johns-wort populations were reduced to less than 1 percent of peak abundance. Success in the more northern states has been variable and patchy, as has also been the case in Australia. Successful regulation on both continents has occurred where the summers are hot and dry. Both larval and adult stages of the beetle feed vigorously during spring and early summer, and the plants are so damaged that they cannot survive a hot dry summer. But if the plants are watered, they will survive (Sweetman,

1958). In cooler, shadier, wetter situations, *Hypericum* can withstand even severe spring defoliation.

Reduction in vitality of the *Hypericum* by the insect herbivores will decrease its competitiveness relative to other elements of the vegetation. But wherever this native vegetation is overgrazed by cattle or sheep, the *Hypericum* will not be as readily replaced.

The attempt at biological regulation of the third plant introduced into Australia has been a failure. The cocklebur, a species of the aberrant composite genus *Xanthium* that probably should be called *X. strumarium*, probably was introduced into Australia from India shortly after 1800 mixed with cotton seed from plants almost certainly derived from the Mississippi Delta region (McMillan, 1971). Because *Xanthium* was first evident at Noogoora Station near Brisbane it is often called Noogoora Burr. *Xanthium* grows readily in disturbed areas and in Australian sheep country is an especial nuisance because the burs catch in the wool and the young plants prove toxic to sheep.

Although many American insects are known to feed on *Xanthium,* and a few Indian species have shifted to feeding on this exotic, the final choice as regulator was an America fly *Euresta aequalis*, the larvae of which feed on the specialized seed of the cocklebur. The seed capsule contains two seeds, one of which will germinate the year after maturation, while the other requires a year's dormancy. The seed fly was liberated in Australia in the early 1940's but has not succeeded in establishing itself well enough to cause any appreciable reduction in the *Xanthium* populations. This appears to be due to the erratic germination of the cockleburs so that the seed is not regularly available from year to year (Huffaker, 1959).

Is it possible that further basic research on the reproductive biology of *Xanthium* would reveal a means by which a continuous supply of seed could be maintained, thus allowing the seed-fly population to reach a sufficient level to ensure its regulatory ability?

A recent paper in *Science* (McMillan, 1971) notes that cockleburs from all latitudes in Australia have the same critical photoperiod for flowering. In North America, however, where cockleburs are found over an even wider latitudinal area, a wide range of photoperiods occurs. The critical day length at each latitude ensures that the seed produced will germinate under generally favorable conditions. Cockleburs transplanted from their normal latitude to another in North America will flower at a day length corresponding to their place of origin. This piece of basic research may contain the germ of a possible regulation strategy, when one other bit of essential information is added. This is the fact that all of these cocklebur populations of Australia and North America will hybridize, and that the hybrids have photoperiods intermediate between those of the parents.

Whether or not this information could be of use in the regulation of *Xanthium* is not our essential point, which is that basic research on any and all aspects of the ecosystem can generate information relevant to the success of a regulation tactic. In the cases of *Opuntia* spp. and *Hypericum* sp. it was discovered only on a trial-and-error basis that many facets of an ecosystem can seriously modify the effectiveness with which even a carefully selected species regulates the target population. It would presumably be more effective to design a systems strategy on a base of knowledge of the entire system; climatology, meteorology, ecology, physiology, and biochemistry are all relevant to the development of the most effective systems strategy for pest regulation.

APPLICATIONS OF SYSTEMS STRATEGIES TO REGULATION PROBLEMS

The three examples of what really must be considered "systems" strategies for pest regulation, although these efforts were not so considered at the time, provide hope that stable biological regulation of pest plant populations can be achieved. Past successes with the biological regulation of insect herbivores of crops and trees similarly provide hope that strategies involving insect predators can be developed. Analysis of the instances of successful regulation of insect herbivores would also suggest that the configuration of the total system within which the regulatory organism operates can strongly modify the effectiveness of the regulation. Some of the most dramatic of these instances of regulation of both pest animals and plants have been achieved when the pests were exotic introductions into the system. The biological regulation of a native species that is unacceptably abundant has proved more difficult. A systems strategy involving simultaneous modification of several elements of the ecosystem might prove effective.

This systems approach has proved useful in the regulation of plant and animal disease agents. Infectious diseases of trees and crop plants are caused by fungi, nematodes, viruses, and bacteria. Most of these live wholly or partly in the soil. Although the biology of soil organisms has proved difficult to study, it is clearly evident that chemically mediated antagonistic and stimulatory interactions between fungi, bacteria, nematodes, protozoa, and the roots of vascular plants control, in large measure, the species composition of the soil. This appears to be a potential arena for systems strategies to regulate plant pathogens and otherwise promote the vigorous growth of beneficial plants. We can do no better than quote the concluding paragraph of the contribution "Progress in the Battle against Plant Disease" by McNew (1966) in the preceding Academy symposium on pest control:

> There are almost unlimited possibilities of controlling the severe destruction from soil-inhabiting pathogens once the forces operative in the soil, especially near to the roots and root hairs, are better understood. The forces involved in the rhizosphere are complex, but progress in the past two decades offers much promise for controlling disease by exploiting natural forces available in the soil.

Two of the most serious infectious diseases of man, malaria and schistosomiasis, are caused by animal parasites with complex life cycles involving passage through an invertebrate animal host or vector. In order for the disease to establish and maintain itself, the configuration of the ecosystem in which the human population operates must be such that populations of man, parasite, and vector come into close contact with reasonable frequency. That these populations do make such contact is attested by the local severity of these diseases. A systems strategy could involve modification of the behavior of the human population or altering the spatial patterning of the system so as to decrease the frequency of contact between the elements of the parasite's life cycle. The other approach is to decrease the population of the vector. The dramatic reduction in malaria since World War II has, of course, been through reduction of the populations of the mosquito vectors through the release of chemical intoxicants into the system.

Schistosomiasis, or bilharzia, is now considered by the World Health Organization to be one of the most significant diseases in developing nations. Freshwater snails are a vital intermediate host. In some areas attempts are being made to modify the behavior of the humans so as to decrease passage of the parasite between humans and snails. Removal of the human population from the banks of rivers and lakes and provision of piped water for bathing and drinking lowers the rate of infection of humans by the larval flukes; disposal of feces and urine other than directly into the rivers and lakes will lower the rate of infection of the snails. However, in many parts of the world where schistosomiasis is prevalent, efforts to reduce the snail populations are appropriate.

Basic research into the biology of a family of dipteran flies, Sciomyzidae, the larvae of which feed upon freshwater snails suggests a possible approach to regulation of certain vector snails. The distribution and prey preferences of more than 100 species of these snail-killing flies have been studied (Berg, 1953; Berg and Neff, 1959; Neff, 1964). The information thus revealed has an obvious potential for biological regulation of vector snails that should be evaluated as part of a systems strategy for the alleviation of the severity of schistosomiasis.

These examples illustrate the broad range of application that the systems approach can develop from basic research. Active programs of basic research in biochemistry, physiology, and ecology, including chemical ecology, are vital to the development of regulation strategies. However, there is a need for active research in other disciplines as well. Basic systematic and evolutionary studies of the great diversity of plants and insects should be actively pursued. New information might be discovered in this way, and new systems strategies might then be implemented.

REFERENCES

Berg, C. O. 1953. Sciomyzid larvae (Diptera) that feed on snails. J. Parasitol. 39:630–636.

Berg, C. O., and S. E. Neff. 1959. Preliminary tests of the ability of sciomyzid larvae (Diptera) to destroy snails of medical importance. Bull. Am. Malacol. Union 25:12–13.

Huffaker, C. B. 1959. Biological control of weeds with insects. Ann. Rev. Entomol. 4:251–276.

McMillan, C. 1971. Photoperiod evidence in the introduction of *Xanthium* (cocklebur) to Australia. Science 171 (3975):1029–1031.

McNew, G. L. 1966. Progress in the battle against plant disease. *In* Scientific aspects of pest control. NAS-NRC Publ. 1402. Washington, D.C.

National Academy of Sciences. 1968. Weed control. Publ. 1597. Washington, D.C.

Neff, S. E. 1964. Snail-killing sciomyzid flies: Application in biological control. Proc. Intn. Assoc. Theor. Appl. Limnology 15:933–939.

Sweetman, H. L. 1958. The principles of biological control. Brown & Co., Dubuque, Iowa.

Whittaker, R. H., and P. P. Feeny. 1971. Allelochemics: Chemical interactions between species. Science 171 (3973):757–770.

Implementing Pest Control Strategies

Edward H. Smith

A review of the literature shows that little attention has been given to implementation of control strategies. We seem to have operated on the assumption that knowledge, once acquired, will flow to the site of need, but I believe this matter needs critical review. We need to establish base lines and some reasonable expectations of progress, although the investigator is ordinarily reluctant to predict the outcome of his efforts.

In seeking to compare the mood of the 1966 symposium entitled "Scientific Aspects of Pest Control" with the present, I sense less optimism today (National Academy of Sciences, 1966). At that time I spoke of "standing on the verge of a renaissance in our concepts of pest control," whereas now one feels a tinge of frustration and impatience to convert promising concepts to practical programs. Perhaps the time scale of expectations is unrealistic.

Pest control is but one of many current problems that arise from the conflict between science, technology, and human values. Lessons learned here can have wide application to questions of environmental quality that will occupy us in the years ahead, for it is largely in the context of environmental quality that the public considers pest control. Progress can be made in only two ways—by invoking new methods of control or by getting greater returns from methods presently available. I will deal here chiefly with the latter. Recent practices and outlook have been largely dominated by commitment to chemical control; with some three decades of experience it is to be hoped that retrospection can provide a basis for sound guidance in the future.

The elements that influence the implementation of pest control strategies are knowledge resources, public attitude, and educational resources. These three factors are interrelated. We will examine these relationships in some detail.

PEST CONTROL AND THE PUBLIC

Man's plight in the latter half of this century is his dilemma of burgeoning numbers against a pattern of uneven distribution of men and material—extremes of wealth and poverty and an approaching crisis in the quality of the environment. Any optimistic projection for the future includes intensified efforts in control of human population and the pest populations that threaten his food, fiber, health, and comfort. Pest control involves technical considerations not fully understood by the man in the street, but his support is required if the new developments are to be made to place pest control on a sounder basis.

Recognizing our dependency on new knowledge, the public's attitude to scientific research is of special concern. These are not good times for science. The public that supported it lavishly through two decades following World War II has become disenchanted with science and technology. Science, which once seemed to offer such great hope in the solution of man's problems, has become suspect. Computer technology is viewed as threat of invasion of privacy. Science and technology seem to have become captives of the military–industrial complex. As the gross national product rises, the quality of life seems to decline. The new challenge is to apply science to the solution of the great social problems that relate to the quality of life and survival of man.

This disenchantment with science and technology manifests itself in a number of important ways, including decline of student enrollment in biological sciences and decreasing financial support. These winds, unfavorable to science and education in general, are certain to adversely affect pest control. In addition to the spin-off from these major trends of public attitude, pest control is also influenced by its own "image" in public opinion and this, too, is unfavorable at this time. It is inappropriate to pursue further here the important question of science and the public mood, but the insightful publications of Handler (1971), Morison (1969), and McElroy (1971) are recommended.

This means that we undertake the task of implementing the strategies of pest control in the face of waning support for science and technology and a crisis of confidence in the use of chemical pesticides. These are formidable odds at best and impossible odds unless we accurately assess their implications for our efforts. In this context the following conclusions seem justified:

1. Support for scientific research has peaked and is on the decline.
2. The appeal of science as a challenging career for youth is waning.
3. Scientific effort will be directed increasingly toward social and environmental problems.
4. The public mood will have greater influence on how science fares in the future.
5. The social consequences of new technology will come under increasing scrutiny.
6. The problems of pest control are intensified by a public credibility gap, based on the public's assessment of pesticide use of the past.

Unless we can close the credibility gap, it is highly unlikely that satisfactory progress can be made. We need, therefore, to understand the factors that created the gap and to develop strategy for overcoming it.

After more than a decade of debate, the public has concluded that pesticides constitute a serious threat to the environment. Earlier the concern was for human health, and this broadened to include environmental concern. The indictment of pesticides by the public was reached despite assurances of regulatory agencies, the pesticide industry, and those responsible for recommendations. The issue focused chiefly on DDT and the evidence of its role in environmental pollution formed the basis for the controversy. The conclusions drawn about DDT have tended to become generalizations about pesticides. The results of this crisis of confidence is a mandate for change in organizations and regulations. The public concern over pesticides has been joined with other environmental concerns. Citizen leadership in the environmental cause includes not only the classical "nature lovers" but a new breed of activists well versed in the workings of the political structure seeking to invoke new concepts of law and individual rights. It is in this setting that the pesticide controversy must now be resolved.

Surely in the prolonged and intense controversy over pesticides there must be some valuable lessons for those of us who bear responsibility in pest control and public education. Let us seek to identify some here.

It seems evident, in retrospect, that the conventional wisdom underlying the practice of chemical control was unsound. We failed to properly assess some characteristics of the pests and of the molecules being directed against them. Thus, our errors in judgment were in the area of both biology and chemistry.

The initial successes of chemical control with synthetic organic insecticides led to a sense of false security and oversimplification in assessing the nature of the problem. We failed to anticipate the development of insect resistance, the resurgence of target species, and the disruption of nontarget species. These are all matters that relate to general biological concepts—genetic variability, natural selection, population dynamics. With respect to the molecules themselves, we failed to anticipate the practical significance of stability, lipid solubility, and broad spectrum toxicity. The interrelatedness of the biosphere was not fully appreciated as the mobility of DDT and its accumulation in living organisms disclosed.

In seeking to make objective assessments of the pesticide problem, we have had the benefit of a number of comprehensive studies. Two of these seem particularly pertinent: *Report of Secretary's Commission on Pesticides and Their Relationship to Environmental Health* (Mrak report) (U.S. Dept. of Health, Education and Welfare, 1969) and *Restoring the Quality of the Environment* (Tukey report) (President's Science Advisory Committee, 1965).

In both cases, groups of competent professional leaders have provided incisive reports giving specific recommendations that are essentially sound. We should find these very helpful in seeking to evaluate our progress. The "task force" approach to studying problems and arriving at recommendations is a well-recognized technique among both governmental and private agencies. Unfortunately, there is rarely any systematic follow-up to consider implementation and evaluation.

Considering the fragmentation that characterized our pre-DDT biological training, it is understandable that we might have failed to anticipate the response of living organisms to chemical selection. What is harder to understand and justify is our slowness to recognize our miscalculation and to make appropriate changes. Even now many individuals working in the field of pest control are fighting a rearguard action in defense of DDT. (Evidence of the continuing controversy can be found in letters to the editor of *Science* and *BioScience.)* This continuing controversy, usually waged without reference to specific data, is most unfortunate and tends to harden the polarization within the ranks of the professionals and between the professionals and the public.

While we can acknowledge errors in judgment in the underlying strategy of chemical control, we should cite a positive feature—the catalytic influence of these errors on our quest for knowledge. These setbacks led to new thinking about the biochemistry of the species, genetic mechanisms, behavior, and a host of other matters that relate to population dynamics and pest management. Hopefully, we can, as pest control specialists, acknowledge mistakes of the past and then without burden of guilt or defensiveness direct our energies in support of positive programs. The public is not very discriminating once its confidence has been shaken; loss of confidence in one segment of the science–technology complex is likely to spread to related areas. We have an obligation to set our house in order, not only because of the direct impact on pest control, but also because of the implications to the broader application of science and technology.

KNOWLEDGE RESOURCES—THE STRATEGIES

Strategies refer to the choice between alternatives and to skill in the art of making choices. From the experience of the past three decades have emerged new concepts of pest control. The prevailing concept at this time centers on population dynamics as the framework of our considerations. From an understanding of the factors regulating pest populations we seek to invoke a number of practices that, in total, reduce the population to tolerable levels. Thus, within the totality of the principles of population dynamics can be fitted in an additive way control measures required to manage the population. This concept de-emphasizes primary reliance on a single method and the objective of complete control or eradication. The success of such a concept depends on the completeness of our knowledge of population dynamics, the strategies which can be invoked, and the degree of population management required. Obviously we are a long way from the knowledge resources required for the synthesis of programs to cope with numerous major pests. The concept is sound and we should keep the goal before us, but in the meantime we must meet the day-to-day control problems with the best strategies available now.

We are indebted to the National Academy of Sciences for a comprehensive inventory of strategies of pest control. The series of publications to which I refer is "Principles of Plant and Animal Pest Control" and it includes the following volumes: *Plant-Disease Development and Control* (1968a), *Weed Control* (1968b), *Insect-Pest Management and Control* (1969), *Control of Plant-Parasitic Nematodes* (1968c), *Vertebrate Pests: Problems and Control* (1970).

Because of the importance of insects and our extensive experience with insecticides, let us give special consideration to Volume 3, *Insect-Pest Management and Control*. This is an impressive volume of some 500 pages. In reviewing the comprehensive list, one is impressed by the length of the list, the many principles represented by the various strategies, and the gap between strategies listed and those in common usage. Despite the numerous controls that qualify for listing, chief reliance is still on chemical control.

Another striking feature is the weight that must be accorded the time scale in the development and application of strategies. Some strategies considered "new" actually represent the work of several decades, for example, the sterilization principle. This principle was established as a practical means of control in the mid-1950's with the eradication of the screw-worm fly *Cochilomyia hominivorax* (Conquerel) from the West Indian island of Curaçao by the release of males sterilized by exposure of the pupae to gamma radiation. This success was preceded by some 20 years of research directed toward that specific goal. While some success has been achieved against tropical tephritid fruit flies in the South Pacific islands and along the Mexico–California border, the initial success of the mid-1950's has not been followed by a pattern of similar successes. Thus, 35 years and millions of dollars invested in research have stimulated high hope and new thought, but have not greatly altered the pattern of pest control practices. This is not to detract from the breakthrough, but to inject an element of sobering reality in terms of our expectations. Similar examples could be cited with some other strategies such as integrated control, sex attractants, hormones, chemosterilants—sound in concept but years from application in the solution of some pressing problems.

In seeking to adapt other methods of control and reduce the pressure for chemical treatment, we need to reassess market standards. Some of them impose exacting specifications that have no significance in terms of nutrition, sanitation, or aesthetic value. On what basis are standards established for the number of insect fragments that constitute contamination in a food product?

It is difficult to assess our strategic posture when such variables exist as the time scale of new development, the public mood, and changes in the dynamic biological systems of which the pests are a part. We seem to publicize "promising new methods" very well. Unless we distinguish better between prospect and promise, the public may form unrealistic expectations that end in disillusionment.

Recognizing the gap between theory and practice and that chemical control must remain our first line of defense, we should give special consideration to it. It is ironic that chemical control has fallen into public ill-repute although we are still dependent on it. The respectability of chemical control must be re-established in the public mind, and, second, the pesticides industry needs to be convinced that investments in pesticide development are justified. The first, public acceptance, is a prerequisite for the second. It has become a national pastime to belittle chemical control. Companies are reluctant to combat adverse public opinion and to hurdle the registration obstacle course, in addition to all the other calculated risks involved in the multimillion dollar investment in development of a single pesticide. The combination of adverse factors has left the pesticide industry in a demoralized state. Many laboratories are operating on a standby basis and it will be some years before the impact of this disruption is fully felt.

There is, of course, another alternative—that of shifting the cost of development from private industry to public agencies. These matters will be considered by other speakers, including Carlson and Brady. The economics of pesticide development and use have received little study, and it is heartening that the subject is featured in this program. Reports such as those of Carlson and Headley are certain to stimulate re-examination of some presently accepted concepts of cost–benefit relationships.

We can censure the pesticide industry for being long on salesmanship and short on education, but let us not make censure the knockout blow. The responsibility should be shared. The pesticide industry delivered the recommended materials and generally operated within the prescribed regulations. Inherent unsoundness in either concept or regulation was likely to produce backlash for the industry. That is what happened, but these events must not obscure the role envisioned for chemical control in the future. The new respectability expected of chemical control is well stated in the recent publication of the National Academy of Sciences *Insect-Pest Management and Control* (1969, p. 456–457):

> Each control technique has a potential role to play in concert and harmony with the others, and a major technique such as the use of pesticides can be the very heart and core of integrated systems. Chemical pesticides will continue to be one of the most dependable weapons of the entomologist for the foreseeable future. . .There are many pest problems for which the use of chemicals provides the only acceptable solution. Contrary to the thinking of some people, the use of pesticides for pest control is not an ecological sin. When their use is approached from the sound base of ecological principles, chemical pesticides provide dependable and valuable tools for the biologist.

While one seeks to make a case for an essential and responsible role for industry, he encounters setbacks from industry itself. The present controversy over a pilot program for cotton scouting in Arizona illustrates the point. An editorial in the *Arizona Republican,* Phoenix, Arizona (February 27, 1971) includes the following:

> No one can say with certainty how far American universities should go in conducting research projects as opposed to teaching students. But the Arizona Agricultural Chemical Association is vigorously opposed to a federal grant for the University of Arizona to determine insect ecology in the cotton fields of Pinal County. . . .
>
> If the Department of Agriculture's Plant Pest Division wants a study of Arizona's pests, it seems to us the best source would be the commercial companies which sell pesticides and deal with farmers all the time.

Until industry extends its cooperation in ecological studies of the type proposed, the partnership called for eludes us.

The role and responsibility of the pesticide salesman has been a subject of controversy. Van de Bosch (1970), in a penetrating study of ecological considerations in the use of pesticides, has identified a conflict of interest as follows:

> The salesman is the key to the system, for he serves as diagnostician, therapist, and pill dispenser. And what is particularly disturbing is that he need not demonstrate technical competence to perform in this multiple capacity. In other words, the man who analyzes pest problems, recommends the chemicals to be used, and effects their sales is neither required by law to demonstrate (by examination) his professional qualifications (as do medical doctors, dentists, lawyers, veterinarians, barbers, beauticians, realtors, etc.), nor is he licensed. Yet this person deals with extremely complex ecological problems and utilizes some of the most deadly and ecologically disruptive chemicals devised by science.
>
> The chemical industry has made some effort to upgrade the quality of its field man, but the men remain salesmen, and merchandising is their real charge.

trol specialists as well as members of the general public.

The need for safer pesticides has long been recognized; indeed, many have been discovered and placed on the shelf. The primary basis for selecting pesticides has been optimum effectiveness against pests, with secondary consideration given to safety to humans and to the environment. The disenchantment with DDT has resulted in a shift to less persistent organophosphorous insecticides, chiefly parathion, with its high mammalian toxicity. This is a paradox. There are safer analogues to both—methoxychlor for DDT; fenthion, dicapthon, and others for parathion—but these have not been substituted.

With our impressive fund of knowledge of toxicology and pesticide chemistry, we can virtually have what we want, but the urgent need to replace the toxic-persistent-cheap molecule has not been reflected in the business between the pesticide and agricultural industries. This gives rise to growing criticism of two industries essential to public well-being and invites increased regulation of both.

To summarize, while our strategies are many, the proven ones are few. The public, having been exposed to glowing releases on new methods, has developed expectations that are not being met. The realities force us to settle for chemical control as the core. While tremendous possibilities for progress in chemical control exist, the sequence of tradeoffs has resulted in only modest gains. The pesticide industry, vital to progress, suffers under public ill will without obvious positive programs designed to improve its public image. For those seeking to implement strategies it is evident that the human component often presents a more difficult challenge than the pests.

EDUCATION FOR IMPLEMENTATION

A tremendous gap exists between knowledge and its application in matters pertaining to the well-being of the individual. Herein lie some of the paradoxes of our time and the sources of great frustration. For instance, the cure for iron anemia has been known for over a half century. It is a simple cure, but the disease is still rampant among segments of our population. The medical profession proclaims that cigarettes are harmful, but we have achieved only modest success in reducing smoking. In the area of pest control some of our oldest strategies, proven sound by the test of time, have not been generally implemented. We continue with highly toxic compounds such as parathion, accepting the annual increment of casualties although alternatives are available. For a nation so committed to education, it would seem that we could do far better. One can hardly imagine a more challenging opportunity for progress through education than in the area of pest control.

The Public

Considering the implications of pest control for human welfare, it is imperative that public understanding be achieved. But pest control as viewed by the layman is merely part of the threatening complex of science and technology. Pest control must be viewed in its sociobiological context, which relates to environmental quality, which, in turn, relates to the quality of life.

A reasonably optimistic outlook for the future requires that science be projected to the public as something that can be understood by ordinary men and regulated to their service rather than becoming their master. In addition, it is necessary to involve the public in choices of alternatives. Unless options are understood and the people believe they have a voice in making decisions, distrust will continue. We are all familiar with cases in which projects are undertaken following perfunctory hearings. In the eyes of the public, it seems that an announcement is simply made that a dam will be built, an atomic energy plant constructed, or an area control program launched. The present controversy over fire ant control programs is an example. The public seeks information on options, and time to consider in the absence of coercion. This may impose inefficiencies from the vantage point of those charged with pest control, but it's the way the public will have it.

To gain public participation in decision-making is a complex process. Biologists will need all the help they can get from the social sciences in gaining the necessary public participation in decision-making on biological issues. The brief treatise by Hahn (1970) "Citizens in Local Politics: Nonparticipation and 'Unrepresentation' " serves to identify some common myths about citizen participation. No matter how complex the process, it is clear that the decade of the 70's marks a turning point in the public's attitude to practices that affect the environment. A recent illustration of this fact is the defeat of the proposal to build the supersonic transport (SST). Headlines in the *New York Times* (March 28, 1971) state "The public cast the deciding vote."

We have had some encouraging experiences in Rockland County, New York, with respect to the public decisions and gypsy moth control programs. The elements of these successes were (a) community interest in educational programs; (b) community leaders committed to the process of orderly decision-making; (c) an effective corps of educators (Cooperative Extension Agents) who enjoyed public confidence; (d) citizen participation in choice of options; (e) the good fortune of effectiveness from the option selected.

Morison in his article "Science and Social Attitudes" (Morison, 1969) has provided an incisive analysis of the need for more effective education of the public in matters of science and technology. He states:

> As for less formal methods for presenting science to adults, we should devise some analogy that would do for the general public what agricultural extension courses have done for the farmer and his wife. The average successful farmer, although he is far from being a pure scientist, has an appreciation for the way science works. Certainly he understands it well enough to use it in his own business and to support agricultural colleges and the great state universities that grew out of them.

Some helpful guidelines emanated from a Conference on Environmental Education for the Adult Public, sponsored by AIBS (Coon and Fleet, 1970). In considering the target audience and educational roles, the following is offered.

> . . .the target should not be "the public" in its broadest and amorphous sense, but people in discretely recognizable assemblies—state and local government officials, civic and community-action groups and the like. . . .
>
> The salient aspect of the proposal is. . .the development of a kind of "public educator," a hybrid of a county extension agent and Peace Corps worker. Not envisioned as a walking encyclopedia of the information gathered or even generated by the centers, nor as an advocate of a particular position, a potentially impossible role at that, this person is proposed to be a catalyst.

While the blueprinting of a comprehensive plan for public education on the scientific aspects of pest control is beyond my purview, I believe the following points can be made:

1. The implications of biology for the well-being of mankind are so great that training in biology should be intensified as an essential component of public education.

2. The public's acceptance of pest control is influenced in a large measure by its general outlook on science and technology.

3. Effective pest control in the years ahead cannot be achieved without better public understanding than now exists.

4. Matters of pest control become part of the total concern the public has for environmental problems.

5. The public has the capacity to acquire the kind of understanding required, as illustrated by groups that have had advantage of special opportunities.

6. Efforts undertaken by biologists without regard for sociobiological dimensions of the problems are not likely to succeed. Inputs from the social and biological disciplines are required.

The Land-Grant Universities

One of the great educational resources that can be brought to bear on the pest control problem is the land-grant university. It grew out of the concept that its mission included the acquisition, transmission, and application of knowledge. This concept has been best exemplified in agriculture where each component—teaching, research, and extension—provides a catalytic influence on the other two. This concept also brought farm families into a sense of partnership with the university. Knowledge from university laboratories flowed freely to problem-solving situations on the farm. The farm families in turn raised new questions for the scientists. And their children sought training at the sources of this knowledge.

It is in the land-grant universities that much of the pest control research has been conducted and from this base that recommendations have flowed through the extension arm for implementation in the field. While the land-grant universities draw on a distinguished tradition, they will not meet the challenge currently posed by pest control without undergoing considerable change.

The Faculties

The primary challenge to the faculties is to make the transition from the traditional disciplinary approach to the interdisciplinary effort required to solve the complex social and environmental problems. This is a major undertaking. There can be no strong interdisciplinary research without strong disciplines from which to draw. Over the years strong disciplinary loyalties have been forged and these respond slowly to the demand for change. The professional leaders of the disciplines for the most part were trained for disciplinary service and made their greater contributions by this route. Assuming that department heads exercise strong influence on departmental programs, let us consider their "profile." Taking entomology as an example, the average age of department heads is 51 years (range 34 to 67). This means that most of them acquired graduate training in the mid-40's when interdisciplinary effort was not in vogue and when chemical control was approaching its zenith—thus, their training during their most impressionable years was not attuned to the concepts current today.

Our system has not provided innovative educational expedients to update training. Sabbatical leaves are not provided at all of our universities. Where they are provided, the experience frequently involves no marked change from the program conducted at the home base. It is interesting to scan the journals and note the shift in emphasis in the research program of individual investigators over the years. In many cases, particularly in applied research, the concepts and procedures change relatively little over the years. Some rework their Ph.D. theses for years. Despite some changes in rhetoric, many working in economic entomology "think chemical control" in the old pattern.

The professional meetings, bowing to specialization, are arranged by sections on the basis of common interest, and this tends to reduce the interchange between disciplines or between branches of the same discipline.

Occasionally, innovative workshops are held to promote the infusion of ideas among established research staff members. These are good as far as they go, but they don't go far enough. The exposure is too short and the dilution factor too great once the individual returns to the old routine.

It is not easy for these individuals to grasp new research concepts of computer science, population dynamics, etc., "on the run." It can be a frustrating experience that drives him back to what he knows best or into administration. Administration is no safe haven because he is expected to administer programs involving concepts that are not integrated into his disciplinary rationale, thus setting the stage for a generation gap between the administrator and the young scientist. If these observations are correct, it would seem desirable to study the most likely methods of updating the training of established investigators and invest in it, at least on a trial basis. We can point to very few examples of innovation in these areas.

Assuming the faculty is receptive and qualified for interdisciplinary research, some problems remain. The disciplinary base must be maintained to ensure the rigorous specialization required for effective contribution to an interdisciplinary team. The individual needs to maintain contact with a "critical mass" of disciplinary interest. In a sense, he needs to be on "detached service" from his discipline, returning to it frequently enough to keep his battery charged. The proper balance here poses problems for department heads with a sense of territorial imperative and for task force leaders who feel the need for something they can call their own.

Research

The matter of research will be adequately dealt with by other speakers. I would like, however, to comment on research as it relates to implementation of strategies. It seems evident that many of the promising leads to which we had looked with high hopes will not materialize in the immediate future. This dictates that chemical control will continue to be our first line of defense. If this assessment is correct, it would seem desirable to reassess priorities and concentrate research efforts on improvements that can be made in chemical control.

In the growing climate of fiscal austerity and disenchantment with research, some reappraisals will be made. I believe that these will identify a need for a more effective continuum of effort from basic research to actual practice. The links between the various levels of research going from basic to applied are weak. We can hardly afford research on plant resistance, for instance, that has as its goal the establishment of differential susceptibilities without regard to the testing of these findings in practice. This in turn requires joint effort by seedsmen, producers, marketers, consumers. The components that form the full spectrum of this effort are frequently not assembled and directed to the goal. The results are progress reports, satisfying to the individual but without impact on practice. It is an easy way out to work in compartments of basic and applied research and extension without ensuring that bridges are built between them. A team effort that identifies the total spectrum of effort from laboratory to field can overcome this "falling between the categories." Just as interdisciplinary research is required, so is there a need for close integration of the various steps from discovery to utility.

The two most obvious needs in chemical control are (1) pesticides that are effective with minimum disruption of the ecosystem and (2) pesticide use in minimum quantities required to bring about the desired reduction in pest population. While the first of these goals requires research, it has already been pointed out that implementation lags far behind knowledge. The second need relates to knowledge of pest populations as a basis for pesticide use. Experience has shown that if reliable information on pest populations is available, growers will make appropriate adjustments in pesticide applications. There is the incentive of economy; to treat only as needed is a simple, old concept. The deterrent to its wider implementation is the knowledge of economic thresholds and seasonal information on population levels.

Another strategy which should have great appeal is in allocating research resources to attack the major offenders. While control principles and population dynamics are essentially the same for all pests, a few major ones account for greatest pesticide usage and improved methods for their control would provide the greatest returns.

Perhaps the most obvious fact is that the funds in support of research on nonchemical methods of control are inadequate for the intensive effort required to make reasonable progress.

Extension

Cooperative Extension is the educational partnership between the USDA, the land-grant university, and the local community (the county). The system provides for program and financial resources from the USDA and the university, with programs being conducted at the county level. In most states, local financial support and program direction is also provided.

The middlemen in this successful educational enterprise between the university and the farm have been the county agents, including those working in home economics and youth development–4-H. They have cultivated the capacity to be on the wavelength of the research scientists on the one hand and the producer on the other. It was the agent who was able to provide educational techniques for gaining almost immediate acceptance of recommendations developed from research findings. The agent had a key role in the implementation of chemical control. He witnessed its excesses and he is in a unique position to provide educational leadership in the readjustment required to place chemical control on a sounder base. There are approximately 10,000 cooperative extension agents in the United States, serving in agriculture, youth development–4-H, and home economics. This provides an impressive nucleus through which to work in educational programs relating to pest control, environmental quality, and the quality of life.

The key role of this group and the need for refresher training was emphasized in the Report of the Environmental Pollution Panel, President's Science Advisory Committee (1965, p. 286–287).

> No matter how effective techniques of pest control may be, if they are to be adopted and used successfully by growers, it will be necessary to educate extension specialists and county agents in their use. Effective use of integrated control will require the reorientation of almost all extension personnel to enable them to change from a relatively simple concept of a single factor approach to the more sophisticated approach based on principles of applied ecology. This change is necessary if the producer is to be furnished with detailed recommendations and consultation on pest population management. Periodic refresher courses on new developments in pest control should be made available to extension personnel and professional consultants.

The 4-H youth program, with its emphasis on learning by doing, has a tremendous educational potential. However, emphasis has tended to be on "dead" entomology—insect collections and preserved specimens that lend themselves to exhibition and competition—rather than live projects. This is unfortunate when insects offer such convenient and fascinating animals for study. For a child to observe that bean plants are cut down by larvae arising from insignificant-appearing armyworm eggs teaches concepts of energy dynamics in a convincing way. Simple ecosystems can be established to create interest in the principles of ecology and pest control.

When we note the tremendous expenditures in such areas as research or area control programs, it appears that there is an imbalance between them and efforts directed to implementation. We badly need support for pilot programs that, if successful, could be implemented widely through the extension system. Going further, it would seem that there is some appropriate ratio between dollars for new knowledge and dollars for delivery.

Similarly, when undertaking comprehensive research programs, the extension component should be built in rather than brought in after the fact. It is encouraging to note that in a recent proposal on environmental quality submitted to NSF for consideration under the new program of IRRPOS (Interdisciplinary Research Relevant to Problems of Our Society) the suggestion was made that a public education component be built in from the start.

Extension Pesticide-Use Specialists

The USDA recognized the urgent need for additional research and extension on pesticides and obtained funding in the 1965 appropriations passed by the Congress. This provided $2,100,000 in payments to states for expansion in educational programs on the proper use of pesticides. This program was undertaken when the primary concern was for human safety in the use of pesticides. The specialists recruited to direct these programs were well versed in these aspects of the problem. They have concentrated on the safe use of pesticides in the context of human safety and have achieved a high degree of success, but, as concern has broadened to include environmental considerations, there has been relatively little change in emphasis in the program. In addition, emphasis has been more on practices ("Read the label," "Store pesticides wisely") than on underlying concepts of pest control. It is also evident that the program has tended to be operated as a separate program without involvement in the research program.

The effectiveness of this cadre of specialists could be greatly enhanced by intensive training in the broader environmental aspects of pesticide usage. This cannot be done by the occasional short-term workshop–swapshop-type training that has been followed. When everyone has the same kind of marbles, there isn't much point in swapping. An intensified course given by a high-ranking university and offered on a regional basis involving field trips to successful programs and interchange with related disciplines would be of value.

To assess the state of these programs, one needs only to examine their publications. Our chief activity has been in eliminating publications that include recommendations no longer approved. We have not yet provided replacements reflecting the ecological considerations in the use of pesticides. This deficiency makes us vulnerable to the critics of chemical control and gives us little to go on in our newer educational programs.

Growers have had to be very sensitive to biological principles; their production units—plants and animals—are regulated by them. Despite the growers' ability to assimilate this information, we have given them relatively little. To gain their compliance through a logical rationale is far better than compliance based solely on following regulations. Growers who are smart enough to survive in modern agriculture can readily grasp the practical significance of DDT's mobility, solubility, and toxicity, but our programs have not concentrated on this kind of education. Such knowledge would enable the grower to understand the rationale for his own practices and to understand the views of some groups who have concerns and seem to be in opposition to him. This would tend to reduce the polarization over the pesticide issue.

Teaching

Because teaching has such profound influence on the future implementation of pest control strategies, we will briefly consider teaching as a part of the complex with which we are dealing, recognizing that comprehensive treatment of the topic must be left to others.

Specialization is a requirement of a technological society, but it is being more fully realized that unless specialization is superimposed upon an educational base that provides sensitivity to human need, we may be producing a technology at odds with the well-being of individuals within the society. We, therefore, need to establish the goal of education. I like the statement offered by Perkins (1966): ". . .the threefold purpose of liberal education is to learn to know nature, society and ourselves; to acquire certain skills, such as clear expression and a grasp of the scientific method and discipline; and finally, to embrace certain values, such as intellectual honesty, tolerance and the capacity for wise judgments."

We in the biological sciences should assume some responsibility for the education of the whole man. The implications of pest control for the well-being of mankind are too important to place this specialized knowledge in the hands of leaders who lack the "capacity for wise judgment." To sense the complexity of the issues, one has only to reflect on the use of DDT to control malaria and thereby save human lives as opposed to withholding its use because of environmental considerations. The role of the biologist has assumed new importance as his knowledge has enabled him to manipulate growth processes and control the living organisms in his environment. To provide the knowledge for biological manipulation without accompanying knowledge in the area of moral responsibility renders the biologist unprepared for the new role that will surely be thrust upon him.

Assuming that we have made a case for liberal education, let us turn to specialization. The major issue here is whether to provide an undergraduate major for the student who plans graduate study. I shall offer comment on the field of my specialization, entomology (presumably, the principles apply to other biological disciplines). I would counsel against the undergraduate major. To acquire an appropriate foundation for graduate study requires so many hours in the physical and social sciences that there is little time left for specialization in entomology. Undergraduate training should provide the base upon which specialization is built at the graduate level.

The chief justification for the undergraduate major is to provide the student an "academic home." The student needs identity and protection from the impersonalization inherent in the "multiversity." A good system of counseling can help achieve this. In providing him a home, we must not do so at the expense of broad exposure to the intellectual climate of the university. We are seeking the best of two worlds, and this is not easy to achieve.

For the terminal B.S., the course is not so clear. He needs technical knowledge to qualify as a practitioner, but, even so, his applied training should be superimposed on a framework of biological principles. Considering that his technical knowledge will obsolesce more rapidly than knowledge, we should opt for the fundamental and hope that he recognizes the necessity of "self-renewal" for professional survival in the midst of a knowledge explosion.

All of our speculation on teaching of entomology needs to be viewed in the context of the recent past in which our mistakes were more in the area of general biological concepts than in inadequate technology.

RESPONSIBILITY OF THE SCIENTISTS

In the final analysis, implementation means making science and technology serve society. This work role will create many conflicts of interest. What is the responsibility of the scientist in these matters as they relate to pest control?

Pest control is serious business because it plays an essential role in providing the food and fiber necessary for life, because its practices affect the viability of the environment, and because it divides people in their views and thereby precludes orderly efforts to provide solutions to problems.

The pest control scientist is presumably a responsible citizen. The society he serves reveres and subsidizes education and grants its scholars sanctuary in the form of tenure. Does his possession of specialized knowledge carry with it a responsibility to speak out on matters of public interest? I believe that it does. The challenge of this new decade is for the scientist to become more engaged in matters that affect mankind. His increased activity in this role will carry new calculated risks, but there is no alternative. If those who possess the specialized knowledge do not make it available, to whom do they default? It is obvious, too, that the scientist must accept the risk of projecting opinions even though all the factors in the equation are not known. Lead time will have been dissipated by the time all the factors are known. Having offered his professional opinion as it relates to alternatives, the scientist must then be willing to let the decision-making process of a democratic society run its course.

In addition to individual responsibility, there is the question of collective responsibility. One of the frustrating features of the insecticide controversy has been the inability to determine the position of organized entomologists. While they were debating whether they had one, some important decisions were made without benefit of their influence. In September 1970, a policy statement on insect control was provided by the Entomological Society of America. This statement pledged the society to encourage

1. Continued research and its application leading to safe, effective and ecologically sound methods and programs of insect pest control;
2. Cooperation with all agencies concerned with the various aspects of insect pest control, including environmental quality;
3. Improved means of educating both the directly-involved and general public in appropriate and safe use of insect control measures.

We were a long time in taking up a position on this matter and we provided a statement with which few could quarrel. We are not likely to find many opposed to continued research on safer methods, cooperation with all agencies, improved methods of educating.

With respect to the scientist's responsibility and the control of technology, the editorial in *Science* by Solandt (1969) is of interest. He says, in part:

Society must so organize itself that a proportion of the very ablest and most imaginative of scientists are continually concerned with trying to foresee the long-term effects of new technology. Our present method of depending on the alertness of individuals to foresee danger and to form pressure groups that try to correct mistakes will not do for the future. A rational institutional framework that will assign a formal responsibility for this critical task to a well-selected, well-organized, and well-financed group of scientists is urgently needed. Clearly this agency must also have strong representation from the social sciences, including law, and close links with political leaders and with "the man in the street." Its task is too important to be left to scientists alone, but scientists must supply the leadership.

In our professional deliberations, we must better identify the role our leaders should play and we must engage our ablest leadership in it. It is clear that vacillation in these matters has muted our voice in these important issues.

SUMMARY

There is continued pressure for more effective methods of pest control to provide for expanding populations and to do so by methods more compatible with environmental quality. Improvement is possible in two ways: by employing new strategies, and by making old strategies more effective. Public support is a prerequisite in either event, and the present public mood is unfavorable. The public's support for scientific research is on the wane, and chemical pest control has encountered a credibility gap.

The novel strategies that have been viewed with such high expectations have not yet moved to general use, and adjustments in the time scale of expectations are required. Such an assessment indicates that chemical control is still the bulwark of pest control, but chemical control must regain respectability in the public eye and the pesticide industry must invest in development of new materials. The challenge is to make chemical control respectable, to restore vitality to the industry, and to promote research on methods that in time will replace chemical control. It is likely too that public agencies will be required to broaden their role in conducting research and development that provides little economic incentive to industry.

Renewed efforts are needed to maximize the positive features of chemical control. New regulations controlling their sale and use, together with intensified programs based on economic thresholds and population levels, can provide impressive gains. In all of these efforts, intensified educational programs will be required.

In the larger context, pest control is viewed as part of the growing conflict between science and technology and the quality of life. These problems call for new roles of leadership within the scientific community and modifications in our educational programs. These will provide a deeper appreciation of the social, economic, and biological factors that must be reconciled to produce a viable environment. Pest control in the 1970's must deal with all of the age-old biological complexities inherent in the host-pest relationship. In addition, it must reconcile matters of environmental quality and the sensitivities of the individual, including his right to a voice in policies affecting him.

REFERENCES

Anonymous. 1971. Fire ant control under fire. Science 171(3976):1131.

Coon, D. W., and R. R. Fleet. 1970. The ant war. Environment 12(10):28–38.

Entomological Society of America. 1970. Entomological Society of America. Bull. Ent. Soc. Am. 16(3):170.

Environmental education: The adult public. Report of a workshop conference sponsored by the American Institute of Biological Sciences. Washington, D.C., April 1970.

Hahn, A. J. 1970. Citizens in local politics: Nonparticipation and "unrepresentation." J. Community Develop. 1(2):63–74.

Handler, P. 1971. The federal government and the scientific community. Science 171(3967):144:151.

McElroy, W. D. 1971. Science, the public, and the new realities. BioScience 21(2):59–61, 70.

Morison, R. S. 1969. Science and social attitudes. Science 165(3889):150–156.

National Academy of Sciences. 1966. Scientific aspects of pest control. Publ. 1402. Washington, D.C.

National Academy of Sciences. 1968a. Plant disease development and control. Publ. 1596. Washington, D.C.

National Academy of Sciences. 1968b. Weed control. Publ. 1597. Washington, D.C.

National Academy of Sciences. 1968c. Control of plant–parasitic nematodes. Publ. 1696. Washington, D.C.

National Academy of Sciences. 1969. Insect-pest management and control. Publ. 1695. Washington, D.C.

National Academy of Sciences. 1970. Vertebrate pests: Problems and control. Washington, D.C.

Perkins, J. A. 1966. The university in transition. Princeton University Press, Princeton, N.J.

President's Science Advisory Committee. 1965. Restoring the quality of our environment. Report of the Environmental Pollution Panel. The White House, Washington, D.C.

Solandt, O. M. 1969. The control of technology. Science 165(3892):445.

U.S. Department of Health, Education and Welfare. 1969. Report of the Secretary's Commission on Pesticides and their relationship to environmental health. Parts I and II. Washington, D.C.

Van den Bosch, R. 1970. Pesticides: Prescribing for the ecosystem. Environment 12(3):21–25.

The Role of the Food Industry in Solving Pest Control Problems

H. L. Wilcke

One of the main objectives of the food processing industry is to ensure that insects and pests are kept out of the food supply that is presented for consumption. The law requires that this objective be met. However, this objective departs from the philosophy expressed by some entomologists and ecologists who have expressed the necessity for coexistence with pest populations or for regulating them at defined levels; it places more emphasis on control programs.

With regard to the interest, the role, and the responsibility of the food processing industry in pest control problems, the following assumptions are made:

- That everyone wants a plentiful, clean, and wholesome food supply at a reasonable cost;
- That as long as man is not the only species inhabiting the earth, there will be competition for his food supply;
- That insects and other pests will continue to multiply rapidly under favorable conditions;
- That these pests must be kept out of our food supply, at least as well as we are doing at present;
- That the food industry will produce, process, store, package, and deliver the type of food the consumer wants in good, wholesome condition;
- That the food industry, by the very nature of its business, is deeply involved in pest control problems.

The industry is concerned with food, starting with the selection or development of proper strains or varieties, through its growing harvesting, storage, processing, packaging, and ending with its delivery to the ultimate consumer. Since food products are most susceptible to damage from pests during the production, storing, and handling of ingredients and raw materials (the processing and the packaging phases), these areas will be considered in more detail.

The initial choice of varieties of food crops to be grown is determined by the preference of the ultimate consumer for certain characteristics in specific foods. The various types of peaches, tomatoes, sweet corn, potatoes, and other foodstuffs sold on the retail market illustrates this. When these characteristics have been determined, the grower is faced with a choice of the specific varieties, within these constraints, which will yield the greatest return in marketable product and, of course, return on investment to the grower himself. Frequently, however, no single variety meets all of these criteria satisfactorily. Varieties that are high yielders are very often susceptible to diseases and parasites. Although some attention has been given to the development of disease-resistant varieties, very little has been given to pest-resistant varieties because the chances of success of such breeding programs are less promising. In addition, one of the principal drawbacks to the development of resistance through genetic improvement is that it is an expensive and a long, drawn-out program. When insects or diseases strike, there is need for immediate control, particularly in cases of severe infestation. This has been the case in attacks by grasshoppers and in the recent corn blight. In addition, the pathogen itself (or the insect pest) may develop certain changes in susceptibilities, or it may become adapted to other environments before the genetic program can be completed. Therefore, genetic programs must be directed toward those problems that occur frequently and persistently and that are predictable. As a usual practice, it is cheaper and quicker to protect a crop by chemical means than it is to take the longer route of breeding for resistance. But, as more and more restrictions are placed on the use of chemical treatments (as seems to be the trend), it is anticipated that the genetic route will be used more often.

One of the devices that has been used to improve quality in the production of crops may also be beneficial in controlling pests. This is the breeding of crops for uniform maturity, so they may be harvested all at one time. This program has been used with tomatoes and other vegetables and with fruits. This practice permits cleaning up the growing area so that the food supply of the pests may be removed and the cycle broken. Crop rotations are frequently used to break the cycle of disease and insect pests. This is particularly effective in controlling nematodes, which affect specific crops, and the same would apply to the corn rootworm. In a variation of this practice, it may be entirely possible to produce seedlings to a more-resistant stage in one area which would be free of diseases that affect the plants in another area and where the organisms or pests may be resident in the soil.

The time of planting may be quite effective in controlling certain types of diseases or pests. The plants in this case may attain sufficient size or maturity to withstand the attacks of the pests before weather conditions permit the development of the pests to the stage at which they may be particularly harmful. This practice may be useful in fighting the cotton boll weevil, if we may classify cotton as a food crop now. If a shorter growing-season variety of cotton may be developed, it may be grown farther north where the insects would be subject to winterkill, or the cotton may mature before the weevils develop.

These cultural practices produce uncertain effects on the total insect pest population affecting our food crops. Therefore, we must develop other methods of controlling these pests. At the present time, the chemical pesticide and herbicide products are by far the most practical and economical and they can be very effective in controlling insect populations and fungi. The problem, then, is the proper use of these products.

The subject of pest control is complex. Madsen and McMullen (1971) accurately described the situation as follows: "It is extremely easy to become lost in a morass of semantics when dealing with the subject of integrated control, pest management, or whatever you will call it, because it involves the sciences of ecology, population dynamics, biological control, horticulture, and applied toxicology. Some of these have become notorious for their confusion of terminologies."

They have also stated some of the broad fundamentals of effective pest control programs as follows:

1. Detailed knowledge of the biologies of the pests and beneficial arthropods and their relationships in the environment.

2. Methods of reliably estimating the population levels of the organisms concerned.

3. Knowledge of economic threshold population levels of the pests.

4. The ability to predict increases and declines of pest populations as influenced by beneficial organisms and abiotic factors such as weather.

5. Knowledge of the influences on the pests and beneficial organisms of all the pesticides that might be used on the particular crop.

6. In certain cases, cultural practices such as irrigation, cover crop management, harvesting procedures, etc., may be of concern.

Not all food processors can be entomoligists or bacteriologists–they must depend on public agencies such as extension services and the U.S. Department of Agriculture and the Food and Drug Administration for information on biologies of pests and of other predatory insects or animals. In many cases, this type of information is also supplied by reliable chemical manufacturers and by their trade associations. The production supervisor must learn which products are effective and which will control his particular problem. The specific pest or disease situation must be identified positively. First of all, good management practices must be used to prevent depredations by pests insofar as possible. Second, production practices must be used that will minimize losses due to pests. Third, we must ensure that pesticides and other methods of controlling pests are used properly. Some of these uses and the need for them have been described by Pratt (1970).

When populations of pests become large enough to be visible, it is often too late to prevent serious damage to the crop. Therefore, some means of early detection must be used. For this purpose, the populations may be monitored by means of sticky traps, black light traps, and CO_2 traps, depending on the insect species involved. This type of information, of course, is most useful in combating insects that appear in large numbers sporadically, rather than against those that are a constant menace or whose behavior is predictable. These data may be used to determine the type of pesticide to be used, the time of application required, and the duration of the treatment that will be required to control the pests. This would be considered proper use of a pesticide compared with the indiscriminate usage that has prevailed at times in the past.

From the food processor's point of view, the control program starts with the construction of the plant. One of the first considerations is to use construction materials capable of resisting the attacks of large rodents, because these rodents will not only carry small insects in with them, but will certainly make openings in the building that will readily admit the smaller insect pests. Specifications for this type of control are usually available through health departments. Careful consideration must be given to the capabilities of the invader, and all openings such as entries for conduits, for electrical wiring, pipes, etc., must be closed very carefully. When the building is planned, it must be carefully engineered to provide good sanitary facilities. It must provide for simple and effective maintenance for housekeeping and sanitation. Food plants must be constructed with good ventilation and lighting. They must be planned for the proper installation of equipment, so that cleaning may be done with a minimum of effort. Self-cleaning equipment or equipment that can be readily disassembled for complete cleaning should be selected. Buildings should have a minimum of corners and sharp angles; this permits more thorough cleaning. In short, the facility should be so constructed as to provide the minimum space for accumulation of dust, food particles, or other materials that might harbor pests.

Cleanliness is one of the most effective means of controlling insect pests in the food processing plant. This applies to materials brought into the plant as well as the housekeeping practices in the plant itself. Probably the most common method for the introduction of pests is through the ingredients that are brought in, that is, the raw materials that are to be processed in the plant. The carriers used to bring the materials to the plant are often the source of introduction. The operators of carriers such as trucks, railroad cars, intermediate warehouse facilities, etc., should recognize the necessity for completely clean carriers. Some of the older-type railroad boxcars and many of the older-type truck bodies are impossible to clean. Considering the pests that may be on the food products as grown, those that may attack the food during harvesting and loading on vehicles of transportation to the processing plant, and those that may attack en route, the incoming products are often a source of introduction of pests to the food plant. This may result in single insect infestation or there may be considerable numbers. Thus, the microscope has become a very important tool in detection of minor contamination.

Processes have been developed to remove these foreign materials from food products as they come into the plant. As stated by Kurtz and Harris (1970),

> Machinery of many kinds has been developed to remove such extraneous materials—sifters, separators, washers, air blowers, etc. Different agricultural crops demand different cleaning techniques because of differences in their

physical makeup—size, shape, texture, weight, etc. Almost always, a combination of cleaning processes must be used, each designed to solve a different problem. Considering the situation—including the fact that they are using air and water systems which are themselves imperfect—the machines are extremely effective.

After the foods have been processed there is still the matter of packaging to protect the product from contamination after it has been packaged. The U.S. Department of Agriculture, Agricultural Stabilization and Conservation Service, Transportation and Warehousing Division (1970, 1971), has issued specifications on containers for dairy products which provide adequate protection from insect pests. They have also specified certain treatments for the CSM products (a food made from corn, soybean flour, and milk, with vitamin and mineral supplements) that are to be purchased for foreign donation programs. In addition, certain methods of sealing packages have been prescribed, including the type of materials which may be used on closures to prevent entrance of insect pests.

Good housekeeping around the plant must include elimination of breeding places for insects, both externally and internally. This means that adequate waste and trash disposal programs must be followed. Proper and adequate cleanup procedures must be carried out to eliminate pest attractants.

The preventive programs that have been described have been based largely on sanitation, good housekeeping, and measures to prevent access of pests into the plant. However, no matter how carefully these practices are carried out, there will be cases of insect pest infestation in food plants. Therefore, pesticides become necessary, not as a substitute for good management, but as an adjunct to it. The food industry does not pretend to be the expert on pesticides. This information must come from government agencies (state and federal) and from the chemical industry, which develops compounds that will be effective and safe, with minimum effect on the environment. The food industry must use pesticides properly and safely, and they must support the program of educating the public on the proper use of pesticides.

The chemical phases of the control program, as far as the food industry is concerned, must be thoroughly understood and properly implemented. All food plants are subject to inspection by either the U.S. Department of Agriculture or by the Food and Drug Administration of HEW. If insect pests or rodents are found in the plant and product, the plant may be closed by these federal authorities or by state authorities. If pesticide residues, insects, or rodent excreta are found in significant amounts in the food products, they may be seized and removed from the market by the processor or by the Food and Drug authorities. Therefore, pest control programs must be used properly by the food industry. There is a very narrow line between failure to control insects and overuse of the pesticide.

The role of the food industry in solving pest control problems may be summarized as follows:

1. The food industry must emphasize necessity for protecting our food supply from fungi, rodents, insects, and other pests.
2. The industry must identify and define the problems in producing a clean and wholesome food supply.
3. Continual emphasis must be placed on proper production practices, sanitation, and general good management practices to prevent infestation of the food supply by pests.
4. The industry must keep informed on the best methods of pest control. There is more latitude in types of programs in production than there is in the food processing plant.
5. When chemical methods of control are indicated, reliable information must be obtained on appropriate products to be used, methods of use, concentrations, and timing.
6. The food industry must insist that proper regulatory agencies establish standards for pest control and extermination companies regarding uses and safety of pesticides and the ability to carry out effective programs.
7. The food industry must cooperate with the chemical pesticide and other pest control industries in developing the proper parameters for the evaluation of pest control programs.
8. The food industry must use all proper and practical means to ensure that chemicals are used properly, and it must defend the right to use all properly approved pest control measures.
9. The food industry must base all of its practices, including pest control programs, on sound, valid information. Its responsibility is to provide food in good wholesome condition, in sufficient quantity to meet the needs of the population, at minimum cost.

REFERENCES

Kurtz, O'Dean L., and Henton L. Harris. 1970. A search for food purity, food product development. Arlington Publishing Co., Chicago.

Madsen, Harold F., and R. David McMullen. 1971. Pest management—It's here to stay. Farm Chem. 134:42–43.

Pratt, J. J., Jr. 1970. Protecting our foods from environmental intrusion. The problem of pesticide and animal matter as non-intentional additives. Food Tech. 24(999):59–64.

U.S. Department of Agriculture. A.S.C.S., Announcement CSM-2, July 31, 1970.

U.S. Department of Agriculture. A.S.C.S., Announcement DP-1, Jan. 15, 1971.

II

Systems Approach to Pest Control

Economics of Pest Control

Gerald A. Carlson and Emery N. Castle

Economists have long recognized that an autonomous market system may not always operate in a fashion consistent with the criteria of a social optimum. One of the classic textbook conditions leading to market failure has been pollution or degradation of a common environment. Until recently this was considered an exception to the general case, and in-depth analysis and empirical treatment has been lacking.* Although recent efforts have added significantly to our knowledge, a comprehensive literature that can be readily applied to current social problems does not exist.

This paucity of economic literature on various facets of environmental quality can be traced to three primary factors:

1. Environmental quality has not been considered to be an important social problem until recently. Only economists who had an interest and available resources were able to make contributions. While some people within a discipline or profession will anticipate social needs, it is apparently unrealistic to expect a massive disciplinary ef-

*Pigou's (1932) classic work on welfare economics is generally regarded as the first treatment in depth of "externalities." Pollution is an example. Much later Bator wrote two classic articles that develop a rigorous framework for judging the performance of an economic system (Bator, 1957, 1958). However, until 1960 there was relatively little empirical work done on the subject. After that a pioneering effort by Kneese (1962) stimulated considerable work in water quality. A rather substantial research program in the economics of water quality was launched at Oregon State University at about the same time (Castle, 1965, 1966).

Paper No. 3456 of the *Journal* series of the North Carolina Agricultural Experimental Station.

fort in the absence of a statement of social need and social support.

2. Economic activity is generally considered to be in opposition to environmental quality. It is not generally recognized that one of the products it is legitimate to expect from an economic system is the level of environmental quality desired by the participants in that system. As a consequence it has not occurred to many people that the economist may have a contribution to make to the problem solution.

3. The most highly developed analytical techniques of the economist lend themselves to treating partial rather than total solutions. While many environmental quality issues can be analyzed by the use of these techniques, the environmental problems often exist because only partial effects have been taken into consideration by decision-makers.

Increased effort is now being made to analyze the production of public goods and environmental quality (U.S. Congress, 1969). The economics of crime, health, insurance, and pollutants generally is receiving more attention. The economics of pest control is a part of this increased total effort. Examples of this increased effort include

1. Production economists analyzing the substitution of other inputs for agricultural chemicals.

2. Welfare economists attempting to evaluate the performance of market and nonmarket institutions with particular reference to problems of environmental quality.

3. Mathematical models, often formulated by a multidisciplinary team, being developed for the purpose of better understanding interrelationships and interdependencies.

In this paper we attempt to present (1) some indicators of benefits and hazards of pest control, (2) a discussion of depletion of and investment in pest control research, (3) economic alternatives in pest control, and (4) a social strategy for pest control.

INDICATORS OF BENEFITS AND HAZARDS OF PEST CONTROL

Estimating Benefits

Evidence of the benefits of pest control come from (1) people's willingness to pay for the controls and control research, (2) increased crop yields, or (3) value of resources released for use elsewhere in the economy (U.S. Congress, Senate, 1969; Robb and Guthrie, 1970).

In 1969 about $851 million were expended for 929 million lb of pesticides for foreign and domestic use compared with $849 million for 960 million lb in 1968 (U.S. Tariff Commission, 1970). Exports were valued at $241 million for 1968 (U.S. Department of Agriculture, 1953–1969). Domestic value of pesticide sales have been increasing rapidly until the last few years. Annual rates of increase for the 1952–68 period have been about 14 percent. However, the pounds of material purchased in the United States have not been increasing as rapidly as the often cited production figures due to the shift from lower to higher priced compounds and the rapid increases in exports. Figure 1 shows the slow rate of growth in pounds of pesticides purchases in the United States. The exception is herbicide use for which a large part of the recent increase is overseas military use [2, 4, D production fell by 90 percent and 2, 4, 5–T output dropped by more than 99 percent between 1968 and 1969 (U.S. Tariff Commission, 1970)]. The quantity of pesticides purchased rather than the value of pesticides is a better figure for examining pesticide hazards. It is doubtful the domestic use of insecticides will double, and the use of herbicides will more than double by 1975 as some have predicted (U.S. Department of Health, Education and Welfare, 1969).

It is difficult to measure the demand for pest controls other than pesticides. Entomologists and plant pathologists at U.S. colleges of agriculture represented about 1,615 man-years of investment annually (U.S. Department of Agriculture, 1967). In analyzing records

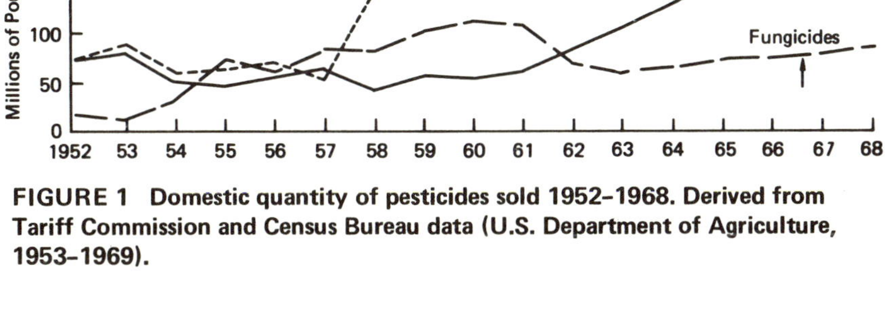

FIGURE 1 Domestic quantity of pesticides sold 1952–1968. Derived from Tariff Commission and Census Bureau data (U.S. Department of Agriculture, 1953–1969).

of these investments, Peterson found that there was a close, positive relationship between nonfarm (rather than farm) income of a state and its expenditure for entomology and plant pathology research (Peterson, 1969). He projected that the number of entomologists and plant pathologists would increase to 1,906 by 1974. Nonfarm state legislators may recognize that lower crop losses due to disease and insects will mean lower food costs; the explanation may also be in state fiscal and political considerations.

Part of the difficulty in evaluating the benefits of chemical pest controls is the diverse nature of the uses. Most people associate pest control with agriculture, yet the most recent farm survey data show that only the following proportion of domestic pesticides were applied in agriculture in 1966 (Eichers *et al.*, 1970):

Fungicides	Herbicides	Insecticides	All Pesticides
27	55	57	51

However, much of the most toxic and persistent pesticides are used in agriculture.

The increase in farm productivity is the indicator that is usually cited as a benefit of pest management. Plant scientists are familiar with the rather substantial percentage increases in yields from pesticide treated crops report in field trials (U.S. Congress, Senate, 1955; American Phytopathological Society, 1957–1969). The major shortcoming of these trials comes in attempting to extend the results to other locations where pest populations differ from those in the trials. Large acreages of crops go untreated, though not all untreated acres will have yield losses less than control costs. The following proportions of these U.S. crop acreages were not treated with these pesticides in 1966 (Fox *et al.*, 1968):

	Fungicides	Herbicides	Insecticides
Corn	99	53	77
Wheat	99	72	98
Soybeans	99	73	96
Cotton	98	48	46

An economic indicator of the productivity of pesticides is the value of the crop that could be produced from an additional dollar spent on pesticides. Headley, using 1963 data, estimated that U.S. agriculture could produce an average of $4 of additional product for each pesticide dollar expended (Headley, 1968). The figure given for England is $5 (Strickland, 1970). One of the authors has estimated a figure of $2.25 for U.S. cotton farms, but only $.95 for cotton farms larger than 100 acres (Carlson, 1970b). Each of these figures, except the last one, indicates that additional pesticide expenditures would be profitable for farmers. The limitation of these economic studies is that they do not take into account the large annual variation in pest infestations. Clearly, an analysis is needed that combines the spatial scope of the economic studies with the detail on pest densities from the pesticide field studies.

Resources released from food production because of pest control is another measure of economic benefit.* The mechanization revolution of the 1930's and 1940's has been augmented since 1945 by a biological revolution in terms of fertilizer, pesticides, and genetic stock. Table 1 indicates the relative shift in farm inputs in the 1952–1968 period. Pesticides and fertilizer are displacing some of the labor and land in our agriculture production process. Part of the value of the manpower that has shifted to other pursuits could be attributed to more efficient pest control. Land has been replaced by farm inputs that raise yields per acre. Expensive acreage reserves or diversion payments are the instruments that have removed about 40 million acres from crop production, but it is pesticides and fertilizers that have made these reserves possible. Hunters and conservationists benefit from the increased use of these reserves for recreational purposes.

There are costs associated with the resource adjustments of labor to cities and land to reserves. Some laborers have moved to cities without jobs. This migration has contributed to the population de-

TABLE 1 Changes in Index of Selected Agricultural Inputs: 1952–1968[a]

	1952	1968
Labor	100	53
Real estate	100	109
Machinery	100	130
Fertilizer and lime	100	292
Domestic sales of pesticides	100	269
All farm inputs	100	110

[a]U.S. Department of Agriculture (1953 and 1969; 1953–1969).

*Note that the value of resources formerly used in pest control is an *alternative* measure to yield increases and cost reductions and we must be careful not to double-count pest control benefits.

cline and demise of many rural communities. This movement of resources from pest control to other economic activity has been sizable and there is some evidence that it represents a benefit to society. However, this is a difficult and, as of now, unanswered question.

Costs or Hazard Indicators

The hazards of pesticides to man, wildlife, and the biosphere have been enumerated in recent congressional hearings, professional meetings, and in the news media (U.S. Congress, Senate, 1969; Rabb and Guthrie, 1970). Costs include resources expended to discover, screen, produce, and distribute the control measures. There are also substantial monitoring costs in determining when control is needed and which control to use.

Our market system is one of ownership and exchange. However, our system of property and contractual rights is not sophisticated enough to capture all exchanges of rewards and detriments. Economists have used the term *externalities* to classify such nonmarket effects. [For an introduction to the forms externalities take in the natural resource area, see the paper by Castle (1965).] In the case of agricultural pesticides we can recognize several forms of uncompensated costs and returns.

Case A: One farmer uses an insecticide to control insects on his crop and some of the toxic material drifts over his boundaries and destroys honeybees or forage of an adjacent farmer. One farmer's production input has retarded the productivity of another farmer. The same physical phenomenon (insecticidal drift) may be a positive externality when it kills the pest in the neighbor's fields or reduces the spread of a contagious disease or mobile pest. Alert pest control farmers of an area may actually take upon themselves the responsibility and costs of controlling mobile pests in neighbor's fields to increase their own yields. When the source and recipient of such production externalities are easily identifiable, then often damages are collected either through court litigation, local agreements, or federal indemnities. In other cases, government agencies provide for mandatory pest control.

Case B: When crops are produced with pesticides, minute quantities of the chemicals may find their way to consumers' bodies. Food consumers are bearing a "cost" in the purchase of certain foods if these minute quantities ever prove to be harmful. This form of externality involves many people so that local agreements or law suits are costly to enact. Federally administered pesticide tolerances for food raise the economic question of the benefits and risks implied by the level of these tolerances. High monitoring costs and unknown food contamination–health relationships may well lead to a situation where incremental social costs (damage from more food contamination) may not be equal to the incremental reduction in the value of production when food is confiscated and discarded (Sosnick and Terlecky, 1969; U.S. Congress, Senate, 1965).

Case C: Another widely recognized externality is the reduction of welfare of farm laborers, pesticide applicators, and other people who may become exposed to pesticides near the production site. (Clearly some of these hazards are reflected in higher wages, but to the extent they are uncompensated they represent externalities.) The magnitude of pesticide occupational hazards is related to the number of hours of exposure and the toxicity properties of the chemicals. The direct costs of prevention, detection, treatment, and rehabilitation, and the indirect costs of loss of productive output via death and disability are real economic factors to be weighed against pesticide benefits. The procedures for analyzing such costs have been applied to several types of illness and air pollution health damage and they await application in the area of pesticide poisoning (Lave and Seskin, 1970; Rice, 1968).

Case D: The reduction in numbers of wildlife, fish, birds and domestic animals; the destruction of vegetation; and the contamination of surface and underground water and soils with farm pesticides are costs that usually are not reflected in our market system. Isolated attempts to evaluate these costs have indicated that they might be low (Edwards, 1969; U.S. Congress, Senate, 1969). However, available data from the National Monitoring Program do not provide adequate information about the environment for these costs to be assessed in general (U.S. Department of Agriculture, 1969; U.S. Department of Health, Education and Welfare, 1969).

Economists have usually treated externalities as minor malfunctions in our market measure of costs and returns. In the case of pesticides we find that they occur in many forms. Before examining the mechanisms for intervening in the market to adjust for externalities, we need a more complete account of the long-range prospects for economical pest control.

AGRICULTURAL PESTICIDES AS BIOLOGICAL CAPITAL

Farmers have typically thought of pesticide expenditures as an annual operating cost similar to expenditures for seed, fertilizer, and labor. Yet, if a long-range view of world agriculture is taken, the effect of agricultural pesticides on agriculture's future productive capacity must be considered. It may be helpful to think of our current stock of pesticide material and use recommendations as *biological capital.* That is, an asset that both accumulates and depreciates when investments are made and services provided change. This capital is distinctive from the production possibilities of fertilizer, hybrids, and resistant crop varieties in that important forms of depletion exist.

Demand

Long-run depreciation of the biological capital occurs when resistant strains of pernicious insects, mites, weeds, or crop diseases develop. The use of the pesticides may temporarily rise as more applications per year are made in an attempt to maintain crop yields. Eventually the demand falls dramatically as growers shift to other control measures. Over two hundred species of insects are known to have developed resistance to one or more pesticides (Food and Agriculture Organization, 1969). R. F. Smith has recently reminded us that as a planning measure we must assume that any arthropod exposed to intensive pesticide pressure will develop resistance (Smith, 1970). The rate of resistance development and its geographical impact appear to be related to percent kill, species mobility, number of generations per year, and other ecological factors. There appears to be a great need to monitor and forecast resistance development as a vital component in determining pest control investments.

Less direct examples of biological capital depletion are entry of foreign pests, secondary pest outbreaks, and institutional restrictions on pesticide use. In each case the productivity of agriculture will fall unless there is an offsetting supply of alternative farm inputs.

Supply

There are several important economic factors influencing the supply of new pesticide capital. Chemical companies prefer to search for chemicals that can be used to control many pest species for many crops. Screening, testing, and production costs can be spread over greater sales volume. However, the nonmarket costs (externalities) of pesticides may increase as the spectrum broadens because of mammalian toxicity and persistence in nature. Because new materials must meet registration requirements, as well as appear to be profitable before investment will occur, few specific chemicals are forthcoming.

A simplified explanation of this investment void can be seen in Figure 2. Figure 2(a) and (b) show the relationships between the breadth of the spectrum of pests that a given pesticide will control and profitability and social costs, respectively. The precise shape of 2(a) will depend upon how fast costs will drop with volume of production and the price of the given chemical relative to other means

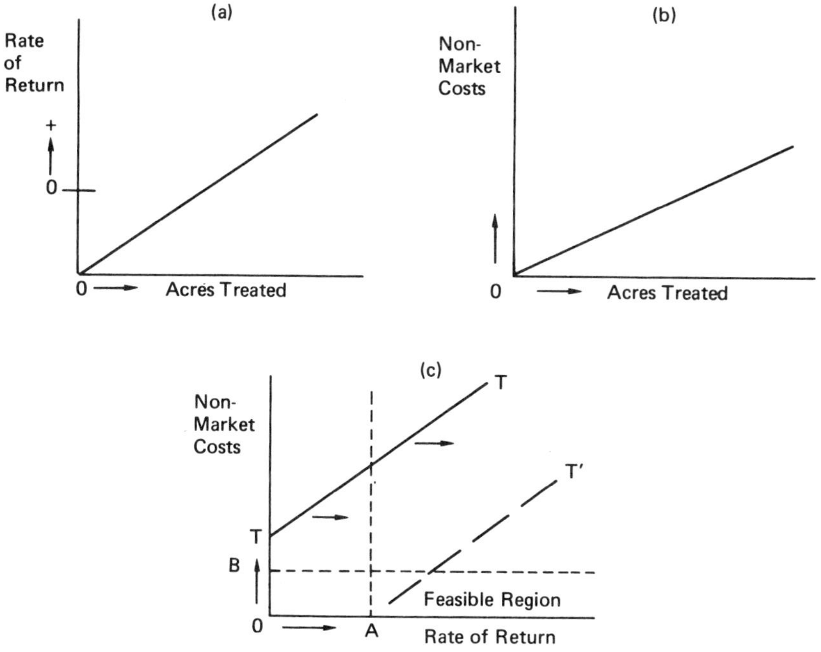

FIGURE 2 Pesticide investment void.

of pest control. The environmental damage potential of a pesticide depends upon other things besides acres treated, but this does not alter the discussion. Figure 2(c) reflects the tradeoff between high rates of return and high nonmarket costs for the existing technology TT. Since the chemical companies have alternative investment opportunities, they will only invest in pesticide search if the expected returns are above A. If the Environmental Protection Agency will only register chemicals whose potential social costs are less than B, then present technology will not produce new pesticides. The only market adjustment which will permit profitable pesticide investments is for TT to shift to the right so that it falls in the feasible region. Such shifts come about when production costs fall or when the price for a viable pest control becomes relatively higher (which also implies higher food and fiber costs).

The pest control demand and supply forces are dynamic and uncertain. The recognition of externalities in registration deliberations may have reduced private pesticide research. Resistance development is continuing and pest control discoveries will become more profitable. The existence of pest control externalities is an economic argument for more public investment in pest control.* The underlying issue is that if we desire the same level of pest control and a clean environment, then someone must pay for it (Headley, 1971).

ECONOMIC ALTERNATIVES IN PEST CONTROL

It is convenient to discuss the many suggested adjustments in pest control in terms of changes by producers, consumers, or spatial adjustments that influence income distributions.

Producer Adjustments

In limited cases, producers can select from among available biological, cultural, or resistant-crop pest control programs. For most of agriculture, use of pesticides remains the least (market) cost-control measure in spite of dramatic and inventive pest control efforts for a few species on a few crops (U.S. Department of Agriculture, 1969; U.S.

*To establish sufficient conditions for public investment, one must show that social returns are higher than private returns and that private returns are for various reasons not high enough relative to other private investments (Castle, 1965).

Department of Health, Education and Welfare, 1969). Those control measures that disrupt agroecosystems in a minor way, such as resistant varieties of crops, are relatively more attractive than before. Substitution of one control program for another must include consideration of all private and public resources expended.

Recently, there has been considerable discussion of substitution between pesticide types. Are substitutes available? How effective are they per dollar? An interesting example is the shift from the relatively persistent organochlorine insecticides to the more toxic organophosphorous compounds. This shift appears to be motivated by pest resistance to the organochlorines and removal of registration of the use of certain organochlorines on selected crops.

The farmer is concerned with the quantity of organophosphates he must substitute for each pound of organochlorines to maintain his crop output. The Pesticide Group of the Economic Research Service of USDA estimated substitution rates for the major organochlorine-using crops (Davis *et al.*, 1970). Table 2 shows that, on the average, about three fourths of a pound of organophosphorus or carbamate insecticides are needed to replace a pound of organochlorine insecticide. By assuming this would hold true and with known relative prices it was estimated that a 75 percent reduction in the use of organochlorines was possible on these crops (which account for over 75 percent of farm organochlorines) at an additional cost of $26.7 million or $2.23 per acre treated. Some economists and entomologists have suggested other rates of substitution (.3–.4 pounds of organophosphates per pound of organochlorine replaced for a 50 percent reduction range) and others have argued that yields may suffer from phytotoxicity or rapid resistance development to the organophosphates (Edwards, 1969; Smith, 1970; California Agricultural Experiment Station, 1969). The net social benefits of such shifts must include the differences in the nonmarket costs.

TABLE 2 Substitution Rates, Organochlorine Reductions, and Additional Production Costs for Organochlorine-Using Crops[a]

	Cotton	Corn	Peanuts	Tobacco	Total
Substitution rates	.68	1.17	.61	1.32	.76
Organochlorine reductions	87%	33%	100%	100%	76%
Additional production costs	$3.12	$1.23	$2.90	$4.22	$2.23

[a]Davis *et al.* (1970).

A similar analysis of the replacement of phenoxy herbicides on the eight major herbicide-using crops leads to much higher replacement costs (Fox *et al.*, 1970). Substitute herbicides, additional cultural practices, labor, and land were needed to maintain output. Net additional costs were estimated to be about $290 million, or $4.64 per treated acre. This would be about a threefold increase in present weed control costs.

There are other inputs which can be used to replace chemicals. As Table 1 indicates, farmers have been substituting pesticides for land and labor. Is it economically feasible to reverse this process? Headley has suggested releasing land from our acreage reserves to replace insecticides. He estimates that a 1 percent increase in cropland would decrease total insecticide use by about 6 percent (Headley, 1970). Considerable refinement is needed in such estimates, but they do help us think about the magnitude of resource adjustment involved in maintaining output with restrictions on pesticide inputs.

One of the authors has found that many precautionary fungicide applications could be avoided if regional crop loss forecasts are developed and used (Carlson, 1970a). For example, it was estimated that only 1 of 4 routine fungicide applications was needed and a 15 percent increase in grower profits would be possible in California peach production if a simple forecast was followed. This approach to pest control involves the substitution of information services and management for pesticides.

The extent of the use of pesticides as insurance against possible infestations appears large (Carlson, 1970b; University of California, 1965). What contribution to reduced pesticide use can crop insurance have? In the 1961–67 period, use of all-risk crop insurance rose 190 percent (U.S. Department of Agriculture, 1968). Sweden has experimented with crop insurance without precautionary pest controls, while the U.S. Federal Crop Insurance Corporation has a policy of requiring normal precautionary pesticide treatments before indemnities will be paid for pest damage. Investigations of the possibilities for replacing pesticides with various crop loss pooling arrangements might identify some acceptable arrangements for raising net social benefits.

Consumer Adjustments

Consumer tastes change over time and people adjust their diets when one commodity becomes more expensive relative to others. We would expect the consumer to pay less for pest-damaged food (or food with insect parts) and more for pesticide-free food to the extent that he can recognize these quality differences. Table 3 provides estimates of the average value of various products that were produced per pound of active pesticide material applied in 1966. For example, only about $19 worth of cotton is produced per pound of insecticide applied compared with about $500 for alfalfa. This "pesticide load" per dollar expenditure varies considerably between products. It gives a rough estimate of the impact of any attempt to impose the full cost of pesticides on the consumers of pesticide-produced products. This approach follows the work of W. Leontief (1970) who has shown how all types of pollution generation can be related to final consumer products by extending input–output analysis.

Recognizing the presticide load of a commodity reminds us of the possibility of altering our consumption patterns. Synthetic fibers are replacing cotton and may proceed more rapidly if pest control costs go up. A tax on insecticides (see p. 95, *Tax on Externalities*) would make soybean oil less expensive to consumers than peanut or cottonseed oil.

The Jensen Commission, along with others, has concluded that public demand for attractiveness in fruit and vegetables has invited excessive use of pesticides (U.S. Department of Agriculture, 1969). In view of this finding, a downward revision of federal grades and federal marketing order quality standards appears to be in line with efforts to have consumers share the cost of using fewer pesticides.

TABLE 3 Average Value of Various Commodities per Pound of Pesticides Applied, 1966[a]

Commodity	Insecticides	Herbicides	Fungicides
Cotton	$ 19.25	$ 688.55	$896.98
Apples	23.16	[b]	19.00
Peanuts	49.14	93.72	16.44
Potatoes	208.54	274.12	175.52
Corn	215.19	103.19	[d]
Alfalfa	502.84	1,397.19[c]	[d]
Soybeans	683.27	245.31	[d]
Wheat	2,322.13	259.54	[d]

[a] Eichers *et al.* (1970); U.S. Department of Agriculture (1967).
[b] Mostly petroleum.
[c] Includes herbicides used on all hay.
[d] Fungicide totals not available.

TABLE 4 Quantity[a] of Insecticides Used per Acre on Selected Crops by Region, United States, 1966[b]

	Northeast	Appalachian	Southeast	Delta States	Corn Belt	Lake States	Northern Plains	Southern Plains	Mountain	Pacific	48 States
Cotton	–	5.03	16.58	8.9	.80	–	–	2.66	10.92	1.71	6.27
Corn	.03	.07	.03	.02	.53	.16	.37	–	.25	.27	.36
Soybeans	.05	.31	1.17	.03	.01	–	–	.18	–	–	.09
Alfalfa	.31	1.07	–	.02	.02	–	.04	.37	.30	.06	.12
Potatoes	2.85	.60	.3	–	6.18	2.73	1.09	.06	1.03	2.94	1.99
Apples	15.50	10.55	–	–	33.87	12.82	–	1.00	3.91	4.67	12.58

[a]Pounds of active ingredients per acre.
[b]Eichers *et al.* (1970).

Spatial Adjustments

E. J. Mishan, a British economist, has argued in his recent book, *Technology and Growth: The Price We Pay*, that spatially isolating pollution-producing activities from other activities may be a less costly adjustment than subsidies, taxes, or input regulations (Mishan, 1970). The essential point is that neither the pesticide user nor the damaged party can increase his claim on damages by increasing his productive or consumption level. This raises the question of what types of spatial shifts might reduce the social costs of pest control.

Table 4 gives pounds of insecticides used per acre for 10 regions of the United States. The figures change somewhat from year to year (the insecticide use on apples was about 15 lb for all the major areas in 1964), but they do reflect the differences between areas in insecticide use and pest densities. For example, cotton production in the Southeast and Delta states utilizes many times more insecticides per acre than the important Southern Plains and Pacific regions. Table 5 gives a more complete breakdown of the relative insecticide use patterns on cotton for 1964–66. The two surveys show a difference in the use on cotton in the Pacific region for 1966. The Table 5 results are probably more representative because of the better sampling coverage (5,000 farms) of a given crop.

There are institutional and political constraints on geographical shifts of agricultural production to where pest densities are lowest, even when other costs of production and crop value are equivalent. Transferability of acreage allotments across the state lines appears desirable for reducing pesticide use. It may be possible for the United States to import high pesticide-utilizing products such as cotton and apples. These products may be grown in regions of the world with low pest populations. An important unanswered political question is whether we want to allow regions to choose lower environmental quality to maintain agricultural and industrial development. How are the inter-regional effects to be balanced against the values of decentralization and local choice?

A SOCIAL STRATEGY FOR PEST CONTROL

The case has been made elsewhere that a given economic organization may fail in its ability to reflect the relevant costs and returns in decision-making (Castle, 1965). When evidence exists that this is oc-

TABLE 5 Average Pounds of Insecticides Applied per Acre of Harvested Cotton[a, b]

Production Region	1964	1965	1966
Southeast			
Southern Piedmont	22.44	16.78	20.26
East Coastal Plains	24.30 }	31.62	32.42
Southern Coastal Plains	26.68 }		
Limestone Valley	3.86	8.22	19.28
Clay Hills	8.58 }	16.78	20.26
Black Belt	11.12 }		
Brown Loam	6.68	6.88	9.30
South Central			
Mississippi Delta	13.70	16.82	14.46
Northeast Arkansas	4.30	5.02	6.72
Black Prairie	3.80	5.28	5.40
Coastal Prairie	13.44	11.00	8.74
Lower Rio Grande	28.88	26.70	37.32
Rolling Hills	1.72	2.12	2.62
High Plains	.96	.64	.66
West			
South Joaquin Valley	25.72	33.50	24.74
Southern California and Southwest Arizona	27.42	24.34	21.34
Southern Arizona	39.28	17.04	23.26
Upper Rio Grande	27.12	17.06	17.24
United States average	11.38	12.40	12.84

[a] Dollar expenditures have been converted to pounds of insecticides at the average price of $.50 per pound.
[b] Starbird and French (1967–1969).

curing, public intervention may be justified. Whether it is justified will depend upon whether the alternative will lead to a better state of affairs. There are three general means of correcting market failure: government ownership or socialization, administrative regulation, and modifying the incentive framework.

Socialization Government control of production and marketing processes may correct a pollution problem. The fact that socialists nations also experience environmental quality problems suggests that the issue is more fundamental than the form of ownership.

Administrative Regulation Administrative regulations generally take the form of a requirement or a constraint being imposed upon the economic system (or individuals within that system) to which it (they) are expected to conform. As necessary as these regulations are, they must be considered as a rather crude and unsophisticated means of correcting failures of economic institutions. They treat symptoms, not causes.

Modification of the Incentive Framework Most economists believe that appropriate economic institutions are not functioning if the existing incentive framework is not yielding the level of environmental quality desired. Economists, therefore, usually will strive to modify the incentive framework when asked to suggest a solution to a problem of environmental quality.

As one considers the modification of the incentive framework as a broad category of social strategies for treating problems of pest control, the question of social goals and objectives logically arises. The decision to rely upon incentives implies that the basic problem is one of relatives rather than absolutes. For example, it implies that absolute safety in the use of DDT, for example, will result in higher food costs. It is suggested that an appropriate objective in the modification of the incentive framework is to reflect to the decision-maker the ratio of benefits and costs in the same proportion as they occur in society. It is obvious that social costs and returns must either be known or estimated with due allowance for risk and uncertainty. A crucial consideration, of course, is that they not be ignored or systematically excluded from consideration. Although space limitations prevent the presentation of a complete incentive-modification program for pest control, examples are presented to illustrate the framework of an incentive modification system.

Tax on Externalities This measure is an obvious attempt to place the increased social cost resulting from pesticide damage on the economic unit using the pesticide. The obvious difficulty is to determine the level of damage and the appropriate tax rate for each type of pesticide. Some argue that such a tax is the purchase of a "license to pollute." In a sense this is correct, yet if the cost of the damage is very high, the effect of the tax is to make this means of pest control very expensive.

Subsidies Subsidies may be of two general types. They may go to those who are damaged. This type of subsidy is inconsistent with the objective stated earlier for the modification of the incentive framework. Another type of subsidy is to make substitutions for the offending pest control more economical and more attractive to those who wish to control pests. Examples are (a) subsidies to the agricultural chemical industry to develop narrow-spectrum chemicals. R. F. Smith (1970) and others have argued that there are many pesticides on the laboratory shelves that have been screened out because they would kill only a few species rather than enough to make them profitable. These are now needed. An arrangement might be worked out in which university laboratories do the pharmacologic studies and a patent is given to a private firm to produce and distribute the chemical, as is now done with hybrid research. Franchises could be arranged between patent holders and formulators who can profitably produce and sell specialized chemicals in small lots. (b) Subsidies to public or private agencies such as the Federal Crop Insurance Corporation to encourage the substitution of crop insurance for pesticide application.

Redefinition of Property Rights by Government Action Because of the public–private mixture that exists in our economic system, it is possible for the government to affect the economic incentive framework significantly by modifying its activities and programs. Examples are (1) permitting acreage allotments to be transferred to areas where pest densities are low, (2) increasing the patent life of narrow-spectrum chemicals, (3) licensing applicators of chemicals, and (4) lowering fruit and vegetable grades, which will have the effect of reducing pesticide use.

Market Allocation of a Restricted Input In the event it is decided the aggregate level of a particular type of pesticide is to be held to a particular level, market forces may be a desirable method of allocating the restricted input. For example, if it is decided that the use of DDT should be reduced by 90 percent rather than banned completely, sales could be made by the government to licensed applicators only (Dales, 1968). This procedure will allocate the DDT to those producers who can use it most profitably. This is similar, of course, to selling licenses for many other kinds of restricted economic activity such as taxi licenses, liquor licenses, and acreage allotments.

In the above material on modification of the incentive framework, an attempt has been made to illustrate a classification system by example. The objective has been to identify those measures that will encourage decision-makers to consider both the direct and indirect social costs and returns likely to be associated with the use of a particular pesticide.

ACKNOWLEDGMENT

Helpful suggestions by J. C. Headley, R. K. Perrin, R. L. Rabb and J. A. Seagraves are gratefully acknowledged.

REFERENCES

American Phytopathological Society. 1957–1969. Fungicide–nematocide tests. St. Paul, Minnesota.

Bator, F. M. 1957. The simple analytics of welfare maximization. Am. Econ. Rev. 47, March 1957.

Bator, F. M. 1958. The anatomy of market failure. Q. J. Econ. 72, August 1958.

California Agricultural Experiment Station. 1969. Pest and disease control program for cotton. Extension Service, Berkeley, California.

Carlson, G. A. 1970a. A decision theoretic approach to crop disease prediction and control. Am. J. Agric. Econ. 52:216, May 1970.

Carlson, G. A. 1970b. The microeconomics of crop losses. Symposium on Economic Research on Pesticides for Policy Decision Making, sponsored by ERS, USDA, Washington, D.C.

Castle, E. N. 1966. Economic and administrative problems of water pollution. American Association for the Advancement of Science, Washington, D.C.

Castle, E. N. 1965. The market mechanism, externalities and land economics. J. Farm Econ. Vol. 47, No. 3.

Dales, J. H. 1968. Pollution, property and prices. University of Toronto Press, Toronto.

Davis, V. W., A. S. Fox, R. T. Jenkins, and P. A. Andrilenas. 1970. Economic consequences of restricting the use of organochlorine insecticides on cotton, corn, peanuts and tobacco. Agricultural Economic Report No. 178, ERS, USDA, Washington, D.C.

Edwards, W. F. 1969. Economic externalities in the agricultural use of pesticides and an evaluation of alternative policies. Unpublished PhD dissertation, University of Florida, Gainsville.

Edwards, W. F., M. R. Langham, and J. C. Headley. 1970. Pesticide residues and environmental economics. Nat. Res. J. 10:4 (Oct. 1970).

Eichers, T., P. A. Andrilenas, H. Blake, R. Jenkins, and A. Fox. 1970. Quantities of pesticides used by farmers in 1966. Report No. 179, Economic Research Service, U.S. Department of Agriculture, Washington, D.C.

Food and Agriculture Organization. 1969. Fifth session on the FAO working party of experts on resistance of pests to pesticides. Rome, Italy.

Fox, A., T. Eichers, P. Andrilenas, R. Jenkins, and H. Blake. 1968. Extent of farm pesticide use on crops in 1966. Report No. 147, Economic Research Service, U.S. Department of Agriculture, Washington, D.C.

Fox, A., R. Jenkins, P. Andrilenas, J. Holstun, Jr., and D. Klingman. 1970. Restricting the use of phenoxy herbicides: Costs to farmers. Agricultural Economic Report No. 194, ERS and ARS, USDA, Washington, D.C.

Headley, J. C. 1968. Estimating the productivity of agricultural pesticides. Am. J. Agric. Econ. 50:13 (Feb. 1968).

Headley, J. C. 1970. Productivity of agricultural pesticides. Symposium on Economic Research for Policy Decision Making. ERS, USDA, Washington, D.C.

Headley, J. C. 1971. The distribution of costs and benefits from environmental protection. Proceedings, Western Agric. Econ. Assoc.

Hyami, Y., and V. W. Ruttan. 1970. Factor prices and technical changes in agricultural development: The United States and Japan, 1880–1960. J. Polit. Econ. 74 (Nov. 1970).

Kneese, A. V. 1962. Water pollution: Economic aspects and research needs. Resources for the Future, Inc., Washington, D.C.

Lave, L. B., and E. P. Seskin. 1970. Air pollution and human health. Science 169 (3947):723.

Leontief, W. 1970. Environmental repercussions and the economic structure: An input–output approach. Rev. Econ. Stat. 72:262 (August 1970).

Mishan, E. J. 1970. Technology and growth: The price we pay. Praeger Publishers, New York.

Peterson, W. L. 1969. The allocation of research, teaching and extension personnel in U.S. colleges of agriculture. Am. J. Agric. Econ. 51:41 (Feb. 1969).

Pigou, A. C. 1932. The economics of welfare. 4th ed. Macmillan and Company, London.

Rice, D. P. 1968. The direct and indirect cost of illness. *In* Federal programs for the development of human resources. A compendium of papers before the Subcommittee on Economic Progress of the Joint Economic Committee, Congress of the United States, Vol. 2.

Rabb, R. L., and F. E. Guthrie (eds). 1970. Concepts of pest management. Proceedings of a conference held at North Carolina State University, Raleigh, March 25–27, 1970.

Smith, R. F. 1970. Pesticides: Their uses and limitations in pest management. Conference on pest management. North Carolina State University, Raleigh.

Sosnick, S. H., and T. N. Terlecky. 1969. An economic analysis of shipping point inspection of California table grapes. Giannini Foundation Research Report No. 306, University of California, Davis.

Starbird, I. R., and B. L. French. 1967–1969. Costs of producing upland cotton in the U.S. Report No. 99 and supplements. ERS, USDA, Washington, D.C.

Strickland, A. H. 1970. Economic principles of management. *In* Concepts of pest management, Proceedings of a conference held at North Carolina State University, Raleigh.

U.S. Department of agriculture. 1953, 1967, 1969. Agricultural statistics.

U.S. Department of Agriculture. 1968. A guide to crop insurance and profitable farm management. USDA, Federal Crop Insurance Corporation, Washington, D.C.

U.S. Department of Agriculture. 1953–1969. The pesticide review. ASCS.

U.S. Department of Agriculture. 1967. Professional workers in state agriculture experiment stations, Agriculture Handbook Series.

U.S. Tariff Commission. 1970. United States production and sales of pesticides and related products, Preliminary, 1969, Washington, D.C.

U.S. Congress. Subcommittee on Economy in Government of the Joint Economic Committee. 1969. The analysis and evaluation of public expenditures: The PPB system. A compendium of papers. 91st Congress, Volumes 1–3, U.S. Government Printing Office, Washington, D.C.

U.S. Congress, Senate. 1965. Interagency coordination in environmental hazards (pesticides). Hearings before the U.S. Senate Subcommittee on Reorganization and International Organization of the Committee on Government Operations, 88th Congress, Appendix 1, U.S. Government Printing Office, Washington, D.C.

U.S. Congress, Senate. 1969. Effects of pesticides on sports and commercial fisheries. Hearings before the U.S. Senate Subcommittee on Energy, National Resources, and the Environment of the Committee on Commerce, 91st Congress, U.S. Government Printing Office, Washington, D.C.

U.S. Department of Agriculture. 1969. Report of Committee on Persistent Pesticides, Division of Biology and Agriculture, National Research Council, Washington, D.C.

U.S. Department of Health, Education and Welfare. 1969. Report to the Secretary's Commission on Pesticides and their relationship to environmental health, Parts I and II.

University of California. 1965. Estimates of crop losses and disease–control costs in California, 1963. Division of Agricultural Sciences, Berkeley, California.

Defining the Economic Threshold

J. C. Headley

Economic entomology as a discipline is concerned with the application of entomological knowledge for the purpose of providing economic benefit. This application may take the form of either preventing insect damage to man and his property or by producing beneficial entomological effects such as pollination and honey production.

Historically, economic entomologists have concerned themselves with preventing damages from insects. From this concern has come the concept of the economic threshold, sometimes referred to as the economic injury level. While the concept has received considerable attention, its definition has not been rigorous and has, therefore, been ambiguous. The purpose of this paper is to develop a definition of the economic threshold that will hopefully be a rigorous one, with a minimum of ambiguities.

SOME PREVAILING DEFINITIONS

Reference to economic thresholds is found frequently in the entomology literature. Edwards and Heath (1964) say that a pest population has reached the economic threshold when the population is large enough to cause damages valued at the cost of practical control. Bierne (1966) suggests that the economic threshold is a critical level of damage above which damages will not be tolerated. He continues to say that damages above this critical level cause the pest responsible to be labeled as a damaging pest and consideration is given or action is taken to try to abate the damage. The Subcommittee on Insect Pests of the Committee on Plant and Animal Pests established by the National Research Council (National Academy of Sciences, 1969) defined the economic threshold as "the level at which damage can no longer be tolerated and, therefore, the level at or before which it is desirable to initiate deliberate control activities." They go on to state that "the definition should probably be amended to consider a more critical density as that where the loss caused by the pest equals in value the cost of available control measures." Their definition is further tempered by including for consideration, adverse side effects which may entail costly ecological consequences. Obviously, the threshold should also include beneficial side effects that might be involved with predators or parasites of beneficial species.

The definitions provided above do not include any statement of the unit to which the definition applies, that is, a field as contrasted to a farm, or an individual producer as contrasted with a community or region. According to the statements cited, the implication is that the economic threshold is one of the components of a decision model for individual producers considering pests one at a time at a given point in time. To the extent that individual producer's actions are not independent of effects felt by other producers or are not independent as between time periods, the aggregate result of decisions based on this concept may not be consistent with expected micro-results.

ANOTHER DEFINITION

The concept of an economic threshold in its simplest form seems to involve about three variables. These are damage, pest population, and, perhaps, time. These variables are related in that damage is a function of pest population and pest population is a function of time.

As the first suggestion, conceive of a damage function at time t causally dependent upon the pest population in time t, such as:

$$D_t = bP_t^2 - A. \tag{1}$$

where D_t = damage in t
P_t = pest population in t
A = constant to define the damage tolerance level

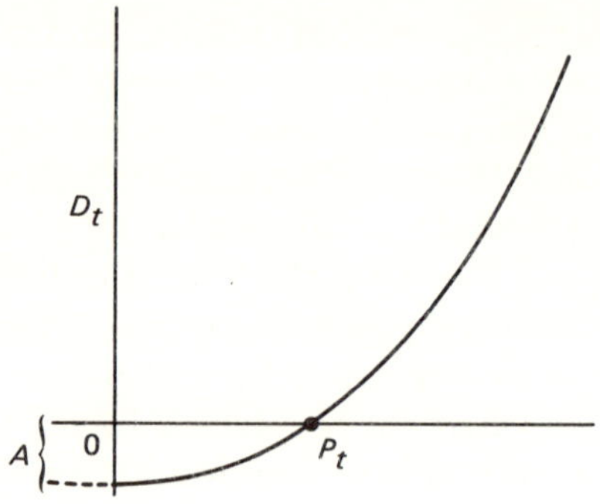

FIGURE 1 Product damage related to pest population.

b = constant parameter which enters into the incremental damage resulting from P_t.

The graph of this function is shown in Figure 1.

Assume that the population, P_t, is related to previous population levels at some point, say, P_{t-n}. That is, at a point n time periods before t, which is entomologically significant, there is a population that will grow to P_t. These populations are related by the function:

$$P_t = P_{t-n}\ (1+r)^n \tag{2}$$

where P_{t-n} = the population n periods prior to t
r = growth rate of the population per time period
$(1+r)^n$ = compound growth factor.

The value for P_t in (2) can be substituted into (1) to derive D_t as a function of P_{t-n}.

$$D_t = b\left[P_{t-n}\ (1+r)^n\right]^2 - A \tag{3}$$

What is present here, then, is a relationship where the population at time period $t\text{–}n$ is the present value of the population at time period t. Therefore, if, through control activities, the population at $t\text{–}n$ is maintained at some level, the population at t is determined and, therefore, the damage in t is determined.

Now, if the product yield is some deterministic biological function of D_t such as

$$y = N - cD_t \tag{4}$$

where y = product yield
c = a constant parameter measuring incremental yield effects
N = a constant measuring yield with no damage, and D_t is defined as before

then, the effects of population on yield can be evaluated. Because by substitution

$$y = N - c\left\{b\left[P_{t-n}\,(1+r)^n\right]^2 - A\right\}. \tag{5}$$

For an individual farm producer who is a price taker, then the yield function (5) can be multiplied by a constant price per unit R to provide value of production as a function of the population at the point $t-n$. This is shown in Figure 2. Note that the value function has

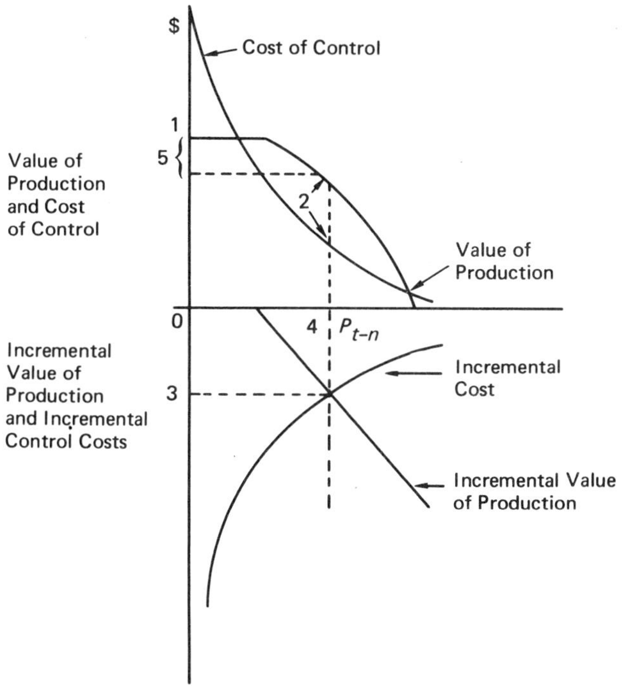

FIGURE 2 Relation of value of production, control cost, and pest population.

the same shape as (5) because of our assumption of constant price per unit. Figure 2 shows that as the population increases, the value of production is decreasing at an increasing rate that is consistent with our assumed damage function (1). When P_{t-n} reaches a value where the expression $cb\ P^2_{t-n}\ (1+r)^{2n} = N + CA$, then y = 0 and the value of production equals zero.

The final element in this model is a function that relates control costs to population, i.e., which tells the cost of control to maintain a population of certain size. Assume that this cost relation is continuous and decreases as the population is allowed to be larger and that this cost decreases at a decreasing rate, which is in keeping with the findings of entomologists that the cost of reducing a population increases at an increasing rate (National Academy of Sciences, 1969).

$$0 = \frac{L}{P_{t-n}} \tag{6}$$

where 0 = the total control cost
L = a constant parameter influencing incremental costs

and P_{t-n} is defined as before. This function is also graphed in Figure 2.

Now is the time to pose the crucial question. What is the economic threshold? A number of points are labeled in Figure 2. Point 1 is the normal producer revenue when the pest population is absent or statistically equal to zero. Point 2 identifies the points where the rate of change of the value of production function is equal to the rate of change of the cost of control function. The value of this rate of change is shown at point 3 on the incremental scale. Point 4 is the pest population level associated with point 2. Point 4, then, is the economic threshold!

This definition can be defended with the following argument: Because of the nature of the damage function, the incremental losses in value of production due to increases in population at low levels are small, while the incremental costs of reducing populations at these levels are high. This is shown by the relative slopes of the two curves in the upper quadrant to the left of point 4. As population increases, the costs of maintaining the population diminishes and the damage mounts, i.e., the value of production decreases. At point 2 on the curves or point 4 on the P_{t-n} axis the cost of reducing the population by one increment is just equal to the incremental increase in the value of production that results. Therefore, it is not advantageous to reduce the population below this level, but rather to sacrifice the value of production shown by point 5, since the costs of obtaining it, in this model, are astronomical.

Therefore, the economic threshold here defined becomes the population that produces incremental damage equal to the cost of preventing that damage. It is not a zero level of damage. Furthermore, the economic threshold population, here defined, also becomes the *optimal* population in an economic sense!

SOME CHARACTERISTICS OF THE DEFINITION

As defined above, the economic threshold has certain properties. First, the critical (optimal) population level will become smaller as the value of the product increases given constant control technology and constant prices for control inputs. This is shown in Figure 3. As the price per unit of the product increases, the slope of the value of production function increases. This merely means that resources can be

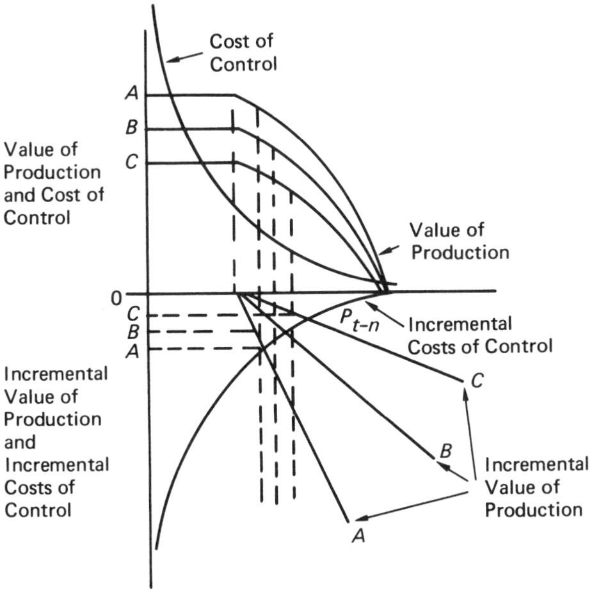

FIGURE 3 Relation of value of production, control costs, and pest population: product prices variable.

justified.to prevent another increment of damage. Therefore, the economic threshold is a function of the product price. Similarly, increases in the unit cost of control methods will shift the control cost curve to the right and also increase the slope of the cost curve, which tends to move the economic threshold toward higher population levels.

Second, the definition given above applies only to an individual producer where the control measures are specific to a particular pest, a particular product, and a particular producer in a particular season. That is, any side effects at a point in time or through time are assumed to be absent. Such side effects could be included in the analysis by altering the control cost function.

Third, this model has assumed that the cost of control function is continuous, i.e., that the incremental costs can be defined at all points on the total cost function. This is not crucial to the findings, however, as long as the general level of control decreases as larger populations are allowed to survive. An example of the discontinuous cost of control function is shown in Figure 4. Here the cost of control function is a step function with a constant rate of change given by the incremental cost function. The increase in costs from level 1 to level 2 brings forth added production value measured by 3. The average rate of change in the value of production over control levels 1 and 2 occurs in the population range that occurs with control level 2 and, therefore, the added production exceeds the added costs. This is evident by the graph showing incremental values. Therefore, the same criterion applies as for the continuous model. The same conclusions regarding product price-changes and control inputs price-changes as in the continuous control cost case also apply.

SOME CONCLUSIONS AND IMPLICATIONS

Based on the definition of the economic threshold given in this paper the following conclusions can be made:

1. The economic threshold is responsive to and determined by prices of the product protected and the prices of control inputs.
2. It is entirely possible that the population level justifying control activities is larger than the population level where yield damage first occurs.

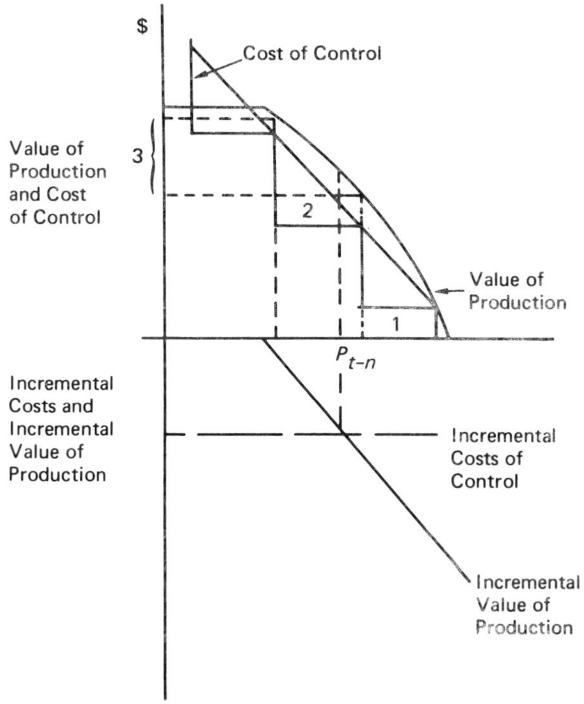

FIGURE 4 Relation of value of production, control costs, and pest population: cost of control discontinuous.

3. The economic threshold, when given this particular *economic* definition also becomes the optimal (net return-maximizing) level of pest population. No smaller or larger population will provide greater net returns above control costs.

4. Unless the incremental costs are less than the incremental value of damages prevented for the entire range of population levels, there is no economic justification for eradication policies under the assumption of time-independence built into this definition. The exception to this would be a guaranteed, once-and-for-all eradication, with a cost of maintaining zero level populations less than the costs of maintaining populations at greater than zero levels.

The implications of this definition of the economic threshold concern the need for research on the topic as a basis for establishing rational control programs and policies. There is a need for the development of damage functions for major pests of important products. Furthermore, control cost functions need to be developed for the same pests for alternative methods of control and combinations thereof.

Following the development of the information mentioned above, damage and cost functions over time are needed to evaluate the longer-term effects of control programs and to design programs that approach optimality over time.

Finally, further information is needed to determine the effects of control programs on overall populations in an area, target and non-target alike. Such research would evaluate strategies for maximizing community returns from pest control and would include information on the expected distribution of damages and benefits as a basis for broadening the support for control strategies and allocating the benefits and costs resulting therefrom.

REFERENCES

Beirne, Bryan P. 1966. Pest management. CRC Press, Cleveland.

Edwards, C. A., and G. W. Heath. 1964. The principles of agricultural entomology. Charles C Thomas, Springfield, Illinois.

National Academy of Sciences. 1969. Insect-pest management and control. *In* Principles of plant and animal pest control. Publ. 1695, Washington, D.C.

Basic Principles of Ecosystems Analysis

J. E. Van der Plank

For convenience of presentation, let me discuss the analysis of ecosystems in terms of plant disease. Basic principles seem much the same in plant pathology, entomology and nematology; but details and emphasis vary so much that it is easier to limit the scope of examples.

The general ecosystem of plant disease is a triangle. There are host plants: wheat, rice, etc. There are pathogens: fungi, bacteria, and viruses, mostly. And there is the environment: the rainfall, temperature, and all the other physical and organic factors that affect disease. To these we may add a fungicide or other chemical. All parts of the ecosystem interact. Thus, a high level of disease may be due to very susceptible host plants, very aggressive pathogens, weather very favorable to disease, or the absence of a fungicide. Models we choose to analyze the interaction must avoid any essential distinction between susceptible hosts, aggressive pathogens, environments favorable to disease, and the effect of chemical agents.

In keeping with the purpose of this symposium the spotlight must fall on that part of ecosystems analysis that concerns the control of diseases and pests.

The general strategy of disease control is either to reduce the amount of inoculum from which the disease starts or to reduce the rate at which it increases after it has started, or, of course, to reduce both. With respect to a seed-borne disease of an annual crop that is planted in the spring and harvested in the fall, control is aimed either at reducing the amount of inoculum available in the spring, i.e., the pathogen carried on the seed, or at reducing the rate at which pathogen builds up later. I shall use the words *initial inoculum* for the inoculum at the start, and *infection rate* for the build-up later. There are, of course, many permutations. Inoculum need not be present in the seed or field at the start, but can come in later from other fields or natural vegetation. But the general principle holds: Disease starts from inoculum, and the pathogen builds up if it is given a susceptible host and a favorable climate.

All methods of control act against initial inoculum or the infection rate, or both. Resistance in the host plants may be of the sort to reduce initial inoculum without reducing the infection rate, or it may act primarily on the infection rate. The environment may be manipulated either to reduce the initial inoculum (e.g., by crop rotation) or to reduce the infection rate. Chemical disinfectants and fumigants generally act against initial inoculum; protectant fungicides applied to the growing crop generally reduce the infection rate.

I have been stating the problem in terms of seasonal epidemics. There is also the problem of endemic disease, especially in perennial crops such as forests and pastures. With long established disease in perennial crops, the concept of initial inoculum falls away; and control is directed at reducing the equilibrium level of disease, i.e., at reducing the level at which the infection rate is zero in the long run.

SOME GENERAL CONCLUSIONS

1. Both the initial inoculum and the subsequent infection rate are determined by the state of the host, of the pathogen, and of the environment, i.e., by the disease triangle or ecosystem.

2. High initial inoculum coupled with a high infection rate means that disease is uncontrolled. Low initial inoculum coupled with a low infection rate means that disease is contolled or unimportant.

3. When the initial inoculum is high and the subsequent infection rate low, reduction of the initial inoculum will usually substantially reduce disease. Control is likely to be trustworthy, in the sense that the good effects of reducing initial inoculum are unlikely to be seriously offset by weather conditions favoring disease and increasing the infection rate in the growing crop.

4. When the initial inoculum is low and the subsequent infection rate high (as when the potato blight fungus can increase a millionfold

or even a billionfold in a single season), the effect of reducing initial inoculum is to delay the onset of a damaging epidemic. Thus, if disease increases 100-fold in a week, reducing the initial inoculum by 99 percent will delay the onset by a week, other things being equal. The relation here between reduced inoculum and the delay is logarithmic. Reducing the initial inoculum by 99.9999 percent would delay the epidemic only three times as long as would reducing it by only 99 percent. Here, control by reducing the initial inoculum is not always trustworthy, because the reduction is easily offset by even a small increase in the infection rate, as when the season's weather favors disease.

5. The infection rate is governed by four factors, all of which need not be present together. I name them in terms of lesions caused by a fungus; the descriptions apply *mutatis mutandis* to other sorts of disease. They are (a) The establishment of infection: The proportion of spores that, having fallen on healthy tissue, germinate and start new lesions. (b) The period of latency: The time needed before a new lesion can begin to sporulate. This has also been called the generation time or, incorrectly, the incubation period. (c) The intensity of sporulation in a lesion, and (d) The period of infectiousness: The period over which the lesion continues to produce spores.

6. Complete control of epidemic disease (apart from the total eradication of initial inoculum) requires that the infection rate be reduced to zero. In this process the period of latency becomes irrelevant, i.e., it does not appear in the equations for the threshold condition for an epidemic. The other factors need not necessarily be reduced to zero; e.g., to stop an epidemic by means of a protectant fungicide does not necessarily require all spores to be killed or stopped from germinating. It is necessary only that the progeny of 1 lesion should, on an average, be less than 1 lesion. All the factors (apart from the period of latency) interact. A spray program that killed, say, 80 percent of the spores before they germinated might be adequate to stop an epidemic if the intensity of sporulation or the period of infectiousness were small, but inadequate if they were large. In turn, germination, the intensity of sporulation, and the period of infectiousness depend on the weather and environment generally, the susceptibility of the host plants, and the aggressivness of the pathogen, i.e., on the disease triangle. The efficiency of a protectant fungicide depends not only on the percentage of spores it kills, but also on the whole disease triangle or ecosystem.

7. In endemic disease, as in permanent pastures or forests, a con-

dition of equilibrium in which the infection rate in the long run is zero does not involve the period of latency. What the level of disease will be at equilibrium depends on the other factors, such as the intensity of sporulation.

8. The period of latency, irrelevant when the infection rate is zero, becomes increasingly important as the infection rate increases. Where explosive epidemics can be expected to occur, one of the first aims should be to increase the period of latency. This can often be done by plant breeding, because a lengthened period of latency is commonly a feature of resistant plants.

These conclusions have been reached with the help of a heuristic model. But they can be reached equally well by simple logic. They are general conclusions and apply equally to control by chemicals, or plant breeding, or the manipulation of the environment.

A DETERMINISTIC MODEL

The following are the equations defining the model used in reaching most of the above conclusions:

$$\frac{dx\,(t)}{dt} = R\,[x\,(t-p) - x\,(t-i-p)]\;[1-x\,(t)],$$

$$\frac{dx}{dt} = rx\,(1-x).$$

Here, x is the proportion of tissue that is diseased (or, for systemic disease, the proportion of plants that are diseased); t is the time; $x(t)$ is the value of x at time t, and $x(t-p)$ the value at time $t-p$, and so on. R is directly proportional to the percentage of spores that, having fallen on healthy tissue, germinate there and cause infection, and also to the intensity of sporulation; p is the period of latency; i is the period of infectiousness; and r is the relative infection rate. These equations were discussed by Van der Plank (1963, 1968). Suffice it to say that the first of the equations relates the increase of disease both to the proportion of tissue that is infectious and to the proportion that is still healthy and available for infection.

The threshold condition for an epidemic to start (from small beginnings), i.e., for $r > 0$, is $iR > 1$.

The model is for continuous infection proceeding without break through day and night. This is unrealistic because infection often occurs discontinuously, e.g., only when leaves are wet with dew or rain. In a model of infection proceeding discontinuously at a particular moment of time each day, R was replaced by a (Van der Plank, 1963, p. 53) and a multiplication rate m (Corsten, 1964; Oort, 1968), which has the same meaning as a. But further analysis shows that a and m are identical with R, and that the modifications needed to cope with discontinuity of infection do not alter the general conclusions that have been stated.

The above equations are essentially for heuristic purposes: to discover relations that might otherwise be missed. There is probably no limit to the number of models that could be used, and readers may be interested in a brave attempt by Waggoner and Horsfall (1969) to design a simulator of epidemics written for a computer.

STATE OF KNOWLEDGE AND KNOWLEDGE GAPS

The principles of ecosystems analysis are the principles of analyzing the interaction of all the relevant organisms (including viruses in this term) between themselves and between them and the environment. Our knowledge of the interactions considered qualitatively embraces much of our knowledge of plant pathology, entomology, and related sciences. It is when one starts probing quantitatively that one becomes intensely aware of ignorance. In plant pathology, particularly, knowledge of even the most elementary relations is vague and inadequate.

For example, possibly the simplest relation we ought to know in the whole ecosystem of plant disease is the relation between inoculum and disease. If 100 spores having fallen on a leaf start N lesions, how many lesions would 200 spores start? Could any question be simpler or its answer more fundamental to quantitative analysis of disease systems? But we cannot answer it, even leaving aside the purely probabilistic complications. Some believe that the relation between inoculum and disease is arithmetic; others that it is logarithmic; and there is a theory of a numerical threshold of infection. Data are not lacking, but many are pure artifacts. There is, for example, a substantial literature on the relation between the concentration of virus in infected plant sap and the number of lesions produced when this sap is rubbed onto leaves. The literature exists because rubbing sap on suitable leaves is one of the standard methods of assaying virus concentration. But the number of infections per unit area of leaf surface is sometimes millions of times greater in these assays than in natural epidemics of virus disease. This literature of assays is quite irrelevant to our problem. After a detailed search of the literature of fungus, bacterial, and virus diseases, I am prepared to assert that at low concentrations of inoculum, disease is proportional to inoculum. But I am not always prepared to define the limit to what I can call a low concentration. In other words, I do not know how far along the course of an epidemic I am entitled to assume proportionality between inoculum and disease.

APPROPRIATE GOALS

For ecosystems analysis we need appropriate models, estimates of parameters and variables to use in these models, and extensive observations in fields and forests.

The necessary information about parameters and variables is fortunately needed only in groups. For example, the whole process from the time a spore settles on a healthy leaf to the time the resulting lesion itself starts dispersing spores is made up of germination time, penetration time, time to spread through the leaf tissues, and so on, up to the time spores start to disperse. Modern research shows each process to be highly complicated with its own complex of enzymes; and even a process like germination is really a multiplicity of processes. But all this detail is not ordinarily needed in ecosystems analysis. What is needed is the total time up to the start of spore dispersal: the period of latency. It is this period, and not all its constituent parts, that normally enters models.

As to models, we can claim a little knowledge of what goes on in the early stages of an epidemic, until disease reaches, say, 5 percent. Beyond this, information becomes increasingly empirical. It is a fine goal to get working models of the whole epidemic process, but we must not delude ourselves that it will be easy.

As to field observations, increasing attention is being given to quantitative epidemiology in plant pathology and population dynamics in entomology. But we want still more measurements—measurements of disease multiplication, measurements of disease spread, measurements from the ground, measurements by aerial photography, measurements of the whole epidemic or population dynamic process from start to finish. All the models, all the equations, all the parameters result in meaningless analyses unless there is something to analyze.

MAGNITUDE AND RELATIVE IMPORTANCE OF PROBLEMS

Since the topic of this symposium is control of diseases and pests, a discussion of ecosystems analysis should therefore relate to control.

Overriding all is economics. With high-priced crops (generally horticultural crops), control is commonly by chemicals. The cost structure supports them. Chemicals dominate the ecosystems, and analyses of the ecosystems must mainly relate to the use of the chemicals. The forecasting of disease outbreaks and hence of the need to start applying chemicals is a case in point. But with out great grain crops, forests, plantations, and pastures we have to work more economically and attend more to the ecosystem, especially in choosing a suitable environment and resistant host plants.

These problems, somewhat interrelated, are especially important: the forecasting of outbreaks of diseases and pests, the determination of losses from diseases and pests, the timing of chemical applications, and the combination of host resistance with chemical control.

Forecasts of disease involve either the initial inoculum or the infection rate, or both. When the pathogen increases greatly during the season, the infection rate is closely watched. For example, the potato blight fungus can increase a millionfold or more during a season, and forecasts of blight are based primarily on the weather, which controls the infection rate. When the multiplication is less explosive, we attend more to initial inoculum. In a warm climate, we do not plant a susceptible crop in soil known to be infested with rootknot nematode, because we can forecast with considerable accuracy what would probably happen.

Because weather forecasts have been inadequate, disease forecasts based on weather have generally been based on past weather. That is, disease is forecasted *after* a favorable weather period. With better weather forecasts, disease will be forecast on weather to come. When that happens, disease forecasting will rank high in the scale of priorities in ecosystems analysis.

One acts to control a disease only if one believes that losses from disease justify the action. When loss is due to reduced yield (as distinct, say, from unsightly surface blemishes), one must know the history of disease throughout the season, i.e., one must know the disease progress curve. Onto this history one must graft details peculiar to the disease in question; thus, with wheat leaf rust, one watches for rust on the flag leaf and its immediate neighbors. When we can forecast the disease progress curve and when we can interpret the curve accurately in terms of loss, we will have gone far toward bringing efficient control measures to the extensively planted, relatively low priced crops.

When economic factors severely limit the number of applications of a fungicide or other chemical that can profitably be given in a season, the applications must be properly timed. If the chemical is applied too early, it may be wasted; if too late, the damage may already have been done. Good timing depends on good forecasting, good knowledge of the disease progress curve, of how an application of the chemical will alter that curve, and of how an alteration of the curve will reduce loss from disease.

In plant pathology, there is too much of a dichotomy of interest. One is too apt to believe in the control of a disease by chemicals or by host resistance, each to the exclusion of the other. (In entomology, if one substitutes biological control in general for host resistance in particular, there is a more balanced favoring of integrated control—in theory, if not always in practice!) Perhaps this has come about because research workers are usually concerned either with chemical control or with host resistance, rarely with both. At all events, chemical control and host resistance are supplements, not rivals. If chemicals and host resistance both reduce the percentage of spores that germinate and establish new infections, then the more the resistance the less the amount of chemicals needed. The beneficial interaction of host resistance and chemicals has long been known empirically; one would like now to see a more detailed analysis and application.

RESEARCH NEEDS, NEW OPPORTUNITIES AND DIRECTIONS

The trend in plant pathology has been toward analysis rather than synthesis, toward cellular and molecular biology rather than the study of populations of pathogens in ecosystems. Plant pathology departments have more biochemists than mathematicians and place more stress on chemistry than mathematics in the courses for a higher degree. I do not deplore this attention to chemistry; I merely point out an imbalance. In the physical sciences there is, at the extreme of smallness, nuclear science; at the extreme of bigness, cosmology. To be interested in the one is not to deplore interest in the other; in any case, they help one another. Agriculture too has its cosmology. We call it here the analysis of ecosystems.

The analogy with cosmology is, I believe, apt. In the analysis of ecosystems one must think big, in the sense of trying to understand what goes on in the wide stretches of fields and forests and whole agricultural regions, rather than small, in the sense of studying what is under the electron microscope or digesting in a test tube. Again, there is no mutual antagonism; what one finds out in the test tube often helps one to understand what goes on in the field. But the tools are somewhat different.

Consider the ecosystem of wheat stem rust in the Mississippi Valley. With a sequence of pictures from satellites; with aerial photographs on special film; with ground surveys recording data quantitatively; with accurate weather forecasts; with knowledge (obtained both in the field and laboratory) of how environmental factors and changes affect the pathogen, the host, and the infection processes; with knowledge of air movements and the various patterns of spore dispersal by air and water; with knowledge of the host plants, their resistance, how weather influences them, and how they influence the ecoclimate; with maps showing the distribution, shape, and size of fields; with knowledge of the pathogen, its races, their abundance, virulence, and aggressiveness in the appropriate environments—with all this, with models to coordinate it and computers to digest it, one hopes the ecosystem could be better analyzed, host resistance better evaluated, and applications of chemicals better timed.

THE OUTLOOK FOR THE FUTURE

Ecosystems analysis (embracing the epidemiology of plant pathogens and population dynamics of insects) has been the Cinderella in many departments of plant pathology and entomology. An imbalance between chemistry and mathematics as basic sciences in these departments may have helped to make it so. (I am thinking here of mathematics apart from what is geared to the design of experiments and tests for significance.) Basic research is needed; for too long in some institutions basic research has been interpreted as research not inspired by farmers' needs and without foreseeable application to agriculture. All this I believe to be changing; experience shows it usually takes more intellectual prowess to solve a problem of immediate practical importance than one judged mainly as a way to scientific publications.

Progress in ecosystems analysis is closely linked with progress in other sciences. An obvious link is with weather forecasting. As weather forecasting improves, interest in disease forecasting will increase, and with this interest will come the basic knowledge needed for systems of forecasting. Progress in ecosystems analysis is also closely linked with progress in designing hardware such as computers, and automatic recorders for a variety of purposes. Progress here is swift, and much of it will rub off on ecosystems analysis.

Lastly, I am optimistic for the very reasons that brought this symposium about. The need for more food and the need for getting it without excessive pollution must force attention on ecosystems analysis and encourage research in appropriate directions.

REFERENCES

Corsten, L. C. A. 1964. Een kwantitatieve beschrijving van de ontwikkeling van een schimmelpopulatie. Meded. Landbouwhogesch. Wageningen 64–15:1–7.

Oort, A. J. P. 1968. A model of the early stage of epidemics. Neth. J. Plant Pathol. 74:177–180.

Van der Plank, J. E. 1963. Plant diseases: Epidemics and control. Academic Press, New York.

Van der Plank, J. E. 1968. Disease resistance in plants. Academic Press, New York.

Waggoner, P. E., and Horsfall, J. G. 1969. Epidem: A simulator of plant disease for a computer. Bull. Conn. Agr. Exp. Stn. 698.

Concept of Pest Management

George L. McNew

Before one can speak authoritatively on "pest control," the nature of a pest infestation must be defined. We freely use such terms as disinfection, sterilization, disease resistance, environmental control, as though they were positive and complete. The horrible truth is that this is not so. We have brought pests under total control or eradicated them in so few instances that you can almost count them on your fingers.

The harsh reality of the situation is that we must live with pests—be they insects, mites, snails, worms, fungi, bacteria, viruses, epiphytic plants, allergens, or weeds. Rarely do we eradicate them; the best we can do is to coexist with them. We and they are part of a giant ecosystem, and man is merely trying to shift the balance of power so he can exist in reasonable comfort and security. It is necessary for us to be sufficiently objective to understand this ecosystem in which the role of man is defined as a competitive force, albeit a bit more intellectual and far-sighted than his competitors.

As a general rule man has disturbed the natural ecosystem just enough to give the predators and parasites an upper hand. We set the stage for pests to prosper on the day we began to cultivate the soil so crops would be grown repeatedly in the same area, when we abandoned mixed vegetation and freely moving animals in favor of intensive monoculture and began to breed our crops and livestock to uniform standards of growth and physiology. Either man had to abandon agricultural pursuits or amend them by supplementary treatments. It is these amendments that we refer to here today as strategy for control. Perhaps it would be better to remember that it is our strategy for survival in competition with biological forces that we cannot crush or obliterate. We merely try to outwit their natural propensity for multiplication and to constantly readjust our strategy to their adjustments.

THE NATURE OF PEST INFESTATION

There is nothing natural about our system of repeated use of the same space—be it a field, a barn, or a residence. We do these things because it fits our convenience and pleasure, but we must pay a price for every convenience. Time does not permit us to dwell on each household insect, each disease, and each weed problem, but if you will bear with me, I would like to draw upon a couple of examples from the field I understand best and you can apply the principles to almost any pest your mind's eye may focus upon.

When the white man came to New England and vicinity, there were several fungous and bacterial diseases attacking the plants that were valuable to the Indians. For example, there was on the banks of the Hudson River a bacterial pathogen of rosaceous plants that was to be christened "fireblight" 150 years later and to be ascribed to the attack of *Erwinia amylovora* after another hundred years. The fireblight bacterium was established in the wild hawthorn, among other hosts, where it survived in stem cankers, occasionally killing off a branch and in wet seasons causing a blight of blossoms.

Those trees that were extremely susceptible were killed outright before they could reproduce, so resistance gradually built up but never achieved perfect immunity. The more virulent strains of the pathogens became better adapted to the physiology of the host so the attack continued on the resistant progeny, but if an extremely potent strain of the pathogen arose it dispatched the host forthwith and thereby sealed its own doom because it is ill adapted to living as a free agent in soil in competition with vigorous saprophytic organisms. Thus, over the millennia this bacterium and host came to a coexistence. If a virulent strain arose and killed off a shrub or tree it was not readily spread to other trees because the crabs and hawthorns were dispersed among other species that were immune to the bacterium.

All of this changed in the 1700's when European settlers introduced susceptible apples from England and extremely susceptible pears from France. These were established in orchards where the bacterium quickly became adjusted to its new host and the more virulent strains could readily spread to the massed suscepts. The cultivation of choice varieties of pears such as Bartlett and Anjou was quickly rendered untenable in the colonies of Massachusetts and New York. It was driven ever-westward to western New York, to Ohio, and on to the West Coast. The bacterium moved along with its new host, reaching California in 1913. Those orchardists in the East were reduced to growing the poorer table varieties and the Keifer pear, which tolerates the bacterium, while the drier climate of the West restricted the bacterium enough to encourage the industry to establish itself there. The conditions of orchard sanitation, unfavorable ecology, and practice of other control measures have made the industry profitable in the West.

The lesson from fireblight is that a suscept that has never had any evolutionary experience with a crop pest should be introduced into an infested area with greatest precaution. If there is reasonable prospect of an attack, man must telescope the processes of pathologic evolution by locating new sources of resistance and breeding them into the desired crop.

This is not easily achieved, as is witnessed by the history of grape culture. When the colonists brought the fine wine varieties of their native lands into the New World, they soon discovered that they were not "adapted" to their new home and died out within a few years, while the native wild grapes thrived. What they did not realize was that the extremely destructive downy mildew fungus had forced a status quo on the wild grapes. In desperation, the colonists began to select the better strains of wild grapes for table and wine. There emerged the Concord and other distinctive varieties of North American grapes. The progressive grape interests of France saw merit in these American varieties so they introduced them to Europe. They made the mistake of carrying with them the dreaded *Phylloxera* insect. Since the American varieties were resistant to this pest, they returned to America to obtain resistant root stocks. This time they succeeded in introducing downy mildew to southern France. The wine industry was wrecked in less than a decade, and its restoration was achieved only by the fortuitous discovery of the fungicidal value of Bordeaux mixture.

The basic principles of epidemics are clearly illustrated by these three examples. There is an evolutionary process in which parasite and host gradually come into a balanced state of resistance and aggressiveness by a natural process of selection. The mixed ecosystem guarantees that if extremely virulent biotypes arise among the pathogens they automatically eliminate themselves by eradication of host in that sector of the ecosystem, while constant pressure by other strains is upgrading host resistance elsewhere. It is utterly fatal to introduce a new cultivar that has had no evolutionary experience with a particular pest into an established population of parasites or predators. Introducing a new parasite of a given species into an established plantation of one of its cultivars is another highly destructive act.

THE MISTAKES OF THE PAST IN PEST MANAGEMENT

Before devising a strategy for future pest management, it is well to consider the mistakes of the past. They fall into two broad categories—the errors of long-range strategy that have upset evolutionary balances, and the misguided tactics in day-to-day operations that have encouraged pest survival. It is only by learning from these that we can plan a strategy for the future.

We have been guilty of introducing susceptible varieties into an established population of pests. Ordinarily, this is not a major catastrophe since the variety is merely abandoned as unadapted, unprofitable, or inconsistently productive, depending on the seriousness of the maladjustment. However, if it has unique horticultural attributes as a crop or has other desirable attributes, it may be given special care and cultivated at great expense and demands upon ingenuity. The much more serious economic problem is when a parasite or other pest is introduced into an established system where climatic conditions favor its multiplication. History is replete with such examples, and it fully justifies every effort at quarantine despite the inadequacies of such control.

The second great error in strategy has been to resort to intensive agriculture where monoculture of crops and crowding of livestock and man has proceeded apace. Where a pest is introduced under these conditions, it spreads like wildfire by contact or near-contact, poor sanitation, and cyclic recurrence of favorable conditions. The use of crop rotations, careful choice of site, etc., are partial palliatives, but they are inadequate because of the basic crowding of suscepts.

The scientist has compounded these errors of strategy by manipulating genes so heterogeneity of populations is lost. There was a time when the heterogeneity of population was ensured by use of open-pollinated varieties of crop and use of rugged mixed breeds of livestock, but this has been lost by intensive selection, inbreeding, and hybridization. In a pest-free situation or even in a mild infestation, there is much economic advantage in using crops and livestock with genetic purity. The use of a single gene or a specific complement of genes to ensure uniform growth, maturity, and gustatory quality, or desired nutritional value is essential in our modern methods of agricultural management. Farm mechanization and livestock management could never have reached its present state had we not sought this uniformity. However, in so doing, we exposed our wards to an enormous hazard of attack from pests.

Even the breeding of crops for pest resistance increases this hazard. Plant pathologists have had three-quarters of a century of experience in this area; they now know the hazards to be faced, but they still have faith in the potentials of genetic resistance in crops. However, when resistance is bred into a crop, a single gene change in the parasite can render a uniformly resistant variety uniformly susceptible. For example, in breeding into oats the Victoria gene for hypersensitivity to numerous races of crown rust (*Puccinia coronata*), they created uniform susceptibility to a minor pathogen of grasses that came to be known as *Helminthosporium victoriae*. It came perilously near to eradicating oat culture before the Victoria gene could be replaced. Also, incorporation of a gene for male sterility into breeding stock of corn hybrids set the stage for another species of *Helminthosporium* to severely damage the corn crop of 1970. The lessons are obvious; genetic purity is a wonderful thing but it is hazardous.

Among our problem-creating situations, there are many soil management practices that have intensified pest control problems. Force-feeding of plants renders them more susceptible to the attack of highly specialized pathogens such as the downy mildews, powdery mildews, and rusts. An excess of nitrogen is almost uniformly deleterious to disease-escaping and disease-resistant attributes. A proper balance of nitrogen and phosphorus is a strong contributing force to escaping full impact of some of the less highly specialized parasites, such as the root-rot diseases. About the only situation where good crop growth and parasite resistance coincide fully is when potassium supplies approach optimum. The tendency toward feeding soils heavily early in the season so as to establish reserves for the crop throughout the season may create an unfavorable condition for seedling blights and promote early establishment of root-rot pathogens which can do damage later, such as the case of take-all disease of wheat (*Ophiobolus graminis*).

The intensive cultivation of crops and failure to restore organic material creates a hazardous condition by depleting the abundance and diversity of soil microorganisms. There is little doubt that the antibiosis of soil-inhabiting plant pathogens contributes materially to suppression of their activities. This has been clearly demonstrated by adding stimulants for bacterial antagonists such as sugars to soils inhabited by the strawberry root-rot complex or plant tissues to the soils dominated by the Texas root-rot fungus *Phymatotrichum omnivorum.*

STRATEGY FOR THE FUTURE IN PEST CONTROL

Obviously, we have made many mistakes in the past. We should heed their lessons in planning for the future. Research has swept forward so strongly that there are opportunities on a hundred fronts that must be explored.

In general, we now have a good picture of what the major pest problems are and we know a few things about their potential for change and adjustment. We know a great deal about their life habits, reproductive processes, and symbiotic relations. There is reasonable knowledge on the impact of environment on pests. A modest start has been made in determining the intricacies of biochemical processes and the response to exogenously applied chemicals, but most of the pattern remains to be filled in. Within the limits of what we now know, we may design a pattern on what seems logical.

The principle that must be faced before getting down to cases is that the human race must face a trade-off between multiple opportunities. There can be no realistic program before we decide how much we are willing to sacrifice. One can grow a perfect apple fruit in its own little biosphere, as the Japanese do, in a paper bag placed over the chosen blossom as soon as it is fertilized. Do we want to invest labor in producing perfect fruit of elegant size, free from blemish and from chemicals, that would probably cost $2 apiece? Possibly, if we persist in having millions of unemployed living on a dole. Both they and society might be better off producing such commodities for the carriage trade.

Or we might reverse the trend of two centuries and restore 75 percent of the urbanites to the rural scene and make them the servants of domestic plants and animals. They could do such things as picking beetles off potatoes and ticks off cattle. For that matter, we might abandon potato culture, for man once lived without this crop before the New World was discovered. These are the extreme alternatives we could trade on; but I doubt whether the public would accept them or even that they could be imposed by martial law.

It would be almost as logical to argue that man made his first mistake when he put a plow in the soil and started an erosion of ecosystems that has grown in intensity with every decade and every expansion of human population, and that the only way to correct the situation would be to eliminate 75 percent of the people and abandon intensive agriculture.

The basic assumptions that seem valid in planning our strategy within which we should plan our trade-offs, are as follows: First, modern intensive agriculture is here to stay and, as long as fuel is available, it will be mechanized. Therefore, monoculture of crops, large livestock feeding operations, and other intensive developments are a necessity if mankind is to be freed for maintaining and expanding his industrial empire and practicing the creative arts. Second, soils are to be managed principally to stimulate productivity and only secondarily as a pest control operation, but they can be modified to avoid hazardous conditions. Third, crop and livestock genetics will have to be manipulated but with the first objective of avoiding pest inducement. Fourth, every weakness in the pest's biocycle must be exploited to suppress its survival, reproduction, and multiplication. Fifth, the environment must be managed so as to suppress or destroy the primary pest or to alleviate its attack by altering ecological conditions where possible or by using properly designed chemicals when necessary.

Proper Managerial Practices These must be observed to suppress pest populations. The crop rotation has gradually been shortened as overhead costs in special equipment increased and food processing plants have forced concentration of production in specific locations. However, some diversity of crops in an area must be maintained to ensure suppression of some pests to tolerable levels by starvation, to permit some replenishment of organic matter in the soil so as to stimulate antibiosis or competitive symbiosis against pests, and to create a conservation bank for nutrients and moisture against the inevitable stress periods. The use of fertilizers must be continued and expanded, but to maintain steady, pest-resistant growth and avoid environment impairment by translocation, the application practices must be changed. However, it is essential that controlled availability of fertilizers be achieved, either by encapsulation of readily available elements or by more extensive use of spaced application such as side dressings.

Breeding of Crops This is still a profitable necessity for reasons outlined above. However, the emphasis on objectives must be shifted so tolerance or, preferably, resistance to pests should be given prior consideration. This means that every breeding stock must be catalogued for resistance to the gamut of pests before time and money have been invested on a massive crop or animal breeding program. We have barely begun to meet this requirement, because all of the biotypes of each pest must be included in testing.

The breeding for pest resistance must be based on three new concepts, one of which is quite obvious and the others essential but more imaginative. The first is use of multigenic sources of resistance. Monogenic resistance is so fragile that it can be circumvented by mutation or reshuffling of genes in the pest. Rarely has any disease-resistant crop variety provided perfect protection decade after decade. The single factorial resistance is the easy one to transmit and to incorporate into a commercially-favored variety, but it is also the easiest to overcome. The theory of multifactorial forms of resistance is that while it may be relatively simple for a single mutation or recombination of genes to occur in a given pest, the simultaneous appearance of two such changes by pure chance increases exponentially.

The basis of inherited resistance must be determined by intensive biochemical–physical research. We must know why a crop is resistant so we will know what we are seeking. Once we understand the chemistry of resistance, we can seek methods of inducing it artifically or of locating materials like it during the attack on the plant. For example, resistance to some plant diseases is incited by a parasitic establishment that induces a phenolic toxicant (known as a phytoalexin) without necessarily flooding the whole plant with a potentially undesirable toxicant. The goal of genetic resistance should be positive and not merely passive (if at all possible), and this connotes development of a toxicant so it is absolutely essential that we understand the chemistry of resistance.

The Introduction of Synthetic Genes by means of the new skills of molecular biology is the third potential on the horizon. To a certain extent this has been achieved by chance improvisation in using irradiation or other mutagens to create molecular realignment in the normal cell genome. However, it is now possible to synthesize nucleotide chains and to simulate small fragments of DNA and RNA. This is important because there is an excellent working hypothesis in plant pathology that there are matching genes for resistance in the host and for virulence in the pathogen. Disease resistance for a highly specialized parasite such as a rust fungus occurs when the host has factors not matched by the parasite or, conversely, when the parasite has a capability not blocked by the host's genome. It is none too early to start unraveling the coding system of pests to see whether we can dislocate normal gene functions. It would seem that the attack in this area should be launched with RNA components since we have at hand tremendous knowledge on transmission of such molecules as plant viruses and they are not likely to be disruptive of the central control system that resides in the DNA of the host or to present such potential hazard to the consumer of produce.

Disruption of the Capabilities of Pests for Reproduction and Attack
This is an essential part of any strategy for pest management. There are any number of practices to escape pest-infested areas, suppress inoculum potential, interfere with survival over winter, etc., that come under good crop and livestock management. These include sanitation, plowing under infested refuse, encouraging antibiotic forces in the rhizosphere or elsewhere, etc. However, there are many more positive direct effects that can be developed.

The use of male-sterile techniques for suppressing insect populations is now an established reality. It can be made to work on insects other than the screw worm and will expand as methods of rearing, sterilizing, and dispersal of insects are improved. The initial investment is tremendously high, and the method is most effective when populations are modest or at a low ebb. However, it is extremely effective in eliminating the residual population that is so difficult to control by ordinary eradicative measures.

The use of hormonal treatments to delay or prevent sexual maturity falls into a similar category and offers much hope for the future. The question may be raised as to why these advances in the field of entomology have not found a counterpart in control of plant diseases and weeds. It is an area of unexplored possibilities that should prove fruitful in weed control. Some of the more difficult problems could be solved better by eliminating seeds than by attempting the eradication of seedlings if suitable methods could be developed for interfering with gametogenesis, pollination, or embryo development. Most of the plant pathogens, however, propagate primarily by asexual devices so that less opportunity exists, but some limited experiments have been made on suppression of conidia formation.

Most of the pests could be controlled if we could get at them with the weapons we have available. However, their populations are so widely dispersed and they multiply with such explosive violence that the methods cannot be applied. If these populations could be manipulated or moved at will, their destruction could be achieved. A good example is the many species of bark beetles that attack and destroy our forest trees. These tiny beetles spend 95 percent of their life cycle, or more, securely embedded in the bark of the trees where they feed, cohabit, and produce a new generation secure from man's anger. For over 100 years we have had to be satisfied with cutting down the dead and dying trees and removing them from the forest—usually after the new brood has already emerged and flown to a fresh victim. It is only within the past five years that we have detected the population-aggregating pheromones that guide them to susceptible trees in sufficient numbers to ensure that they will overwhelm the tree's resistance. By isolating and identifying these pheromones produced in the hindgut of the pioneering beetles, we have been able to synthesize them, and when they are supplemented with monoterpenes of their hosts, we have been able to direct their flight to predetermined spots. Once we have been able to fight them on our terms rather than theirs, new vistas have been opened for attracting them to traps, encouraging attack on trees or blocks of trees that are scheduled for removal, or directing their flight to areas where their predators and parasites abound. Interestingly enough, the clerid and ostomid beetles that prey upon the bark beetles have learned to seek out these chemical messengers so we can now upset the balance of nature in favor of the predator, which becomes a much more effective ally in our battle.

A comparable use has been made with lures for certain Diptera such as the Mediterranean fruit fly where a combination of chemical toxicants and trapping has proved very effective. The use of sex attractants as lures has proven effective for many of the Lepidoptera and Orthoptera insects so one of the sexes can be trapped and removed from the reproductive activities.

The strategy of the future must inevitably use all the natural enemies of the pest. There are many weeds that are partially debilitated by disease and insects and there is demonstrated ability of some beetles to provide promising control. Likewise, substantial progress is being made with population suppression of moths and caterpillars by highly potent viruses. The bacterial, fungal, and nematode pathogens of insects have had modest success, but they and their toxins may be be expected to contribute a measure to population suppression techniques.

The Modification of Environment This has always offered considerable promise either by changing the ordinary factors of climate so reproduction and growth is suppressed or by adding toxicants designed by man. The fundamentals of housefly control still rest upon sanitation so the larvae cannot develop. There are many plant diseases that can be tolerated if the crop is placed on the proper soil type, or where air circulation is good and exposure to light is such that the fungus spores cannot germinate. By proper spacing and pruning of orchards, the severity of infestations can be suppressed, if not fully controlled. A crop such as sugar beets can escape leaf blight caused by *Cercospora beticola* if the beets are cross-checked and spaced so drying air can circulate around the lower leaves while the same crop sowed in drills that soon fill the aisles is rapidly blighted.

Chemicals These have always been the ultimate insurance against pests. There is no need to dwell upon the rapid expansion in this area since 1941 when the organic insecticides, fungicides, and herbicides were discovered and it became literally possible to design a molecule to fit a specific need. Safety for crop, man, and livestock was increased; pests that had never been suppressed by any device were brought under effective control.

These materials are going to be used extensively in the foreseeable future. As in the past, and in light of all the progress outlined above for other measures, there probably will be needed not only a larger volume of such chemicals but a greater diversity of materials. The various biological and managerial control devices outlined above are fundamental to sound agricultural practices but they depend on a normal condition for their effectiveness, i.e., on a stable population of pests, on a reasonable climatic condition, on proper timing of routine operations, on a healthy condition for antagonistic biological forces, be they parasites, predators, or simple antagonists. Therefore, the normal fluctuations in every biosphere will create crises from time to time. The use of chemicals to disinfest, to suppress, to destroy the pest, or to permit the crop to escape the most extreme ravages will remain a potent force.

Without this form of crop insurance, most agricultural practitioners cannot afford to invest in expensive equipment, precise fertilizer components, and tailor-made crop varieties. All of these things can be afforded only when the farmer takes out crop insurance of several sorts, including the suppression of pests. These chemicals of the future are going to be more expensive because they must be more specific so as to avoid adverse side effects, must meet rigid standards of environmental safety, and must be distributed and used with better control than at present.

There is no reason why these chemicals cannot be used effectively and be accepted by the general public as an essential and safe component of an industrialized urbanized society. In the current demands for environmental protection that often reach hysterical proportions, we became acutely aware that fear is bred from ignorance and a sense of insecurity. Toxicological methods have never been so perfect that they reassure the layman who has a hard time in grasping the significance of a dosage-response curve. Nearly every toxicant known for any form of life can have an effect on other forms. The question is, How much and how frequently can each accept the chemical? From the experience of the past 30 years, one can safely say that safe chemicals can be designed to meet the most exacting demands for crop and consumer safety. We are now in the process of adding a new dimension to our standards—safety for nontarget areas of the environment.

Analyzing the experience since 1940 objectively, we find that three strategic errors were made in designing molecules in light of this new demand. They were the following: the search for lipophilicity to promote rapid permeation of the living cell, unconcern for adsorption to soil colloids and organic matter, and a conscious effort to choose wide-spectrum activity so as to secure as large a market as possible. There is every reason to believe we can correct these errors.

Lipophilicity of a molecule is not absolutely essential, as can be shown by the toxicity of the silver, copper, or mercury ions, not to mention many hydrophilic organic molecules. The pesticide dissolved in body fat is so immobilized that it presents no great hazard to health as shown by feeding experiments with DDT on man and domestic animals. As long as the intake is held at nominal levels, the material partitions into the fat at a constant level and the remainder is detoxified and excreted. The immobilized molecules are outside the pool of toxicant except under two circumstances. If starvation diets are suddenly imposed, a toxicological manifestation could be visualized. Furthermore, as we all know, there can be a magnification of effects when the carrier (fat) is consumed in a food chain in which the end consumer is a cold-blooded animal or fowl, as occurs in certain forms of predaceous wildlife.

The tendency of a pesticide to adsorb onto and into clay particles may lead to its survival in lakebeds and sluggish streams. An organic phosphate that is normally either hydrolyzed or oxidized may adsorb into the alumina sheet of a montmorillonite clay where the silica sheets to either side protect it from direct action by microbial decomposition since enzymes cannot reach it. One may properly question whether we should be concerned with such hidden reserves but they do show up in analyses as residues in soil sludges.

The general pesticide will probably give way to the more specific types as demands for safety gain prominence. The greater cost can be justified only by increased security and less chance of hazard to unspecified target subjects. There are many unresolved pest problems especially among plant diseases that can be satisfied by specially designed molecules that simply were not profitable to search for and produce under past standards. If the demands for pest control continue to the point of such perfection and the demands of the public for absolute safety for the environment are pushed to the extreme, these high-cost chemicals will be inevitable. The public will have to pay the bill, but only time will prove whether it is through higher-priced commodities or taxes to provide the subsidy.

SUMMARY

There are no general rules that apply to all kinds of pests under all circumstances, but there are a few general principles that can be applied in working out a game plan for the future:

- Most pests are shifty, versatile adversaries capable of adapting to their host, their environment, and our scheme. Even with the great opportunities for research afforded by modern concepts, man cannot expect to achieve perfect control, much less eradication of any substantial number of them. Our strategy calls for learning to live with them.

• There are certain evolutionary schemes of predation and parasitism that will not be denied. Under normal conditions, a crop and a pest such as fungous parasite come into a balanced condition where severity of attack may vary from time to time and as new biotypes evolve; but ordinarily a status quo condition evolves in which the pest and its host live as commensals.

• Man has a propensity for disturbing this balance by his material needs of the moment by introducing a susceptible progeny into an established population of pests, by carelessly introducing a pest into a population of suscepts that have had no evolutionary experience with it, or by changing the environment or population arrangement so the pest can run wild. The result is a disaster that cannot be tolerated in the pest management strategy of the future.

• The first requirement is to use varieties and breeds that have maximum tolerance, resistance, or capability to recover from attack. Even the best germplasm composition possible is inadequate to meet all pests under any circumstance, but if a measure of resistance is present it permits greater latitude in other strategies.

• In breeding new crops, the primary objective should be for pest-resistance rather than produce quality or productivity since these attributes can be superimposed on the more fundamental requirement. Even the best of resistance can be expected to break down as new biotypes evolve in the pest population.

• Management of crop and livestock should be designed to escape or suppress the pests by sanitation, sterilization, choice of sites, etc. Such measures must be designed to be operative under certain normal ecological conditions and can be expected to break down when those conditions deviate excessively.

• Direct action may be taken against the pest by suppressing its reproduction by use of sterile males, interfering with the hormonal balance necessary for sexual maturation, or trapping one sex by use of sex attractants.

• Insect populations can be manipulated or aggregated under conditions where they can be trapped or killed by use of food pheromone complexes.

• The destruction by parasites and predators can be enhanced by drawing a pest into a site where such agents thrive or can be assembled by pheromones.

• After all the foregoing have been put into practice, conditions still may develop when pests develop explosively because of inevitable shifts in the environmental conditions regulating the pest development, changes in physiological resistance, etc. The various biological control devices are based upon well-defined norms of environment, and modest changes from this norm can wreck the best laid plans.

- All the evidence suggests that pesticides will be used extensively in the foreseeable future. They are the crop insurance that makes it possible for farmers to invest heavily in mechanization, soil fertilization, crop adaptability, and other expensive practices of the farm revolution of the past fifty years. The chemicals are part of a production team that must be further perfected if we are to meet the growing demands for food and fiber.
- There will be need for many new types of chemicals in the years ahead. The current situation would seem to demand compounds tailor-made to provide (a) more specificity so there will be less opportunity for unexpected side effects; (b) a reasonable margin of safety for applicator and consumer; and (c) less hazard to a sound environment and its valued wildlife.
- Research on new types of molecules must be based on the assumption that pesticides will be more expensive and the public will have to pay for its greater demands. The errors of the past in the design of molecules will have to be reassessed and the following characteristics avoided: strong lipophilicity to promote cell permeation, since it promotes transmission in the food chain of predaceous wildlife; strong adsorption on and in clay or organic matter where it is not subject to microbial or chemical decomposition; and resistance to microbial decomposition in soil and water. Although microbial decomposition may be promoted by training soil microorganisms to be more aggressive for new classes of molecules, the compounds should be readily susceptible to β-oxidation, hydrolysis, or oxidation by enzymic action.

III
Strategies for Better Chemical Control

Development of Selective and Biodegradable Pesticides

Robert L. Metcalf

Pesticides are of such fundamental importance to the human race, both for the production of adequate quantities of food and fiber and for the preservation of adequate standards of public health, that no responsible authority has envisioned a time when their use will not be required. The Mrak Commission of the Secretary of Health, Education, and Welfare (1969) stated, "our need to use pesticides and other pest control chemicals will increase for the foreseeable future." However, this virtual indispensibility does not continue to give license for using pesticides on an infinitely increasing scope or irresponsibly with regard to the quality of the environment. The volume of the outcry about the side effects of pesticides on nontarget organisms and upon environmental quality provides substantial proof that changes must be made in pest control practices and, in some cases, in the nature of the pesticides themselves. It must be remembered always that pesticides are applied to the environment as purposeful contaminants; consequently, the benefits from their use must greatly exceed any damage to environmental quality.

The importance of changes in the use patterns of pesticides in reducing environmental contamination falls outside the subject of this discussion. Adequate pest control can be achieved in most cases with lower volume applications more precisely timed and placed. The fixed volume of the biosphere cannot tolerate the insult of a total growth rate in pesticide applications of nearly 10 percent per year *ad infinitum*. Even more important is the nature of the pesticide chemicals with which man carries out his purposeful environmental contamination. The theme of this discussion is that we have failed to use adequately our sophistication in synthetic chemistry to produce superior pesticides that are primarily selective against pests and essentially harmless to higher animals, and that are appropriately biodegradable in the environment thereby preventing widespread pollution.

ACUTE TOXICITY OF PESTICIDES

It is disturbing that progress in pesticide design has given us chemicals that are not only more and more effective against pests but also are, in many cases, increasingly more toxic to humans and higher animals. Table 1 provides an indication of the progress that has been made during the last century in producing synthetic insecticides. The compounds available in each decade are rated by their rat oral LD_{50} as to the *most hazardous* and *safest.* For reasons difficult to explain, evolution in the insecticide area seems to favor the widescale use of compounds highly toxic to man, while their more innocuous relatives are relegated to minor applications or back to the shelf. Many of the newer compounds are essentially nonselective as insecticides, i.e., the ratio of their LD_{50} values to mammal versus insect approaches 1.0. For the purpose of this discussion we shall use the female rat or mouse oral LD_{50}/female housefly topical LD_{50} as the mammalian selectivity ratio, or MSR.* This gives a simple and generally practical measure of the relative hazards of acute usage for the various insecticides (see Metcalf, 1964, Winteringham, 1969).

Mammalian selectivity ratios for most of the widely used organochlorine, organophosphorus, and carbamate insecticides are presented in Tables 2, 3, and 4. Great differences in the MSR values exist among each of the three classes of insecticidal compounds. DDT, MSR ca 60 (Table 2), has the lowest ratio of the commonly used DDT analogues and methoxychlor is approximately 10 times more selective. However, from the acute toxicity viewpoint DDT obviously provides no hazards to the human population as proved by worldwide use for a

*This is justifiable not only because the two species are the most commonly employed toxicological test organisms for mammalian and insect toxicology, but also because the largest variety of precise LD_{50} data is available for them. Further, the oral route represents a common one for poisoning in mammals and the cutaneous or topical the common one for insects.

TABLE 1 Comparative Acute Mammalian Toxicity of Insecticides in Use in the United States, 1900 to 1970

Decade Ending	Toxicity (♀ rat, oral LD_{50} mg per kg) Most Hazardous	Safest
1900	Paris Green (100)	Mineral oil (> 10,000)
1920	Nicotine (30)	Cryolite (13,500)
1940	Dinitro-*o*-cresol (30)	Pyrethrins (820)
1950	Parathion (3.6)	Methoxychlor (> 6,000)
1960	Phorate (1.1)	Ronnel (2,630)
1970	Aldicarb (0.96)	Phoxim (8,500)

generation in human habitations for malaria control and eradication without evidence of ill effects. No human fatality has been attributed to DDT.

For the cyclodiene insecticides the record is not unblemished. Chlordane, MSR 72 (Table 2) has been widely used as a household insecticide and for termite control without injurious effects. How-

TABLE 2 Mammalian Selectivity Ratios for Organochlorine Insecticides

	LD_{50} mg per kg		
	♀ Rat–Oral[a]	♀ House Fly–Topical[b]	MSR
DDT derivatives			
DDT	118	2.0	59
DDD	3400	19.5	174
Methoxychlor	6000	9.0	668
Ethoxychlor	300[c]	2.9	104
Methylchlor	3350[c]	21.0	180
Prolan®	4000	4.8	840
Cyclodienes			
γ-Chlordane	430	6.0	72
Aldrin	60	2.25	27
Dieldrin	46	0.95	24
Endrin	7.5	3.15	2.4
Heptachlor	162	2.25	72
Dihydroheptachlor	5000[d]	3.75	1330
Isobenzan	7	1.25	5.6
Thiodan	18	4.15	4.3
Lindane	91	0.85	107
Toxaphene	80	11.0	7.3

[a]Hayes (1963) except as indicated.
[b]Original data from this laboratory (WHO, 1968).
[c]Mouse oral LD_{50}, Domenjoz (1946), Von Oettingen and Sharpless (1946).
[d]Buchel *et al.* (1964).

TABLE 3 Mammalian Selectivity Ratios for Organophosphorus Insecticides

	LD_{50} mg per kg		
	♀ Rat–Oral[a]	♀ House Fly–Topical[b]	MSR
Parathion	3.6	0.9	4.0
Methyl parathion	24	1.2	20
Chlorthion	980	11.5	85
Dicapthon	330	1.5	220
Ronnel	2630	2.0	1315
Bromophos	3750[c]	3.2	1170
Iodofenphos	2000[c]	2.5	800
Fenthion	245	2.3	106
Fenitrothion	503	2.3	219
EPN	7.7	1.2	6.4
Malathion	1000	26.5	37.7
Diazinon	76	2.95	25.8
Coumaphos	15.5	3.2	4.8
Dimethoate	215	0.55	390
Guthion	11	2.7	4.1
Demeton	2.5	0.75	3.3
Mevinphos	3.7	1.5	2.5
Dichlorvos	56	1.8	31.1
Naled	430[c]	1.25	344
Phoxim	8500[c]	2.3	3700

[a]Hayes (1963) except as indicated.
[b]Original data from this laboratory (WHO, 1968).
[c]Kenaga and Allison (1969).

ever, several human deaths have been reported following suicidal attempts (Hayes, 1963). Dieldrin, MSR 24, has been used safely in house spraying for malaria eradication but can be toxic under conditions of repeated exposure as 2 to 40 percent of spraymen have developed symptoms of poisoning in from 2 weeks to 24 months exposure. Endrin, MSR 2.4, and isobenzan (Telodrin®), MSR 5.6,

TABLE 4 Selectivity of P=O Analogues of Methyl Parathion and Fenitrothion[a]

	LD_{50} μg per g			K_e for ChE l/mole/min	
	Mouse–Oral	Housefly–Topical	MSR	bovine AChE	fly AChE
R=					
H	21	2.5	8.4	5.2×10^5	2.9×10^5
CH_3	120	4.3	28	0.73×10^5	7.6×10^5
$(CH_3)_2CH$	> 500	6.5	ca 100	0.22×10^5	22.6×10^5

[a]Hollingworth *et al.* (1967a).

are considerably more hazardous. Mass epidemics of human poisoning have followed leakage of endrin solutions into flour during shipment (Mrak, 1969) and endrin poisoning has occurred in formulators and applicators (Hayes, 1963). Fewer data are available for human poisoning by other cyclodiene compounds in Table 2. However, human deaths have resulted from ingestion of lindane and toxaphene, especially in children. In view of the very high MSR value for dihydroheptachlor, together with its substantial insecticidal activity, it seems regrettable that this selective insecticide has not been marketed commercially (Büchel *et al.*, 1964).

ORGANOPHOSPHORUS INSECTICIDES

The vast majority of pesticide poisonings and fatalities in the United States result from exposure to organophosphorus insecticides, especially parathion. This compound is a general biocide with a low MSR of 4.0 (Table 3). Its analogue, methyl parathion, has a substantially higher safety factor, MSR 20. These insecticides are used very widely in the United States, and the 1966 annual production was parathion, 19,444,000 lb and methyl parathion, 35,862,000 lb (Mrak, 1969). Largely as a result of the widespread use of these highly toxic compounds, Blackbourne (1970) has estimated that there are 2,000 nonfatal organophosphorus poisonings in the United States and 200 fatal poisonings. In Dade County, Florida, alone, of 79 accidental poisonings documented between 1964–1968, 51 percent involved parathion. In Dade County, between 1959–69, there were 56 deaths resulting from organophosphorus poisoning, 28 accidental and 26 suicidal. Eighteen children died from contact with improperly stored organophosphorus pesticides (Blackbourne, 1970). On a worldwide basis, the toll is much greater, especially in relatively unsophisticated societies. Contamination of wheat with parathion in India, and flour in Colombia resulted in 102 and 88 deaths, respectively (Mrak, 1969).

Selectivity of O-P-Esters of Parathion Type

The illnesses and deaths from parathion poisoning are particularly unfortunate because they could largely be prevented if some of the more selective organophosphorus insecticides shown in Table 3 were substituted for the essentially nonselective parathion and methyl parathion. It is a reasonable generalization that the *O,O*-dimethyl phosphorothionates are substantially more selective to mammals versus insects than the corresponding *O,O*-diethyl compounds. This appears to be a reflection of the greater metabolic instability of the CH_3OP- bonds, which provides increased opportunity for mammalian detoxication during the "delay factor" resulting from *in vivo* activation of P=S to P=O. The discovery of the notable selectivity of Chlorthion® *O,O*-dimethyl *O*-(3-chloro-4-nitrophenyl) phosphorothionate (Schrader, 1961), with a MSR of 85, was particularly significant as it demonstrated that a simple molecular change in methyl parathion, i.e., the addition of the 3-chloro atom, increased the selectivity by at least fourfold, although there is a loss of insecticidal potency and in spectrum of activity. Dicapthon, the 2-chloro- analogue has higher insecticidal activity and an even higher MSR of 220 (Table 3). The 3-methyl- analogue, fenitrothion, *O,O*-dimethyl *O*-(3-methyl-4-nitrophenyl) phosphorothionate (Drabek and Pelikan, 1956) combines a high degree of insecticidal activity with a very substantial MSR of 219. The use of fenitrothion as a general insecticide for the control of rice pests in Japan and the Far East in place of methyl parathion has virtually eliminated the serious problems of human poisoning frequently encountered with the latter. It has also proved effective and safe for residual spraying for malaria control in human habitations. Other notably selective and useful compounds in this series include the trihalogenated derivatives ronnel *O,O*-dimethyl *O*-(2,4,5-trichlorophenyl) phosphorothionate, MSR 1315; bromophos *O,O*-dimethyl *O*-(2,5-dichloro-4-bromophenyl) phosphorothionate, MSR 1170, and iodofenphos *O,O*-dimethyl *O*-(2,5-dichloro-4-iodophenyl)-phosphorothionate, MSR 800. Efforts to introduce any of these safer compounds into large-scale usage in the United States as replacements for parathion and methyl parathion have failed because of the very low cost of the parathions and their high insecticidal effectiveness. However, it seems that a small additional cost per pound is a small price to pay for a substantial saving of human life and for a much better public attitude regarding the use of insecticides.

Mechanisms of Selective Action

The selectivity of the methyl parathion analogues provides an appropriate example of the possibilities for the design of better pesticides through understanding the biochemical nature of the selective process. Plapp and Casida (1958) compared the metabolism in the rat and American cockroach of parathion, methyl parathion, Chlorthion®, dicapthon, and ronnel, using radiolabeled compounds. In the rat, hydrolysis of the alkyl phosphate bond was substantially greater with the dimethyl than the diethyl phosphorothionates and this was suggested as a factor in the lower mammalian toxicity of the former (Table 3). The rate of alkyl phosphate hydrolysis for the "safe" compounds such as ronnel, dicapthon, and Chlorthion® was substantially greater at high than at low dosages, and it was concluded that this metabolic pathway was an alternate mechanism for detoxication in mammals at high dosage levels, but was substantially less efficient in the cockroach. Hollingworth *et al.* (1967b) studied the metabolism of methyl parathion and fenitrothion in the mouse and house fly. The patterns were qualitatively similar but quantitative differences related to the much greater amounts of P=O hydrolysis products in the mouse, 60–70 percent as compared to 15–35 percent in the house fly, and to the substantially greater role of desmethylation in the mammal, 26–42 percent, as compared to 10–18 percent in the house fly. Thus the oxidative and phosphate-degrading mechanisms of the insect have a lower total efficiency than those of the mammal and this is reflected in the higher toxicity of both methyl parathion and fenitrothion to the insect. The methyl phosphonate of fenitrothion had a MSR of 7.4 compared with 540 for the parent compound, using mouse LD_{50} data. As the phosphonate is not readily *O*-desmethylated, this pathway is seen as a key process in selectivity.

Selectivity in this class of compound is also related to target site reaction of the P=O analogues produced *in vivo* by microsomal enzymes. Hollingworth *et al.* (1967a) considered the specific details of reaction of the P=O analogues of methyl parathion and fenitrothion with bovine erythrocyte and house fly head cholinesterase. Details of the selectivity of these activated P=O analogues are shown in Table 4. It is apparent that successive substitution of methyl paraoxon in the 3-position with CH_3 and $(CH_3)_2CH$ dramatically decreases the toxicity to the white mouse without substantially changing the toxicity to the house fly thus increasing the selectivity of the compounds. There is a corresponding relationship with inhibition of bovine and fly acetyl cholinesterase in that reactivity with fly AChE increases with introduction of alkyl groups in the 3- position in the order $H < CH_3 < (CH_3)_2CH$. For bovine AChE the effect is exactly opposite and reactivity decreases with $H > CH_3 > (CH)CH$. Hollingworth *et al.* (1967a) interpreted this data to reflect structural differences in bovine and fly AChE, particularly in the distance between esteratic and anionic site. Recent study in our laboratory has fully confirmed this earlier work and has shown that Chlor-oxon and dicapthoxon also have enhanced reactivity for fly AChE as compared to bovine AChE (Metcalf and Metcalf, 1972). It seems clear that the degree of selective toxicity demonstrated by the P=O analogues of the methyl parathion derivatives can be accounted for by specific differences in reactivity between mammalian and insect cholinesterase. The much higher selective toxicity ratio, demonstrated by the P=S analogues of Table 3, results from the combined action of reduced target site reactivity with AChE, and enhanced detoxication through desmethylation as a result of the delay factor provided by P=S to P=O oxidation. Although the biochemical explanation for the high degree of selectivity of the methyl parathion analogues in Table 3 is complex, it provides a rational approach to the development of organophosphorus compounds of high selectivity that could provide safe yet effective insect control.

CARBAMATE INSECTICIDES

The carbamates are the newest class of insecticides to have had large scale development. As seen in Table 5, their acute oral toxicity to

TABLE 5 Mammalian Selectivity Ratios for Carbamate Insecticides

	LD_{50} mg per kg		
	Rat–Oral[a]	♀ House Fly–Topical[b]	MSR
Carbaryl	540	900	0.6
Mobam®	178	17.5	10.0
Carbofuran	4.0	4.6	ca 1.0
Propoxur	116	25.5	4.5
Bux	87–170	65	1.3
Aminocarb	30–50	85	0.35
Zectran	39	65	0.60
Methiocarb	130–135	24	5.4
Carbamult	140	22	6.4
Landrin	178	41	4.3
Butacarb	> 4000	39.0	ca 100
Aldicarb	0.96	5.5	0.175
Methomyl	37	3.65	10
Dimetilan	25–64	9.0	2.8

[a]WHO (1968).
[b]Original data from this laboratory (WHO, 1968).

mammals varies from aldicarb LD_{50} 0.96 mg per kg (MSL 0.175) to Butacarb > 4000 mg per kg (MSL ca 100). Our understanding of the nature of the selective process for carbamates is less well developed than for the organophosphorus insecticides. In general, it appears that the insecticidal carbamates are of low selectivity and it is particularly interesting that the MSL value for carbaryl, a material of low mammalian toxicity, is approximately equal to that of carbofuran, a material of very high mammalian toxicity. However, carbaryl is substantially more toxic to other insects than the house fly, or it would have little insecticidal usefulness. The topical LD_{50} for carbaryl to the honey bee, for example, is 2.3 mg per kg. It seems clear that to systematically develop carbamate insecticides of high selectivity, we need to know much more about detoxication patterns in mammals and in insects and about the relative reactivity of mammalian and insect acetyl cholinesterases with specific carbamate structures (Metcalf and Fukuto, 1965; Metcalf, 1971).

PERSISTENT BIODEGRADABLE INSECTICIDES

DDT, dieldrin, and other organochlorine insecticides have been outstandingly successful in the control of noxious insect pests of public health and agriculture largely because of their persistence in a variety of ecological niches—soil, water, plant surfaces, animal bodies, and habitations. These insecticides are used as purposeful environmental contaminants on a very wide scale and it is only relatively recently that we have become aware of the ecological consequences of their ubiquitous presences as environmental micropollutants. Such substances produce adverse environmental effects at concentrations less than 1 ppm, have high physiological and ecological specificity, are resistant to biochemical degradation, and are sequentially concentrated in organisms of trophic webs, where they have a capacity for the delayed onset of intoxication (Warner, 1967).

DDT is the most widely studied micropollutant. Its undesirable effects result both from its persistence and from its slow degradation in biological systems to form DDE and DDD (Figure 1). These three substances are highly soluble in lipids (to about 100,000 ppm) and extremely insoluble in water (to about 0.002 ppm) so that their lipid/water partition coefficients approach infinity. As a result, they are stored in plant and animal fats and concentrated in higher members of food chains. The rate of concentration is a function of level of intake and time of exposure and is also dependent upon the rate

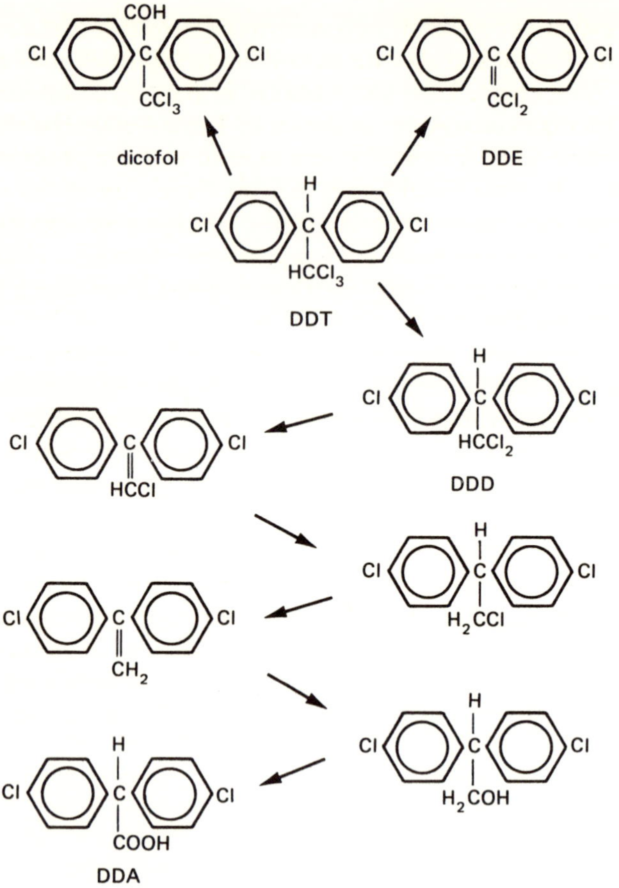

FIGURE 1 **Pathways of DDT metabolism (Kapoor *et al.*, 1970).**

of detoxication and excretion. Over wide limits, the level of storage in fat of laboratory animals and humans is directly proportional to intake (Mrak, 1969), although it appears that a plateau level may eventually be attained. This process of accumulation is especially dramatic in fish and other aquatic animals where rates of detoxication and elimination are very low. DDT and DDE can be absorbed directly from ppb concentrations in water and accumulated to levels of 10^5 to 10^6 those of intake. The rate of accumulation is proportional to the concentration of DDT in the water × time of exposure or CT (Figure 2). Thus the presence of trace quantities of DDT in lakes and estuaries can result in concentrations in fish that can cause reproductive failure and also are hazardous to raptorial birds at the top of the food chain. The concentration of DDT in components of the Lake Michigan ecosystem as shown in Table 6 after Harrison *et al.* (1970) illustrates the potentially serious nature of the problem. In large lakes such as the Great Lakes these pollution problems are amplified by the long water-retention times, averaging 30.8 years for Lake Michigan and 189 years for Lake Superior (Rainey, 1967). Thus, their contamination may be a major disaster for which there is no apparent solution.

It should be evident that there is urgent need to replace DDT and other persistent nondegradable (hard) insecticides with persistent yet biodegradable substitutes. The employment of methoxychlor as a biodegradable substitute for DDT was suggested at the Symposium on Scientific Aspects of Pest Control (Metcalf, 1966) and this compound is now being used as a replacement for DDT in such predominantly environmental uses as the control of blackfly larvae in streams and the elm bark beetle vectors of Dutch elm disease. We have established that methoxychlor, unlike DDT, is substantially biodegradable through attack by the multifunction oxidase or oxygenase enzymes found in living tissues. These are the drug metabolizing enzymes that degrade xenobiotic compounds. Through their action methoxy-

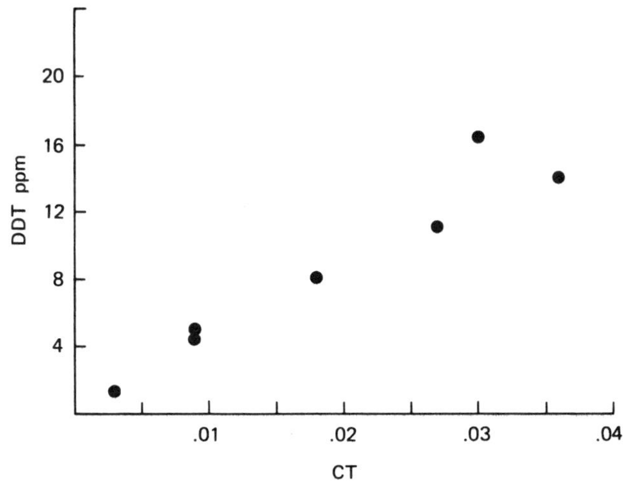

FIGURE 2 **Storage of DDT by green sunfish (ppm) as function of concentration in water (ppm) X days of exposure (CT) (data from Reinbold *et al.*, 1971).**

TABLE 6 Concentration of DDT in a Lake Michigan Food Chain[a]

	DDT in ppm
Water	0.000002
Bottom mud	0.014
Fairy shrimp	0.410
Coho salmon, lake trout	3–6
Herring gull	99

[a]Harrison *et al.* (1970).

chlor is O-demethylated to produce mono- and bisphenols that have much lower oil/water partition coefficients than the parent compound and are conjugated and eliminated from animal tissues (Kapoor *et al.*, 1970) as shown in Figure 3.

The DDT-type molecule can serve as a model substrate for at least four distinct types of multifunction oxidase attack: (1) ring hydroxylation, (2) O-dealkylation, (3) sulfide oxidation, and (4) side-chain oxidation. Typical reactions of DDT analogues as substrates for these enzymes are shown in Figure 4.

STUDY OF BIODEGRADABILITY IN A MODEL ECOSYSTEM

A model ecosystem for the precise evaluation of pesticide biodegradability has been developed in this laboratory (Metcalf *et al.*, 1971b). It consists of a 10 × 12 × 18 in. glass aquarium containing a shelf of washed white quartz sand, which is molded into a sloping soil–air–water interface. The lower portion is covered by 12 liters of "standard reference water" which provides satisfactory mineral nutrition for the growth of *Sorghum* on the terrestrial portion and *Oedogonium* algae in the aquatic portion. The latter contains a selection of plankton, *Daphnia, Physa* snails, *Culex* mosquito larvae, and *Gambusia* fish. The aquarium is aerated and is kept in an environmental growth chamber at 26° C and with 12 hr day-light exposure to 5,000 ft-c. A representation of the system is shown in Figure 5.

When the *Sorghum* plants are a few days old, the leaves are treated with 5.0 mg of radiolabeled pesticide in acetone, applied by micropipette. Ten large *Estigmene* larvae (salt marsh caterpillar) are placed in the chamber and feed until the plants are defoliated. The radiolabeled fecal materials thoroughly contaminate the aquatic part of the system and are taken up in the several food chains, e.g., algae → snail, plankton → *Daphnia* → *Culex* → *Gambusia*. After 33 days, the experiment is terminated and the various organisms are weighed and the radioactivity present is extracted and examined by TLC, radioautography, and scintillation counting.

Experiments of this nature have been conducted with DDT, DDE, and several DDT analogues and some of the results are shown in Table 7. It is evident that the pure *p,p'*-DDT applied to the plants was readily converted to large amounts of DDE and DDD and that the overall concentration of these three compounds from water to *Gambusia* was of the order of 100,000-fold. The situation in the

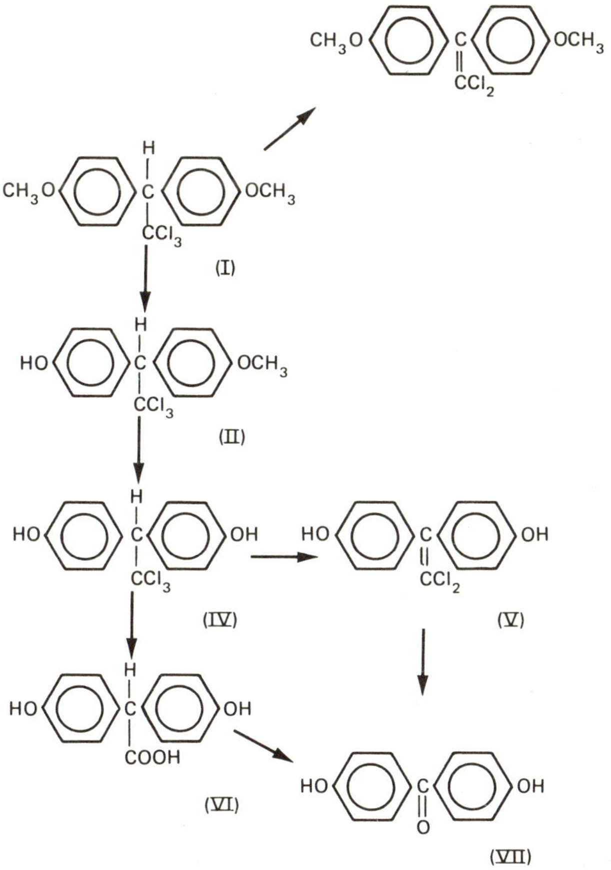

FIGURE 3 Pathways of methoxychlor metabolism (Kapoor *et al.*, 1970).

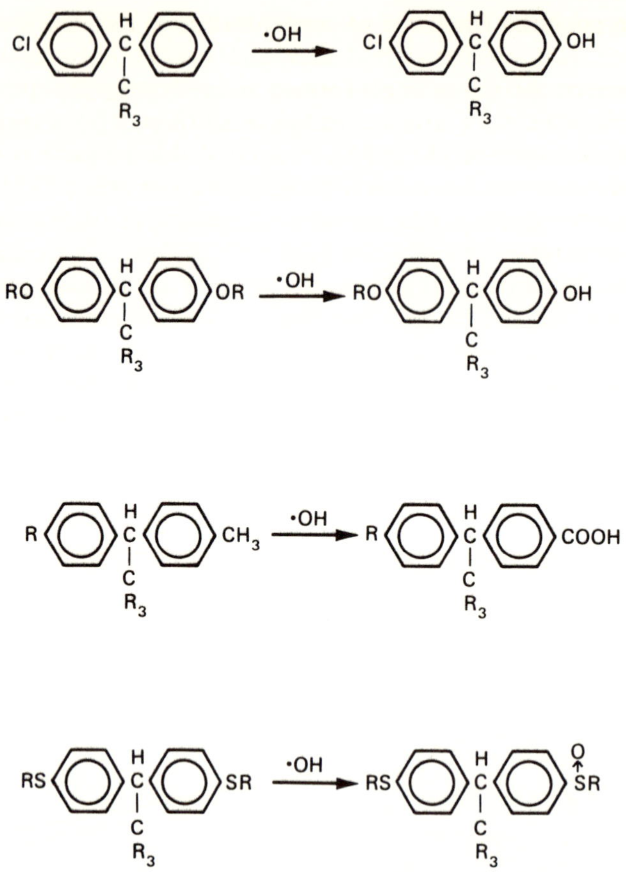

FIGURE 4 DDT analogues as substrates for multifunction oxidases (Metcalf, Kapoor, and Hirwe, 1971).

model system resembled closely that of Lake Michigan (Table 6). When *p,p'*-DDE was the starting material, little further metabolism occurred and storage in the animals of the ecosystem occurred at levels substantially higher than those for DDT. It is evident from these experiments that DDE is responsible for a major portion of the adverse environmental effects of concentration and storage in animal tissues occurring after the use of DDT.

The other three DDT analogues, methoxychlor, methiochlor, and methylchlor, each possess *para*-substituents readily attacked by the multifunction oxidases and, as shown in Table 7, all three are substantially biodegradable. This is shown by the large variety of more polar metabolites identified in the water phase of the ecosystem and in the animal components. The order of concentration in fish was low, 0 for methiochlor, 100-fold for methylchlor, and 1,000-fold for methoxychlor. However, the *Physa* snails formed an interesting exception in that they absorbed very large quantities of all three DDT analogues. Subsequent biochemical study has shown that this animal has a very low titre of multifunction oxidase activity (Hansen *et al.*, 1971).

The model ecosystem data, therefore, suggest that DDT analogues with *para*-substituents readily attacked by multifunction oxidases are substantially biodegradable and do not concentrate through food chains as does DDT. This has been confirmed on a larger scale with methoxychlor, which was not readily concentrated from water or from treated *Daphnia* by green sunfish, *Tilapia*, or guppies, although DDT accumulated to high levels under identical conditions (Reinbold *et al.*, 1971).

DEVELOPMENT OF BIODEGRADABLE ANALOGUES OF DDT

Precise knowledge of the metabolic pathways of DDT analogues which are readily attacked by the multifunction oxidases and the

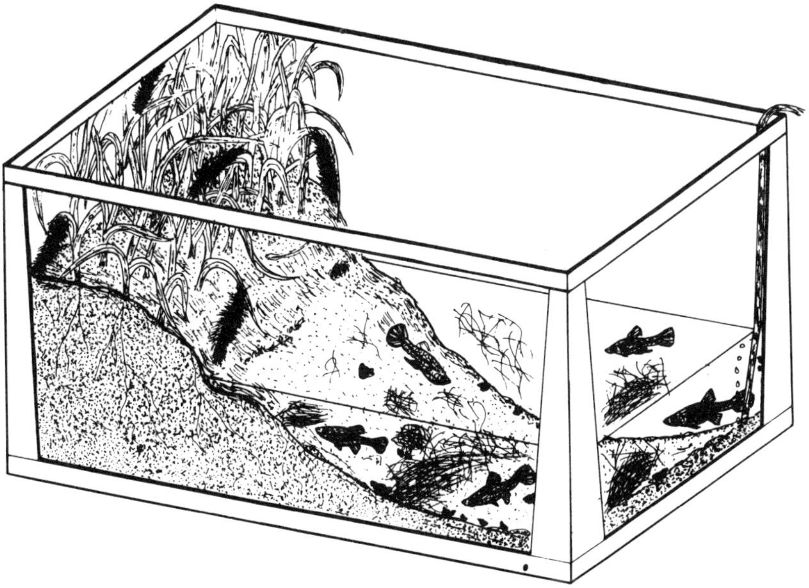

FIGURE 5 A model ecosystem for evaluating pesticide biodegradability and ecological magnification (Metcalf, Sangha, and Kapoor, 1971).

TABLE 7 Distribution of ^{14}C-DDT, ^{3}H-Methoxychlor, ^{3}H-Methiochlor, ^{14}C-Methylchlor and Their Metabolites in a Model Ecosystem[a]

	Concentration (ppm)			
	H_2O	*Physa* (snail)	*Culex* (mosquito)	*Gambusia* (fish)
DDT				
Total ^{14}C	0.004	22.9	8.9	54.2
p,p'-DDT	0.00022	7.6	1.8	18.6
p,p'-DDE	0.00026	12.0	5.2	29.2
p,p'-DDD	0.00012	1.6	0.4	5.3
polar metabolites	0.0032	0.98	1.5	0.8
DDE				
Total ^{14}C	0.008	121.6	168.9	149.8
p,p'-DDE	0.0053	103.5	159.5	145.0
polar metabolites	0.0027	18.1	9.4	4.8
Methoxychlor				
Total ^{3}H	0.0016	15.7	0.48	0.33
p,p'-methoxychlor	0.0011	13.2	–	0.17
p,p'-methoxychlor ethylene		0.7	–	–
$HOC_6H_4HCCCl_3C_6H_4OCH_3$	0.00013	1.0	–	–
$HOC_6H_4HCCCl_3C_6H_4OH$	0.00003	–	–	–
$HOC_6H_4C{=}CCl_2C_6H_4OH$	0.00003	–	–	–
unknowns	0.00009	–	–	–
polar metabolites	0.00125	0.8	–	0.16
Methiochlor				
Total ^{3}H	0.073	1.3	0.39	0.48
p,p'-methiochlor	0.0018	0.539	–	–
p,p'-methiochlor ethylene	0.0006	0.199	–	–
$CH_3SC_6H_4HCCCl_3C_6H_4SOCH_3$	0.0029	–	–	–
$CH_3SOC_6H_4HCCCl_3C_6H_4SO_2CH_3$	0.0207	0.186	–	0.162
$CH_3SOC_6H_4HCCCl_3C_6H_4SO_2CH_3$	0.0092	–	–	–
$CH_3SO_2C_6H_4HCCCl_3C_6H_4SO_2CH_3$	0.0093	–	–	0.059
polar metabolites	0.019	–	–	0.058
Methylchlor				
Total ^{14}C	0.216	101.0	1.0	0.68
p,p'-methylchlor	0.0006	72.16	0.83	0.084
$CH_3C_6H_4HCCCl_3C_6H_4CH_2OH$	–	6.64	+	–
$CH_3C_6H_4HCCCl_3C_6H_4COOH$	0.034	–	+	–
$HOOCC_6H_4HCCCl_3C_6H_4COOH$	0.018	–	–	–
unknowns	0.055	21.34	–	–
polar metabolites	0.118	0.851	0.167	0.60

[a]Metcalf *et al.* (1971); Kapoor *et al.* (1970, 1971).

demonstration that analogues such as methoxychlor, methiochlor, and methylchlor do not readily accumulate through food chains, suggests that there are a number of asymmetrical analogues of DDT which should be effective and persistent insecticides. When absorbed into living organisms these will have weak points for attack by the multifunction oxidases, thus promoting rapid detoxication and elimination as water-soluble metabolites. Such DDT analogues should have many advantages as safe, relatively stable, and potentially inexpensive residual insecticides. As a result of the synthesis and evaluation of a number of asymmetrical DDT analogues having alkoxy, alkylthio, and alkyl groups attached to one or both aryl rings, we have demonstrated the substantial insecticidal activity of the analogues shown in Table 8 (Metcalf *et al.*, 1971a). The asymmetrical analogues are effective insecticides and are generally superior to DDT against the DDT-resistant R_{SP} house fly because of their decreased capacity for dehydrochlorination resulting from the presence of electron-donating groups which reduce the rate of attack of the DDTase enzyme (Metcalf and Fukuto, 1968). The synergistic ratio

TABLE 8 Toxicity of Analogues of DDT[a]

		LD_{50} ppm					
		House Fly–Topical					
		S_{NAIDM}		R_{SP}			
R_1	R_2	Alone	SR[d]	Alone	SR[d]	*Culex*–Larva	Mouse–Oral
Cl	Cl	14.0	2.5	170	4.3	0.07	200[a,b]
CH_3	CH_3	100	5.7	135	4.7	0.081	3350[c]
CH_3O	CH_3O	45	12.8	48	10.4	0.067	1850[c]
C_2H_5O	C_2H_5O	7.0	4.0	11	2.3	0.04	300–325[a,b]
C_3H_7O	C_3H_7O	125	6.8	500	10.2	0.6	200[b]
CH_3S	CH_3S	225	13.2	2800	5.6	0.21	1000
H	H	2900	5.0	> 2500	> 1.0	1.1	2000[b]
Cl	CH_3	62.5	9.6	140	4.0	0.032	
Cl	CH_3O	41.5	5.9			0.058	
CH_3	CH_3O	23.5	4.7	62.5	5.3	0.085	> 1000
CH_3	C_2H_5O	9.0	5.3	27.0	7.1	0.13	1000
CH_3O	CH_3S	32.0	8.0	80	6.4	0.11	1000
C_2H_5O	CH_3S	32.0	11.4	85	7.4	0.07	> 1000
C_2H_5O	C_3H_7O	9.5	2.9	17.5	2.5	0.036	75–100
Cl	H	160	40	175	3.2	0.12	> 1000[b]
CH_3	H	> 500	> 20	> 500	2	0.61	

[a]Metcalf *et al.* (1971).
[b]Von Oettingen and Sharpless (1946).
[c]Domenjoz (1946).
[d]Synergistic ratio = LD_{50} alone/LD_{50} after pretreatment with 50 mg piperonyl butoxide.

(SR values) in Table 8 provide a quantitative measure of the rates of attack of the various analogues by the multifunction oxidase enzymes as these are specifically inhibited by the piperonyl butoxide synergist (Metcalf, 1967). The new asymmetrical analogues (with the exception of the ethoxy-propyl analogue, which is quite resistant to *O*-dealkylation) are all substantially less acutely toxic to the white mouse than DDT as shown in Table 8.

SUMMARY AND CONCLUSIONS

Most chemicals used for agricultural pest control have been selected on the basis of optimum effectiveness and maximum persistence. Relatively little thought has been given to their safety or selectivity with regard to higher animals or humans and to the ultimate quality of the environment. The severe criticism and public disenchantment with the use of pesticides proves that this must change. Safety to farm workers, consumers, and the environment is a criterion that must become equally as important as performance in the developing and marketing of pesticides.

The objective of this discussion has been to demonstrate that selective and biodegradable pesticides can be developed through better understanding of the biochemical processes involved in the intoxication and detoxication processes. Two examples have been emphasized: (1) improving the selectivity of parathion, a widely used general biocide, by developing analogues that retain high toxicity to insects yet are safe to humans and higher animals; and (2) improving the biodegradability of DDT, a ubiquitous environmental micropollutant, by developing persistent yet rapidly biodegradable analogues that are not stored in the lipids of animals or concentrated through food chains.

The principles of selectivity and biodegradability illustrated for parathion- and DDT-type molecules are generally applicable to pesticides of many descriptions. This approach to the development of better pesticides must be emphasized in the pest control strategies for the future if we are to begin to solve the many problems of human ecology arising from pest-control practices.

ACKNOWLEDGMENTS

The research reviewed in this paper was made possible by research grants to the author by the Herman Frasch Foundation–American

Chemical Society, by The Rockefeller Foundation, by the U.S. Public Health Service Grant No. FD00271 from the Food and Drug Administration, and by the World Health Organization.

REFERENCES

Blackbourne, B. D. 1970. Pesticide poisoning–a medical examiner's view. Proc. 22nd Illinois custom spray operators training school, Urbana, p. 16.

Büchel, K. H., A. E. Ginsberg, R. Fischer, F. Korte. 1964. β-Dihydroheptachlor, ein Insektizid mit sehr niedriger Warmblütler-Toxizität. Tetrahedron Lett. No. 33, 2267–72.

Domenjoz, R. 1946. Action biologique de quelques derives DDT. Helv. Chim. Acta 29:1317.

Drabek, J., J. Pelikan. 1956. Metyl-analogy "chlortioner." Chem. Prumysl 6:293.

Hansen, L., R. L. Metcalf, I. P. Kapoor. 1971. Unpublished data.

Harrison, H. L., O. L. Loucks, J. W. Mitchell, D. F. Parkhurst, C. R. Tracy, D. G. Watts, V. J. Yannacone, Jr. 1970. Systems studies of DDT transport. Science 170:503.

Hayes, W. J., Jr. 1963. Clinical handbook on economic poisons. U.S. Department of Health, Education and Welfare, Washington, D.C.

Hollingworth, R. M., T. R. Fukuto, R. L. Metcalf. 1967a. Selectivity of Sumithion compared to methyl parathion. Influence of structure on anticholinesterase activity. J. Agr. Food Chem. 15:235.

Hollingworth, R. M., R. L. Metcalf, T. R. Fukuto. 1967b. Selectivity of Sumithion compared to methyl parathion. Metabolism in the white mouse, metabolism in susceptible and resistant house flies. J. Agr. Food Chem. 15:242, 250.

Kapoor, I. P., R. L. Metcalf, J. R. Coats, Po-Y. Lu, A. S. Hirwe, R. F. Nystrom. 1972. Comparative metabolism and biodegradability of DDT, ethoxychlor, and methylchlor. J. Agr. Food Chem. 20, 1.

Kapoor, I. P., R. L. Metcalf, R. F. Nystrom, G. K. Sangha. 1970. Comparative metabolism of methoxychlor, methiochlor, and DDT in mouse, insects and in a model ecosystem. J. Agr. Food Chem. 18:1145.

Kenaga, E. E., W. E. Allison. 1969. Commercial and experimental organic insecticides. Bull. Entomol. Soc. Am. 15:85.

Metcalf, R. A., R. L. Metcalf. 1972. Molecular basis for selective toxicity of analogues of methyl parathion. In preparation.

Metcalf, R. L. 1964. Selective toxicity of insecticides. World Rev. Pest Control 3:28.

Metcalf, R. L. 1966. Metabolism and fate of pesticides in plants and animals. *In* Scientific aspects of pest control, p. 230. Publ. 1402. National Academy of Sciences, Washington, D.C.

Metcalf, R. L. 1967. Mode of action of insecticide synergists. Ann. Rev. Entomol. 12:229.

Metcalf, R. L. 1971. Structure/activity relationships for insecticidal carbamates. Bull. World Health Org. 44, 43.

Metcalf, R. L., T. R. Fukuto. 1965. Effects of chemical structure on intoxication and detoxication of phenyl N.-methylcarbamates in insects. J. Agr. Food Chem. 13:220.

Metcalf, R. L., T. R. Fukuto. 1968. The comparative toxicity of DDT and analogues to susceptible and resistant house flies and mosquitoes. Bull. World Health Org. 38:633.

Metcalf, R. L., I. P. Kapoor, A. S. Hirwe. 1971a. Biodegradable analogues of DDT Bull. World Health Org. 44, 363.

Metcalf, R. L., G. K. Sangha, I. P. Kapoor. 1971b. A model ecosystem for the evaluation of pesticide biodegradability and ecological magnification. Environ. Sci. Technol. 5:709.

Mrak, E. (Chairman). 1969. Report of the Secretary's Commission on Pesticides and their relationship to environmental health. U.S. Department of Health, Education and Welfare, Washington, D.C.

Plapp, F. W., and J. E. Casida. 1958. Hydrolysis of the alkyl-phosphate bond in certain dialkyl aryl phosphorothioate insecticides by rats, cockroaches, and alkali. J. Econ. Entomol. 51:800.

Rainey, R. H. 1967. Natural displacement of pollution in the Great Lakes. Science 155:1242.

Reinbold, K. A., I. P. Kapoor, W. F. Childers, W. N. Bruce, R. L. Metcalf. 1971. Comparative uptake and biodegradability of DDT and methoxychlor by aquatic organisms. Bull. Ill. State Natural History Survey 30(6):405.

Schrader, G. 1961. Zur Kenntnis neuer, wenig toxischer Insektizide auf der Basis von Phosphorsaureestern. Angew. Chemie 73:331.

Von Oettingen, W. F., N. E. Sharpless. 1946. The toxicity and toxic manifestations of DDT as influenced by chemical changes in the molecule. J. Pharmacol. Exp. Therap. 88:400.

Warner, R. E. 1967. Environmental micropollutants, Bull. World Health Org. 36, 181.

Winteringham, P. 1969. Mechanisms of selective insecticidal action. Ann. Rev. Entomol. 14:409.

World Health Organization. 1968. Evaluation of insecticides for vector control. WHO/VBC/68.66.

Rodent Control Strategy

David E. Davis

The control of rodents has been a major problem for centuries and despite decades of research remains an insurmountable task. The present discussion will summarize briefly the results of research for the past three decades and suggest a strategy for the control of urban rats and other rodents. The Norway rat (*Rattus norvegicus*) and the pine or orchard mouse (*Pitymys pinetorm*) will be used as examples. Many rodents and other mammals produce damage in North America and a multitude of species in other parts of the world cause trouble in cities and in agricultural areas. It is impossible to discuss these species but it is hoped that the strategy for one species can be adapted to another.

The central theme of this presentation is that we now have sufficient knowledge of the principles of populations; hence we should design our control measures according to these principles. The major concepts here are included under the phrase *population dynamics*, which is the study of birth rates, mortality rates, and movement rates. From these data one can measure the changes in the population and also note that the population will be regulated at some number that is a reflection of the environmental circumstances. Another principle that can be used is the concept of diversity, which suggests that populations are more stable if there is greater diversity in the numbers of species and in the types of environment. Given these and other subsidiary principles, it would seem that the appropriate strategy for control is to devise methods of manipulating the environment in accordance with the change desired in the population (Davis, 1949). Unfortunately, as will be indicated in more detail later, our political mechanisms often are unable to administer the control procedures that are dictated by the principles. A corollary of the strategy of working with principles is that research should not continue in clear violation of population principles in expectation that a politically acceptable solution will be found. Let us consider an analogy in the construction of a bridge. The engineer recognizes the law of gravity, designs his bridge in accordance with the law of gravity, and then indicates the costs and how the bridge must be constructed. He does not continue research in a futile attempt to build a bridge that violates the law of gravity. This concept must be recognized in the control of pests, and our research results must be made available to those who can alter the political processes to achieve the objective. This strategy will be mentioned throughout this paper and summarized at the end.

PRINCIPLES OF RODENT CONTROL

Research on rat control had been carried out in a desultory manner for decades but in the latter part of World War II a concentrated program was developed by the Public Health Service in Texas and by The Johns Hopkins University School of Public Health in Baltimore. The results of the work in Baltimore have been summarized (Davis, 1953) and some subsequent papers have extended these concepts to other population aspects (Bentley *et al.*, 1959; Davis, 1969; Gomez, 1960; Strecker *et al.*, 1962; Spillett, 1968). The mechanism that controls the population has been clarified (Christan and Davis, 1955; Davis and Christian, 1956, 1957, 1960) by the development of an understanding of the pituitary-adrenal mechanism that acts as a negative feedback by reducing the birth rate and increasing the mortality rate as the population grows in numbers. While an understanding of the mechanism is extremely important for the planning of rodent control strategy, nevertheless, the details are not important for this presentation.

The history of the research in Baltimore as described in the above papers can be summarized rather briefly. Beginning in 1943, Dr. John T. Emlen developed the census method that permitted rapid estimates of numbers of rats in city blocks. The advantage of this procedure was that many discrete populations could be followed for the results of experimental treatments evaluated. The various treatments, of course, included the assessment of the value of poisons and the comparison of sanitation, slum rehabilitation, and slum clearance. Results of these extensive and detailed studies showed that rat populations will increase to the capacity of the environment and then remain at essentially the same level for an indefinite period of time or until the habitat is changed. Additional studies including laboratory work on rats and mice showed that this level was maintained by the negative feedback mechanism. It was thus apparent that the reduction of rats by predation (poisoning or trapping) had only a temporary effect since the population returned to the capacity in a short period of time.

Data will now be presented that are purposely chosen from research done three decades ago to emphasize the point that these results were established long ago. Thus these principles have had an opportunity to be tested and verified in various ways. One of the basic points is to consider the changes that occur in a population over a period of time after reduction in numbers by some form of predation using either traps or poisons. The history of rat populations in six different blocks for approximately two years (Figure 1) shows that, after poisoning, the population promptly returns to approximately the original level and that it then remains at that level for many months. If the predation by poison, traps, and gas is intensive, then (Figure 2) the population can be kept at a low level but will return to its capacity promptly even after subsequent treatments with poison. In this history, the capacity of the block remained essentially the same for four years. More detailed studies indicated a general relationship first pointed out by Emlen *et al.* (1945). When the population (Figure 3) was reduced 50 to 70 percent, the recovery was rapid. When the reduction was almost 90 percent, the return was slow but, nevertheless, it occurred. The data illustrate the basic aspects of the sigmoid growth curve, which is now recognized as a fundamental feature of populations. The history of rats in another block (Figure 4), which was completely freed from rats and remained free for 2.5 years, showed that once the rats returned the population increased in 6 months from zero to about 150. A spontaneous decrease occurred followed by an oscillating increase.

That a block has a capacity for rats is clearly indicated (Figure 5) by the history of a block whose population was reduced by poisoning then recovered as usual and then again was reduced, but this time by the introduction of rats. When the introduction of rats ceased, the population in each block began to return to the capacity. Additional

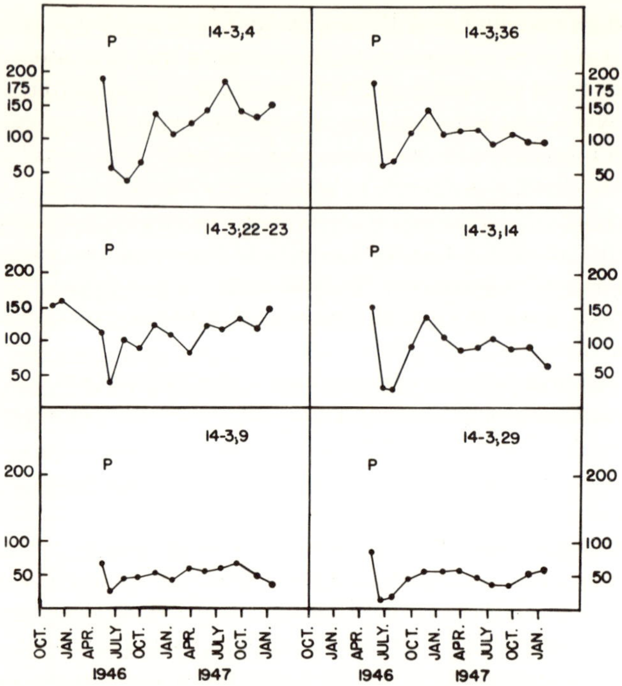

FIGURE 1 Number of rats in six blocks following reduction by poisons. [Note that the numbers returned to approximately the original level and remained there.] (The precision of estimate is about 20 percent.)

studies (Davis and Christian, 1956) indicate that the phenomenon of reduction occurs when rats are simply replaced, either in stationary populations or in growing populations. Thus, for example, the removal of about 20 males and a replacement by 20 males from other areas was followed by a reduction in the population. Results such as these in rats and others in mice stimulated research on social behavior and aggression that led directly to the concept that a self-regulating mechanism exists (Davis, 1949), which has recently been extended to many species (Wynne-Edwards, 1959) The significance for present strategy is the demonstration that a homeostatic population mechanism exists that causes a population to return to a level, either from below capacity or from above it. Thus, the proper principle for control of populations is to use this homeostatic mechanism to adjust the population to a lower level by lowering the capacity.

Predation by traps or poisons merely makes space for more rats to grow. Indeed, a few simple calculations show that a population with capacity of 110 will produce 330 rats a year at capacity and 360 if kept at 75 percent by predation. Predation *could* be effective, as is true of any species, but for rats it rarely works.

An alternative approach is to manage the habitat. In the case of rats, of course, the procedure is to reduce the food and shelter; this can be carried out, in part, by sanitation. The removal of garbage and fences (Figure 6) shows that this procedure reduces the population and little recovery occurs. An additional complication can be seen (Figure 4) in which a block which had oscillated around the asymptote for about 4 years declined to zero when the food supply (–F) was greatly reduced. A more drastic approach is to rehabilitate the housing in an area. This procedure consists of inspections by various city officials and enforcement of the sanitation and housing codes on both the tenant and the landlord. The history of two blocks (Figure 7) that were rehabilitated, while a reference block was not, shows the drastic reduction and maintenance of a low population of rats. At the end of the experiment the population in the reference block declined because the people noted the benefits of rehabilitation in the nearby blocks and spontaneously cleaned up their housing conditions. Ob-

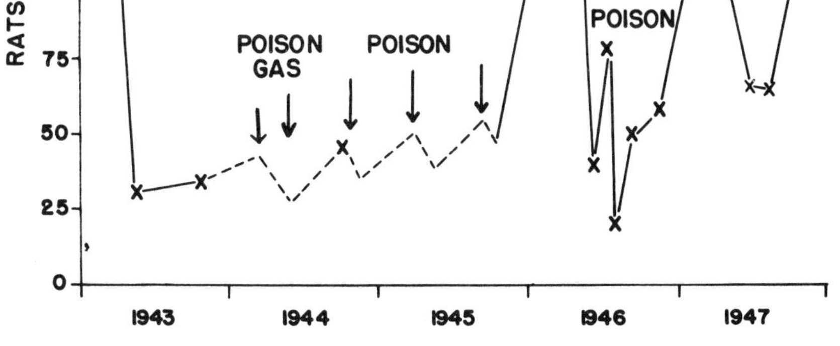

FIGURE 2 History of rat population for 4 years. The capacity remained at approximately 100. The estimates (X) were made at intervals of 2 months.

viously, an even more drastic procedure is the clearance of slums and the complete rebuilding of the structures that will reduce the rat populations drastically.

As part of these studies, estimates were made of rats in the city of Baltimore by counting the rats in randomly chosen blocks in census tracts throughout the city (Figure 8). By this means the rats were estimated in 1947, in 1949, and in 1952 (Brown *et al.*, 1955). An approximation to the number of rats in 1944 was made retrospectively. These data (Figure 9) show that the rat population decreased from about 400,000 to about 60,000, a decline of almost 90 percent. Note that by 1952 the rat population was increasing following the introduction of anticoagulants between 1948 and 1950. The public and the government officials believed that anticoagulants would solve the problem and relaxed their efforts to reduce the food and shelter for rats. No estimates have been made since 1952, and no research has been done in Baltimore since 1959. However, it was clear during years 1952–1959 that the rat populations were perservering at capacity in those blocks that were not being cleaned up. Recently (September, 1970) I visited a number of the blocks and found the rat populations at capacities maintained three decades ago.

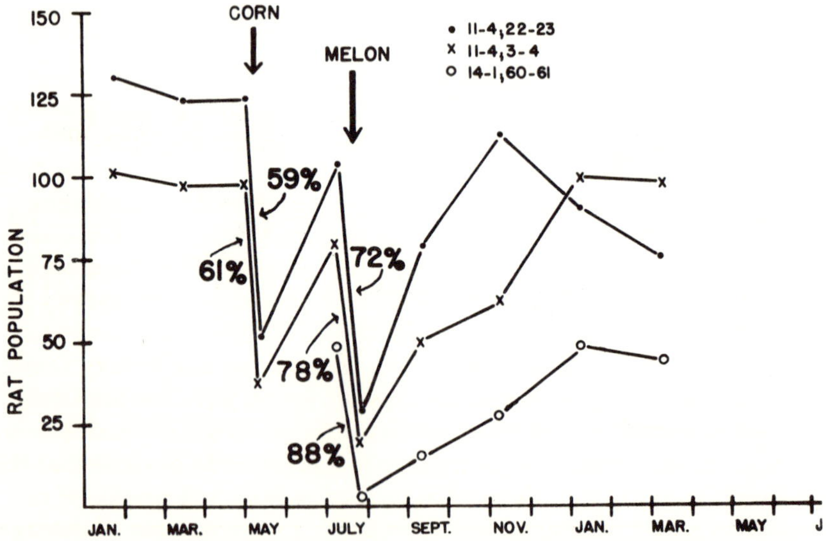

FIGURE 3 Detailed history of numbers of rats in two blocks where different baits were used. The second poisoning operation merely interrupted the return of the population to capacity.

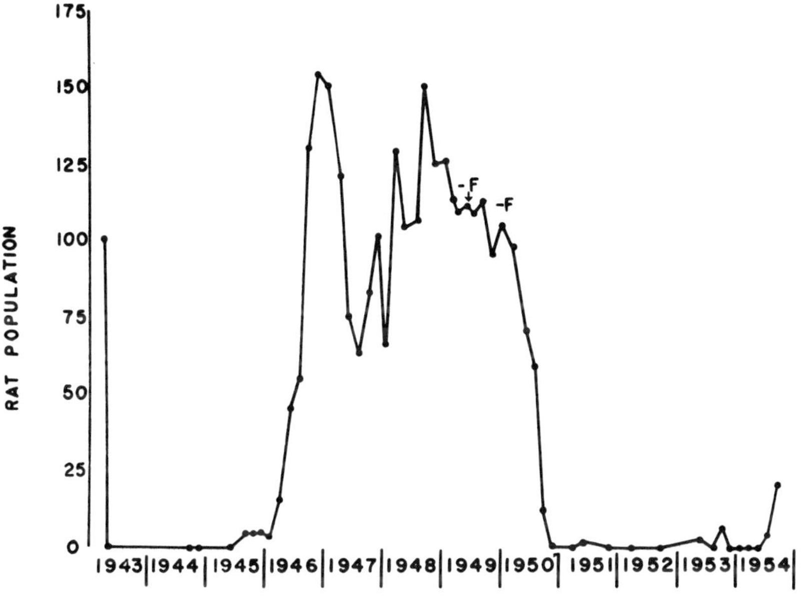

FIGURE 4 In a block where rats were exterminated, no invasion occurred for 2.5 years. Then a few rats entered the population, rapidly increased, and then, by homeostatic mechanism, declined. The removal of the food supply (–F) later resulted in disappearance of rats for 3.5 years.

Many other rodents have been studied and although the evidence is by no means as complete, some species, such as prairie dogs, are clearly vulnerable to poisons because of their habits and can be eliminated from various areas. However, changes in habitat conditions are equally effective. Many other species, especially microtines, are not very vulnerable to poisons and many years of effort have been devoted to finding a better procedure for poisoning these species. A long-term project has been conducted since 1933 in an attempt to find a procedure that will reduce the pine mouse. The history shows that poison after poison and device after device were invented only to fail to produce satisfactory results. Just recently a promising new poison has been again proclaimed, but it is perfectly obvious that the history of nearly four decades of tests of poisons will simply be repeated again. The appropriate approach for the reduction of damage by pine mice and other microtines is to develop a means of altering the habitat condition, which usually is some agricultural situation. Research should be devoted to finding a procedure for harvesting or

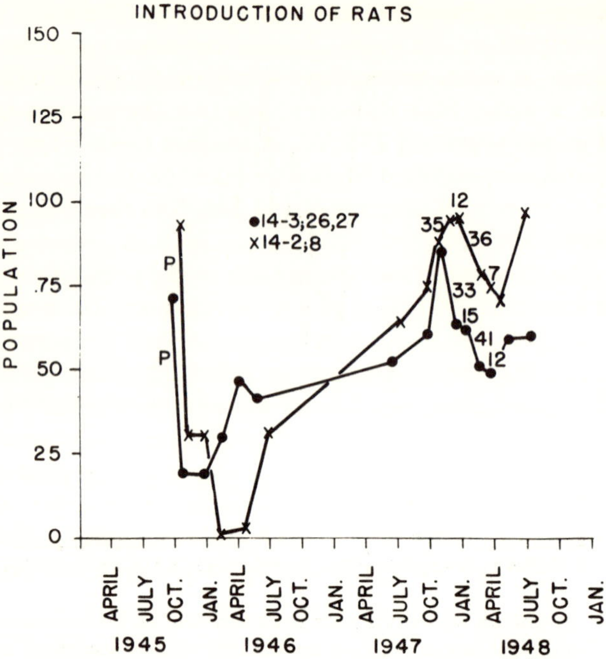

FIGURE 5 The introduction of rats resulted in a decline in numbers, demonstrating that the capacity was exceeded.

managing the crop so that the habitat for pine mice will be reduced or eliminated.

CURRENT KNOWLEDGE–1971

The goals of rodent control are, of course, to reduce the damage below a level of economic interest. It usually is not necessary to eliminate the population from an area. To achieve this goal we now have adequate knowledge about the life history and population principles for the control of urban rats and perhaps for microtines of agricultural interest. The principles are available for many other species such as the bandicoot (Spillet, 1968) and other tropical rats (Strecker *et al.*, 1962). Many details are known for these species and it would seem that control could be achieved if the proper strategy were adopted.

It must be recognized that new information can always be collected. Indeed, the amount of information to be obtained is infinite since local differences and yearly changes can always provide new facts. However, it seems unnecessary to obtain complete information; for example, it makes little difference whether the home range of rats is 150 feet in one town and 175 feet in another town. Thus, for urban rat populations in the United States it would seem unnecessary to embark on a program of obtaining local data. In other places, some new knowledge is desired but it should be carefully oriented to determine the local circumstances in respect to the population principles mentioned above. In all cases it should be related to the goal of manipulation of the habitat to reduce the capacity of the area to support the rodents, so that the damage will be below economic levels.

The quantitative aspects of predation should be explained, since many of the contributions to this symposium deal with the encouragement of predation for insect control. The results of predation depend on the quantitative values for reduction in numbers and on the efficiency of the homeostatic regulatory mechanisms. For many insects and a few vertebrates, these relations are such that predation

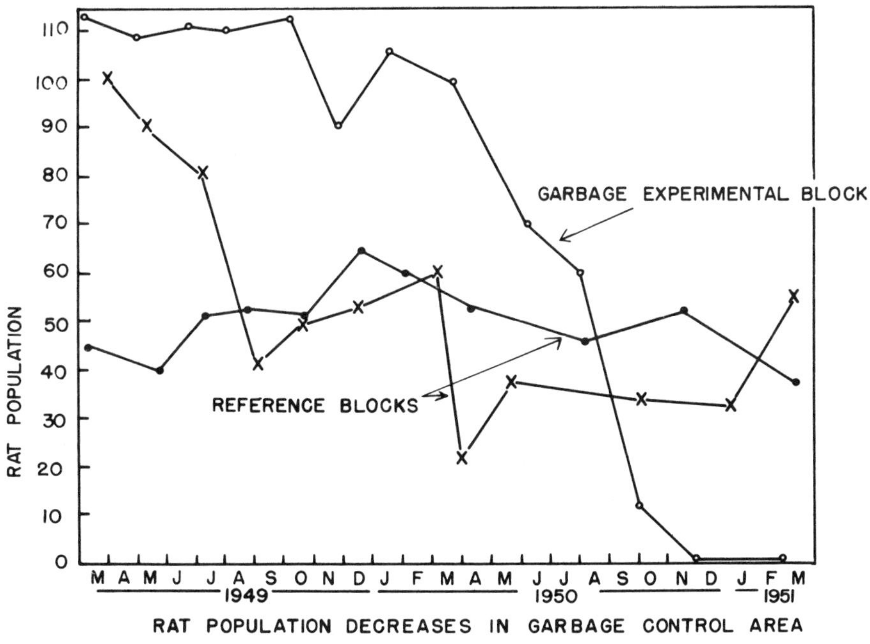

FIGURE 6 Removal of fences and garbage resulted in a reduction of rats as long as sanitation was maintained.

can be effective. But for most birds and mammals these relations render predation ineffective. The reasons are numerous and complex, but a few can be cited here such as long breeding season, flexible birth and mortality rates, social organization, and aggressive behavior controlling emigration and immigration. Thus while predation (the actual means is not important) could reduce rodent populations below economic levels, it rarely does.

In other parts of the world the habitat conditions differ so that barriers may be effective (Telle, 1967; Henschel, 1965; Greaves *et al.*, 1968) and local reductions may succeed, although resistance to anti-coagulants appears (Drummond, 1966).

POLITICAL PROCESSES

The strategy for research in the control of economic damage should be to obtain adequate information for action. Generally there are several levels of research, more or less as follows:

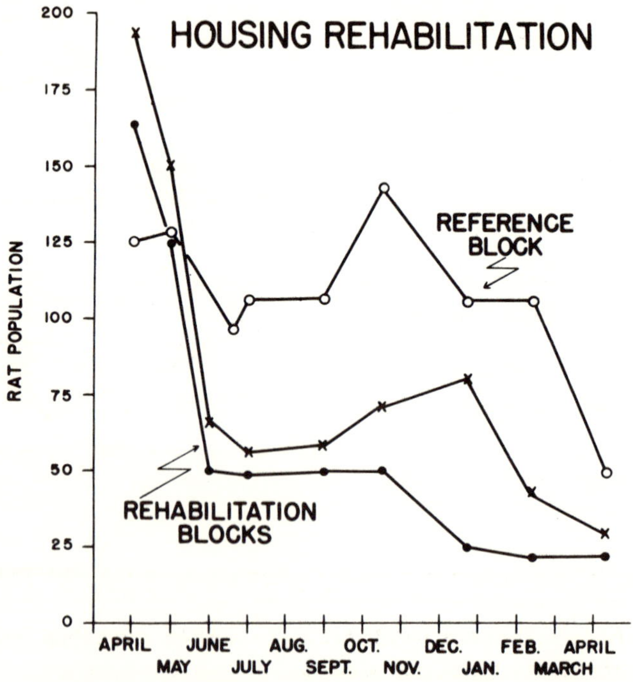

FIGURE 7 Rehabilitation of housing conditions reduced rat numbers.

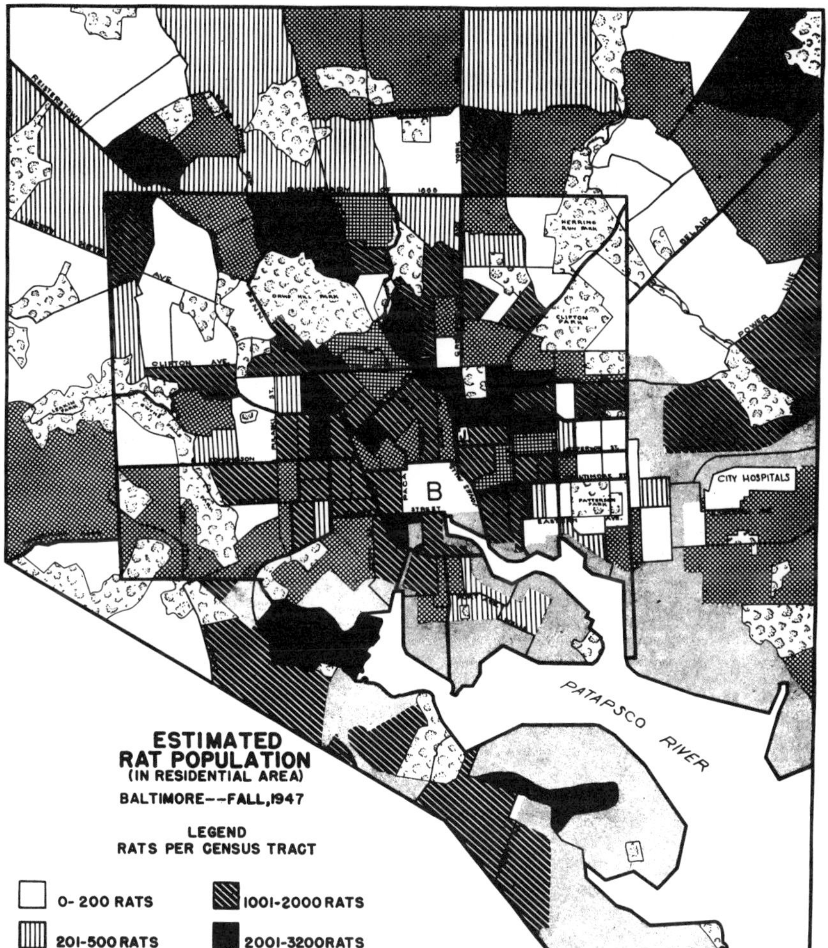

FIGURE 8 A map of the distribution of rats in Baltimore in 1947, when the population was estimated at 160,000.

1. Laboratory work on reproduction, energy consumption, behavior.
2. Populations in laboratory situations or, better, in special field situations such as city blocks or islands.
3. Control procedures on a wide scale in a city or other large area.
4. Appraisal of the results in economic or psychic terms.

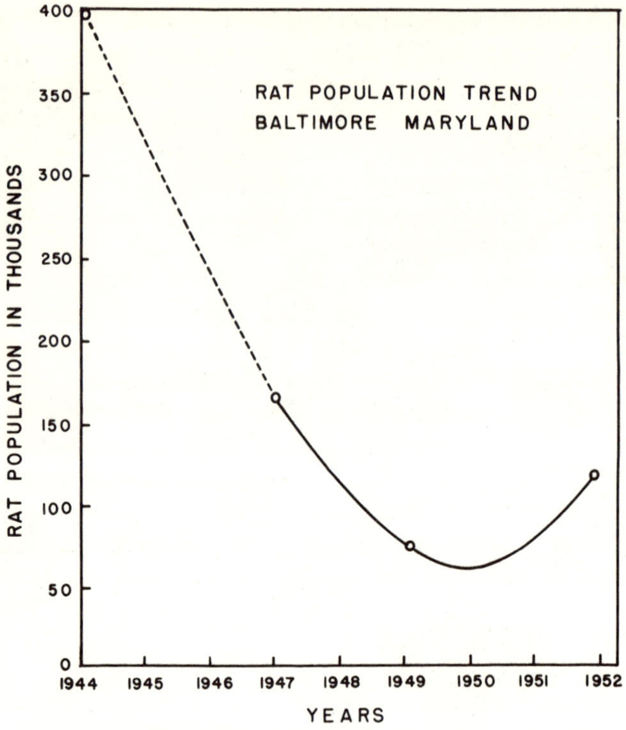

FIGURE 9 The decline of rats in Baltimore during sanitation and rehabilitation programs. The increase followed the introduction of anticoagulants.

It should be noted that, of course, these stages merge one into another but a significant aspect is that different methods will be used and different types of persons will conduct the research. For example, the person who is qualified to conduct technical laboratory research on the physiology of reproduction would not be qualified either by training or temperament to evaluate the economic damage in a whole city. Thus, one aspect in this strategy of research is to recognize that different types of persons, perhaps as many as four, are necessary to carry out the application of a control procedure. At the present time, since the basic principles are known, the biologist needs only to collect certain local data of a survey type to determine the extent of the problem. Then the research should be shifted from the research biologists to the administrator and to the political scientist who should be in a position to see that the action is carried out. The general public may not find that the program is acceptable, at least without ample explanation and education. However, it should not be the duty of the biological scientist to continue his research in the vain hope of finding a plan that is politically acceptable, since basically the biological laws will have to be obeyed.

Many political difficulties are encountered in the actual application of the basic population laws. A multitude of conflicts of interest develop in an action program. Consider, for example, the problem of sanitation and housing rehabilitation. The tenant complains that the landlord does not do his part and, of course, the landlord returns the compliment. Arguments develop over who should buy the garbage can and who should pay for the repairs of the plumbing. On the commercial side various persons wish to profit from the sales of poisons and develop intensive advertising campaigns knowing full well that the poisons will leave plenty of rats for further sale of poisons. Others object to the clearance of slums, recently claiming that persons who advocate such procedures are activists or communists. Even adequate ordinances fail when lawyers obstruct the procedures or the judge ignores the case. These problems are, of course, very real but they are the province of political science and the educator; they are not the province of the biologist. The biologist has the choice (at least theoretically) between continuing scientific research in some topic or changing to action in the political arena. If he continues in research, he uses the talents and training he has developed. If he changes to political action, he uses hidden talents and develops new abilities (or fails). Only rarely can he do both, if for no other reason than available time. Thus, rather than transform biological scientists into political scientists, it is more efficient to turn the job over to political scientists.

A few comments on costs seem desirable. During the past two decades research on rat control has been conducted in a few places. It has consisted of laboratory tests on new poisons, baits, gametocides, behavior, resistance to poisons, and similar topics with some field demonstrations. I have not seen any results that have altered the rat population or problem in the United States except for a temporary alleviation of a local situation. The research has not been intensive—perhaps three full-time persons each year or $75,000 a year or $1,500,000 in two decades. In my view these funds have been wasted because they were spent in a vain attempt to circumvent the basic laws of population regulation. The situation now, when research funds are restricted, is especially critical. We cannot afford the luxury of conducting research in violation of population principles.

The control of damage by pests is, of course, part of the problem of environmental quality. It is now recognized that laws and strict enforcement of these laws will be necessary to alter the assumptions about social responsibility and the role of administration of public affairs. Irresponsible private control will be replaced by public control because of the inexorable action of economic laws concerning private and public gain. Given our political system with many centers for decision, with divided responsibility, and with pressure for profit, the dominant tendency is to favor local, short-term solutions that nullify local pressures on jobs and profits. Unfortunately, rewards for general improvements such as water and air sanitation rarely show visible or direct benefits.

The strategy of pest control research is to describe what can be done under biological laws; the art of politics is to get it done. The biologist must not waste research funds in futile attempts to avoid biological laws.

CONCLUSIONS

From this survey of the development of knowledge of urban rats during the past three decades and from some brief comparisons to other rodents, several conclusions emerge. The strategy for pest control by the biologists should be to uncover the basic laws of populations and also to obtain factual information in local areas for quantitative relationships. The strategy does not include research in the deluded expectation that somehow or other the laws of population can be nullified. At present the laws of population dynamics and some of the homeostatic mechanisms in rodents are sufficiently known that their application should be turned over to persons skilled in the political arena. Political scientists, administrators, lawyers, and educators must become involved in the translating of biological laws into reducing the damage done by pests.

REFERENCES

Bentley, E. W., A. H. Bathard, and J. D. Riley. 1959. The rates of recovery of sewer rat populations after poisoning. J. Hyg. 57(3):291–298.

Brown, R. Z., W. Sallow, David E. Davis, and W. G. Cochran. 1955. The rat population of Baltimore, 1952. Am. J. Hyg. 61(1):89–102.

Christian, J. J., and David E. Davis. 1955. Reduction of adrenal weight in rodents by reducing population size. Trans. N. Am. Wildl. Conf. 20:177–189.
Davis, David E. 1949. The role of intraspecific competition in game management. Trans. N. Am. Wildl. Conf. 14:225–231.
Davis, David E. 1953. The characteristics of rat populations. Quart. Rev. Biol. 28(4):373–401.
Davis, David E. 1969. Appraisal of rat populations. Indian Rodent Symp. (1966):51–58.
Davis, David E., and John J. Christian. 1956. Changes in Norway rat populations induced by introduction of rats. J. Wildl. Manag. 20(4):378–383.
Davis, David E., and J. J. Christian. 1957. Population consequences of a sustained yield program for Norway rats. Ecology 39(2):217–222.
Davis, David E., and J. J. Christian. 1960. Role of density in populations of mammalian plague reservoirs. Bull. World Health Org. 23:417–418.
Drummond, David. 1966. Rats resistant to warfarin. New Sci. p. 771–772.
Emlen, J. T., A. W. Stokes, and C. P. Winsor. 1948. The rate of recovery of decimated populations of brown rats in nature. Ecology 29(2):133–145.
Gomez, Juan C. 1960. Correlation of a population of roof rats in Venezuela with seasonal changes in habitat. Am. Mid. Nat. 63(1):177–193.
Greaves, J. H., L. E. Hammond, and A. H. Bathard. 1968. The control of reinvasion by rats of part of a sewer network. Ann. Appl. Biol. 62(2):341–351.
Henschel, J. 1965. Rat infestation in the Hamburg district before and after the 1963 control operations. Pflanzen-Schutz Nachsch. Bayer 18(4):157–164.
Spillett, James Juan. 1968. The ecology of the lesser bandicoot rat in Calcutta. Bombay Natural History Society and The Johns Hopkins University Center for Medical Research and Training, Calcutta.
Strecker, Robert L., Joe T. Marshall, William B. Jackson, Kyle R. Barbehenn, and David H. Johnson. 1962. Pacific Island rat ecology. Report of a study made on Ponape and adjacent islands 1955–1958. Bernice P. Bishop Mus. Bull. 225.
Telle, H. J. 1967. Die Auswahl von Rodentiziden fur die Rattenvertilgungen und fur die Beibehaltung eines rattenfreien Zustandes. Anz. Schaedlingsk. 40(11):161–166.
Wynne-Edwards, V. C. 1959. Population ecology. The control of population-density through social behaviour: A hypothesis. Ibis 101(3/4):436–441.

Hormonal Chemicals for Invertebrate Pest Control

William E. Robbins

The hormonal regulation of essential physiological processes, such as growth, development, molting, and reproduction, has been shown to occur in representative species from many of the major invertebrate phyla (Charniaux-Cotton and Kleinholz, 1964; Wigglesworth, 1964; Highnam and Hill, 1969). As we expand our knowledge of invertebrate hormones it may eventually be possible for us to develop selective hormonal pesticides and chemotherapeutic agents for use against a variety of invertebrate pests and parasites. However, we now have the requisite information and chemicals to undertake the development and use of hormonal pesticides solely for insects, which include man's most important competitors for food and fiber and the vectors of certain of his most dreaded diseases. These requisites include

1. Basic information on the hormone-regulated processes of insects; a knowledge of how and when the hormones normally act to regulate insect growth, development, and reproduction. This information will aid us in determining how we may best use hormonal chemicals in insect control.
2. A knowledge of the chemistry of insect hormones, gained through their isolation, characterization, and synthesis.
3. The availability of synthetic hormones, hormone analogues, and other synthetic compounds that either have hormone activity or that disrupt the hormone-mediated processes of insects.

The underlying basis for the development and use of hormonal chemicals for insect control is that these chemicals should be safe for man and higher animals since they selectively disrupt physiological and biochemical processes peculiar to insects. Examples of such processes are the hormonal regulation of molting and metamorphosis in insects and in related arthropods and the hormonal control of reproduction and diapause in insects. These processes either do not have a counterpart in vertebrates, or the controlling mechanism in insects is very unlike that operative in higher animals. In addition, the known hormones of insects that control molting, metamorphosis, diapause, and reproduction are structurally different from the hormones of vertebrates. Thus, insect hormonal chemicals offer a considerable potential for the development of safe and selective insecticides. The remarkable progress in insect hormone chemistry during the past six years now makes it possible to attempt to realize the predicted applications of insect hormonal chemicals as insecticides (Williams, 1967; Schneiderman *et al.*, 1970) and to determine how we may best use these potent biological regulators as insect-control chemicals.

Our progress in the development and use of hormonal insecticides should serve as a guide for the similar development and use of hormonal pesticides and chemotherapeutic agents for other invertebrate pests and parasites. Indeed, reports indicating that insect hormonal chemicals regulate or disrupt the development of a number of invertebrates other than insects (Cleveland *et al.*, 1960; Krishnakumaran and Schneiderman, 1968; Williams and Robbins, 1968; Ilan *et al.*, 1969; Krishnakumaran and Schneiderman, 1969; Muftic, 1969; Wright, 1969; Shanta and Meerovitch, 1970) suggest even more direct application of the information gained from insects.

HORMONAL REGULATION IN INSECTS

Hormonal Control of Insect Development, Reproduction, and Diapause

Molting is a characteristic feature of insect growth and development, and three types of hormones control molting and metamorphosis in insects (Figure 1): the brain hormone, the molting hormones (or ecdysones), and the juvenile hormones (Schneiderman and Gilbert, 1964; Wigglesworth, 1964; Highnam and Hill, 1969). The brain hormone, secreted by neurosecretory cells in the insect brain, activates

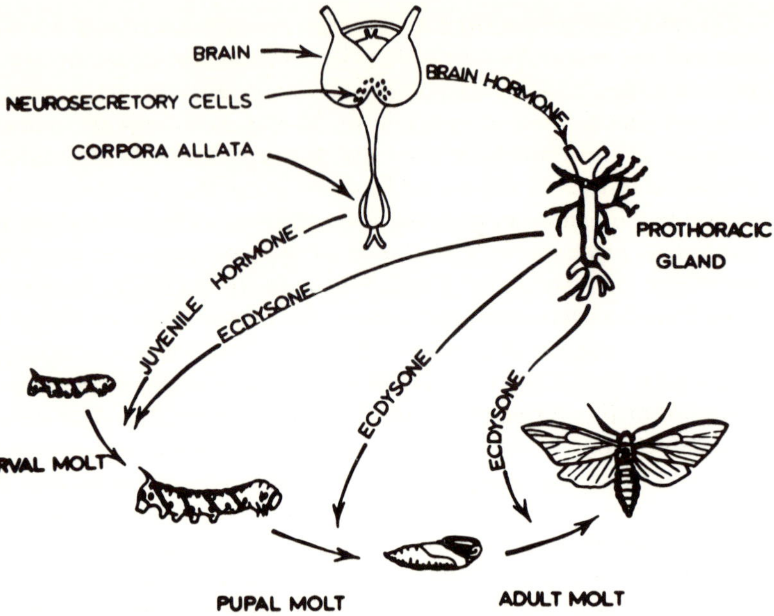

FIGURE 1 Schematic diagram of the hormonal control of molting and metamorphosis in a holometabolous insect, showing the principal endocrine organs and hormones.

the prothoracic glands. The activation of these glands brings about the secretion of one or more closely related steroids, the ecdysones, which cause the insect to molt. The ecdysones, then, are the true growth and molting hormones of insects and possibly of all arthropods (Krishnakumaran and Schneiderman, 1968, 1969).

The kind of molt that the insect undergoes is controlled by the third type of hormones, the juvenile hormones, which are secreted by the corpora allata. The presence of juvenile hormones in the immature insect causes the larva to molt to a larger larva and thus prevents it from differentiating into an adult. The juvenile hormone, then, is an unusual biological regulator in that it permits growth yet prevents maturation. In the absence of the juvenile hormone at the time of molt, differentiation does occur; the nymphs of hemimetabolous insects molt into adults and the larvae of holometabolous insects molt into pupae and then into adults. In summation then, the molting hormones or ecdysones cause the insect to molt and the juvenile hormones influence the type of molt that occurs. In many adult insects, the juvenile hormones also function in postimaginal maturation and are essential for ovarian development, yolk formation, and the maturation of eggs in the females and development of the accessory glands in the males.

In addition to controlling growth, development, metamorphosis, and reproduction, the hormones are also involved in the regulation of insect diapause—a physiological defense mechanism of insects. During unfavorable environmental conditions, such as extended periods of heat or cold, or the absence of the preferred host plant, the insect may enter diapause, a hibernation-like state of physiological arrest that permits it to survive. Diapause results from the temporary cessation of certain hormone-regulated processes: In some lepidopterous pupae, diapause may be terminated by the administration of compounds with either molting-hormone activity or juvenile-hormone activity, while in certain coleopterous and hemipterous insects, compounds with juvenile-hormone activity interrupt adult diapause.

Assay of Insect Hormone Activity

The insect brain hormone appears to be a protein (Yamazaki and Kobayashi, 1969) but its structure still remains unresolved (Berkoff, 1969), so only insect hormonal chemicals with juvenile hormone (JH) and molting hormone (MH) activity will be discussed as potential hormonal insecticides. Since compounds will be referred to as having JH or MH activity, it is desirable that a brief description be made of the standard or most commonly used biological assays for detecting and measuring these insect hormone activities. The most widely used bioassay for JH activity is the *Tenebrio* test (Bowers and Thompson, 1963). In this assay for morphogenetic activity, recently formed pupae of the yellow mealworm (*Tenebrio molitor* L.) are either treated topically or injected with an active compound or an extract, and the abnormal retention of pupal characters in the ensuing adult signals the presence of JH activity in the test material. Quantitation is based on the degree of retention of immature characters, which may range from a near-perfect adult with pupal genitalia to a perfect second pupa.

The universal bioassay for MH activity consists of injecting an active extract or compound into the isolated abdomen of a cyclorrhaphous dipteran larva. The abdomen, which was separated from the source of the MH in the anterior portion of the larva by ligation prior to a critical period, would normally never tan or form a puparium.

A positive response in this assay may range from slight patches of tanning to complete puparium formation in the isolated abdomen. Originally, the blowfly *Calliphora vicina* Robineau-Desvoidy was the principal test insect (Karlson, 1956), but an improved assay using the house fly, *Musca domestica* L. (Kaplanis *et al.*, 1966a), and a modified test with a ligated lepidopteran larva, the rice stem borer, *Chilo suppressalis* (Walker), (Sato *et al.*, 1968) are currently the most extensively used MH assay systems.

Both the JH and MH assays have been used in the isolation and identification of hormones or compounds with hormone activity from insects and other natural sources, as well as to evaluate the hormonal activity of synthetic chemicals. The *Tenebrio* assay, since it measures a disruptive effect on normal insect development and metamorphosis, is also extensively used in the development and evaluation of hormonal insecticides. Since MH assays such as the house fly assay, which employ the ligated abdomens of insects, measure a normal effect of MH compounds on insect development, a number of inhibitive assays have been developed and used to assess the disruptive effects of the ecdysones and related steroids on insect development and reproduction (Hora *et al.*, 1966; Robbins *et al.*, 1968, 1970).

CHEMISTRY OF INSECT HORMONES

Juvenile Hormones and Compounds with Juvenile Hormone Activity

Williams (1956) prepared the first extract containing JH activity from the abdomens of male Cecropia moths, *Hyalophora cecropia* (Linnaeus), and it was not until 11 years later that the structure of the Cecropia JH was elucidated. During the period that elapsed between Williams' discovery and the characterization of the first insect JH, a great diversity of extracts and a number of terpenoid compounds were shown to possess JH activity. Three of the terpenoid compounds in particular provided important advances in our knowledge of juvenile hormone chemistry: farnesol (I), methyl *trans, trans*-10,11-epoxyfarnesenate (II), and juvabione (III) (Figure 2). Farnesol, the well-known sesquiterpenoid alcohol, was the first chemical shown to have juvenile hormone activity by Schmialek (1961), who identified it as an active component in mealworm feces. At the time it was synthesized and shown to have JH activity by Bowers and coworkers

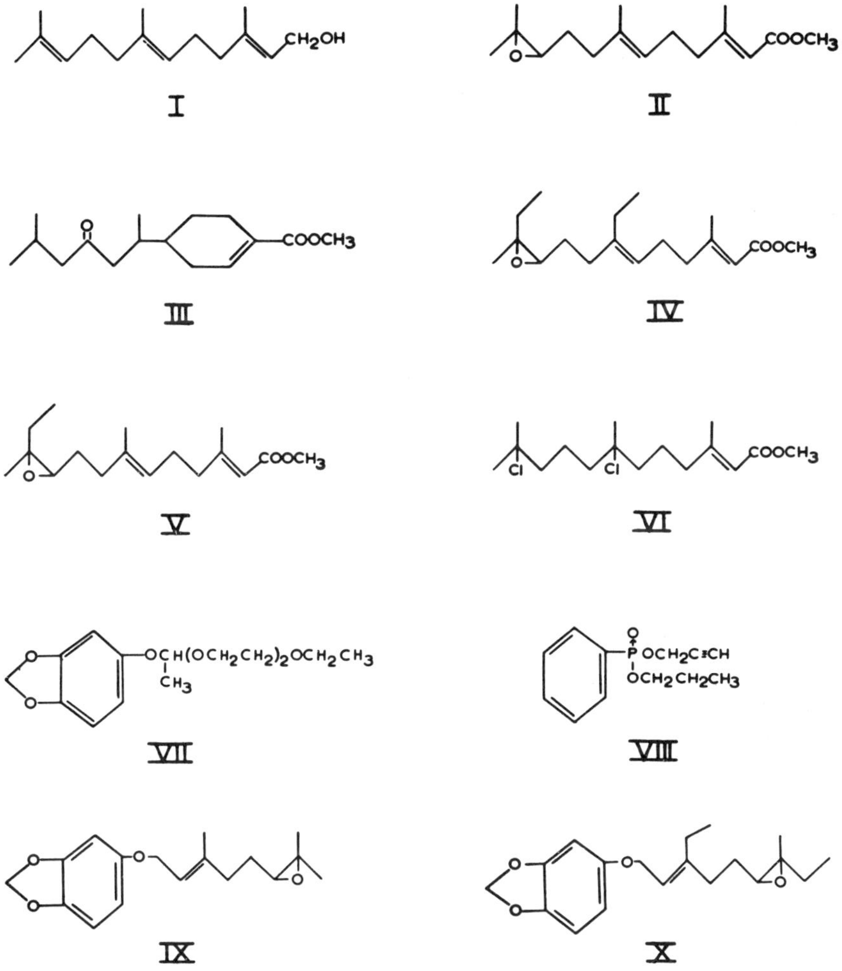

FIGURE 2 The insect juvenile hormones (IV and V) and some additional compounds with juvenile hormone activity.

(1965), the farnesenic acid derivative, methyl 10,11-epoxyfarnesenate was the most active JH chemical known. This compound, which had many of the biological and chemical characteristics of the active component(s) in Cecropia extract, was subsequently found to be a lower homologue of the two Cecropia juvenile hormones, differing from the natural hormones only by either one or two methyl groups. The monocyclic terpenoid juvabione, which was isolated and identified from balsam fir (Bowers *et al.*, 1966), is particularly interesting because of its highly specific JH activity in members of the insect family Pyrrhocoridae (Slama and Williams, 1966a), thus demonstrating for the first time the high degree of selectivity that may be attained with insect hormonal chemicals.

In 1967, Roller and his associates established the structure of the major Cecropia JH as methyl *trans, trans, cis*-10,11-epoxy-7-ethyl-3, 11-dimethyl-2,6-tridecadienoate (IV, Figure 2) using a combination of chemical and spectral analysis and synthesis (Roller *et al.*, 1967; Dahm *et al.*, 1967; Dahm *et al.*, 1968). One year later, Meyer and coworkers (1968) confirmed the structure of the major Cecropia hormone and identified a second JH from Cecropia, methyl *trans, trans, cis*-10,11-epoxy-3,7,11-trimethyl-2,6-tridecadienoate (V, Figure 2), which is a lower homologue of the first hormone, and differs only in the presence of a methyl rather than an ethyl group at C-7. The structure of the second Cecropia JH has been confirmed by synthesis (Johnson *et al.*, 1969).

The list of synthetic compounds with JH activity has continued to expand rapidly and now includes a large number of terpenoid and nonterpenoid chemicals (Bowers, 1971). One compound that is particularly interesting for both its high JH activity and its equally high selectivity for certain insects is the chlorinated sesquiterpenoid, methyl *trans*-7,11-dichlorofarnesenate (VI, Figure 2). This compound was shown by Romanuk *et al.* (1967) to be one of the active components in a highly active reaction mixture prepared by the hydrochlorination of farnesenic acid in alcohol (Law *et al.*, 1966). The report by Bowers (1968) that a number of insecticide synergists, with such diverse structures as sesamex (VII, Figure 2) and propyl-2-propynyl phenyl phosphonate (VIII, Figure 2), possess moderate-to-high JH activity provided us with a new group of nonsesquiterpenoidal JH compounds. Even more important, this discovery served as the basis for the design and synthesis of a number of the most active JH chemicals known to date—the methylenedioxyaromatic-terpenoid ethers (Bowers, 1969). These compounds, exemplified by the 6,7-epoxygeranyl ether of methylenedioxyphenol, *trans*-4-[(6,7-epoxy-3,7-dimethyl-2-octenyl)oxy]-1,2-(methylenedioxy)benzene (IX, Figure 2), and its diethyl homologue, *trans*-4-[(6,7-epoxy-3-ethyl-7-methyl-2-nonenyl)oxy]-1,2-(methylenedioxy)benzene (X, Figure 2), combine the terpenoid geraniol with the methylenedioxyphenoxy group of the synergists. They not only have JH activity in a wide spectrum of insects but in certain species these compounds are active at subnanogram levels (Bowers, 1971).

Ecdysones (Molting Hormones) and Analogues

The first insect-molting hormone to be characterized, *α*-ecdysone, was isolated in crystalline form by Butenandt and Karlson (1954). Eleven years later its complete structure was elucidated by chemical techniques (Karlson *et al.*, 1965) and x-ray crystallography (Huber and Hoppe, 1965) and shown to be the steroid 2*β*,3*β*,14*α*,22R,25-pentahydroxy-5*β*-cholest-7-en-6-one (Figure 3). Shortly thereafter, the structure of *α*-ecdysone was confirmed when its synthesis was independently and almost simultaneously reported by two groups (Kerb *et al.*, 1966; Siddall *et al.*, 1966). Two additional ecdysones have also been isolated and identified from insects: 20-hydroxyec-

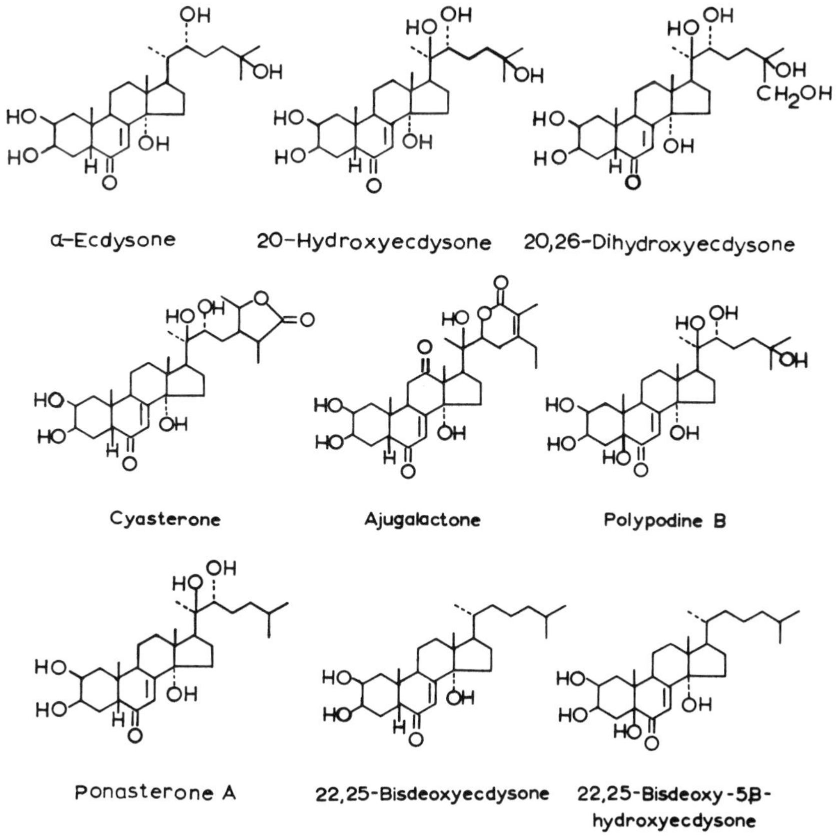

FIGURE 3 The insect ecdysones (molting hormones) and some examples of the phytoecdysones and synthetic ecdysone analogues.

dysone (Hocks and Weichert, 1966; Kaplanis *et al.*, 1966b; Hoffmeister and Grutzmacher, 1966), also known as crustecdysone, ecdysterone, and β-ecdysone, a steroid first characterized from material isolated from a crustacean (Hampshire and Horn, 1966), and 20,26-dihydroxyecdysone, which was isolated and identified from tobacco hornworm pupae, *Manduca sexta* (Johannson) (Thompson *et al.*, 1967) (Figure 3). All of the insect ecdysones possess the steroid nucleus of α-ecdysone and differ only in the number and the position of the hydroxyl groups on the side chain. The insect molting hormones are distinctly different, structurally, from the steroid hormones of higher animals. The insect ecdysones are all C_{27} steroids that contain the intact cholesterol side chain and thus differ from the C_{18}, C_{19}, and C_{21} steroid hormones of vertebrates. The insect steroid hormones contain an α,β-unsaturated ketone group as do a number of steroid hormones of higher animals. However, unlike the steroid hormones of vertebrates that often contain a 4-ene-3-one system in the A ring, the ecdysones have a 7-ene-6-one system in the B ring. The 5β-configuration (A/B ring *cis*) of the insect ecdysones occurs in certain vertebrate steroids; however, in the vertebrate steroids this configuration is more characteristic of the urinary metabolites of steroid hormones or excretory products.

In addition to the ecdysones isolated from insects and crustaceans, 30 or more ecdysteroids, including the two major insect ecdysones, α-ecdysone and 20-hydroxyecdysone, have been isolated and characterized from plants (Okauchi, 1969; Horn, 1971). A complete description or discussion of the diversity of structure of these phytoecdysones is not within the scope of this paper. However, the structures of four of the plant steroids, ponasterone A (Nakanishi *et al.*, 1966), cyasterone (Takemoto *et al.*, 1967), polypodine B (Jizba *et al.*, 1967), and ajugalactone (Koreeda *et al.*, 1970), which will be discussed specifically, are shown in Figure 3. The majority of the phytoecdysones have a steroid nucleus identical to that of α-ecdysone and differ only in the structure of their side chains, as do ponasterone A and cyasterone. However, a number of the plant ecdysones, including polypodine B and ajugalactone, do differ in the structure of the steroid nucleus. Currently there is no information as to the physiological or biochemical functions of the phytoecdysones in plants, yet their concentration in plants may be as much as a thousandfold that found in the best insect source (Horn, 1971). These high concentrations in plants have given rise to the suggestion that in addition to any function these steroids may have in the biochemical economy of the plant, they may also serve as plant protectants.

Added to the numerous naturally occurring ecdysones from arthropods and plants is an ever-increasing number of synthetic analogues of the ecdysones (Hocks *et al.*, 1966; Hora *et al.*, 1966; Thompson *et al.*, 1970, 1971a; Horn, 1971). Certain of the synthetic analogues, when tested in the diet or medium, are far more active inhibitors of insect development and reproduction than the two major insect ecdysones or the phytoecdysones–inokosterone, cyasterone, and ponasterone A (Robbins *et al.*, 1968; Kaplanis *et al.*, 1968; Robbins *et al.*, 1970; Thompson *et al.*, 1971a). The most potent of these synthetic steroids are 22,25-bisdeoxyecdysone and 22,25-bisdeoxy-5β-hydroxyecdysone (Figure 3). The former compound has a steroid nucleus identical to that of the insect ecdysones and differs only in the absence of hydroxyl groups on the side chain, and the latter compound, patterned after the phytoecdysone polypodine B, is the 5β-hydroxy-analogue of 22,25-bisdeoxyecdysone.

INSECT HORMONAL CHEMICALS AS INSECTICIDES

Insect hormones exert their effects in insects at very low levels and thus are extremely potent biological regulators; nanogram to microgram quantities of the insect hormones are readily detected in the standard hormone bioassays. When hormones or chemicals with hormone activity are administered to an insect at an inappropriate (hence, vulnerable) period or are administered in excessive quantities, these compounds disrupt normal development so that death or sterility occurs. To date, quite logically, most of the research on hormonal insecticides has centered around natural and synthetic compounds with insect hormone activity. One reason for this is that the most widely used bioassays measure insect hormone activity since these assays were originally developed and used for the isolation and identification of the insect hormones. However, as we progress in the development of hormonal pesticides, it becomes increasingly apparent that we should investigate all the possible types of compounds that may affect the hormone regulated processes of insects. These include, in addition to compounds with hormone activity, hormone antagonists or antihormones, and compounds that interfere with hormone biosynthesis or metabolism. Recent reports already indicate interesting progress in the latter two areas, particularly in relation to the molting hormones. Koreeda *et al.* (1970) have shown that the phytoecdysone ajugalactone inhibits the MH activity of ponasterone A in the *Chilo* dipping assay (Sato *et al.*, 1968), and a number of rela-

TABLE 1 Lethal Effects of Compounds with Juvenile Hormone Activity When Applied Topically to Insects or Added to the Larval Diet or Medium

Insect	Stage	Compound[a]	Dose per Insect or Concentration	Effect	Reference
Topical application					
Yellow mealworm	Pupa	X	1.0 ng	Pupal–adults	*b*
Confused flour beetle	Pupa	X	1.0 ng	Pupal–adults	*c*
Cotton stainer (*Dysdercus cingulatus*)	Nymph	VI	10.0 ng	Nymphal–adults	*d*
Diet or Medium					
Confused flour beetle	Larva	X	0.5 ppm	Pupal–adults	*c*
Tobacco hornworm	Larva	X	0.5 ppm	Blocks pupation	*c*
Corn earworm (*Heliothis zea*)	Larva	X	1.0 ppm	No adult emergence	*c*
Yellow fever mosquito (*Aedes aegypti*)	Larva	IX	1.0 ppm	No adult emergence	*c*

[a]See Figure 2 for structure of JH compounds.
[b]Bowers, 1969.
[c]Insect Physiology Laboratory, A.R.S., U.S.D.A., unpublished data.
[d]Slama *et al.*, 1969.

tively simple azasteroids block molting and metamorphosis in the tobacco hornworm and several other test insects when present in the larval diet or medium in ppm concentrations, apparently by blocking ecdysone biosynthesis and metabolism (Svoboda and Robbins, 1971; Robbins *et al.*, 1971b). Compounds with comparable activities for juvenile hormones in insects should, in particular, offer a considerable potential as hormonal insecticides.

Compounds with Juvenile Hormone Activity as Insecticides

The JH compounds certainly offer the greatest immediate potential for use as hormonal insecticides. Many of the more active JH chemicals are relatively simple molecules that would make their commercial production economically more feasible. These compounds readily penetrate the insect cuticle and thus are active by contact as are conventional pesticides and certain of them are sufficiently volatile to act as fumigants. The JH chemicals are extremely active compounds, and, under laboratory conditions, certain of them disrupt development and cause death when applied topically to the insect at doses as low as a nanogram or when present in the diet or medium at concentrations of less than a ppm (Table 1).

The principal lethal effect of compounds with JH activity is to interfere with the larval–pupal, pupal–adult, and nymphal–adult molts. The untimely presence of a JH compound during these molts causes the production of either supernumerary larvae or nymphs or intermediate forms such as pupal–adult intermediates. These abnormal forms usually fail to undergo normal ecdysis, are incapable of normal reproduction, and soon die (Bowers, 1971).

A second lethal effect of the JH chemicals is their interference with normal embryonic development, resulting in ovicidal activity. This effect may occur when the JH compound is administered either to newly laid eggs or to gravid females (Slama and Williams, 1966b; Riddiford and Williams, 1967). In the latter case a female chemosterilant action is obtained; with *Pyrrhocoris apterus* (L.) a single application of a dose of less than 1 μg of methyl *trans*-7,11-dichlorofarnesenate (VI, Figure 2) at any time in the reproductive cycle will cause sterility in the female insect for the remainder of its life (Masner *et al.*, 1968). The ovicidal action is obtained only when the eggs are treated during early embryonic development. If treated later a delayed action occurs in that the eggs may give rise to larvae that fail to develop normally during succeeding larval stages or during metamorphosis (Riddiford and Williams, 1967; Riddiford, 1970). With certain of the methylenedioxyaromatic-terpenoid ethers, ovicidal activity has been obtained by treating the eggs either by contact or by fumigation (Walker and Bowers, 1970).

In certain insects that diapause as adults, such as the alfalfa weevil *Hypera postica* (Gyllenhal) and the cereal leaf beetle *Oulema melanopus* (L.), treatment with certain compounds with JH activity terminates diapause, and the insects immediately begin to feed, mate, and to reproduce (Bowers and Blickenstaff, 1966; Connin *et al.*, 1967). The untimely termination of diapause would offer an additional means of using the JH compounds in insect control by exposing the insect to unfavorable conditions in which it cannot survive, such as heat, cold, or the absence of its host plant.

The JH compounds, then, affect a number of developmental stages including eggs, mature larvae or nymphs, pupae, and female adults. However, they do not block the development of or kill the growing larvae or nymphs, which in many plant-feeding insects is the most destructive stage. This is a major disadvantage of JH insecticides. In addition, to disrupt the development of either the egg or the mature larva and pupa, the JH chemical must be applied to or be in contact with the insect at a time when the insect is sensitive to the compound, and these periods may be of rather short duration in certain species. Thus, for effective control, the juvenile hormone must be present in the insects' environment for a sufficient period of time to ensure that large numbers of the insect population are exposed to the hormonal insecticide during sensitive or susceptible periods in their development. To be effective in the field, then, it will be necessary either to use repeated applications of the JH chemicals or to use JH compounds that are quite persistent. However, this persistence may occur in the environment or in the insect, as observed in the delayed action effect when eggs are treated late during embryonic development. In the insect the "residue" may be of a chemical (Willis and Lawrence, 1970) or of a physiological (Riddiford, 1970) nature. Finally, many of the above-mentioned disadvantages for compounds with JH activity might be avoided by the development and use of antijuvenile hormones, which should block development at the earlier larval or nymphal molts. These compounds could provide us with highly active hormonal insecticides that could be used more like conventional pesticides and still retain the safety and selectivity of the hormones.

Ecdysones and Analogues as Insecticides

Compounds with MH activity would be more expensive to produce commercially because of the requirement of the steroid nucleus for activity and the complexity of their structure. They penetrate the cuticle less readily than the JH chemicals and thus are usually taken up by the insect via ingestion. However, cuticular penetration may be effected through their application in certain volatile or nonvolatile solvents (Sato *et al.*, 1968; Sato, 1968; Williams and Robbins, 1968) and some apolar or organosoluble derivatives of ecdysones and their analogues penetrate the cuticle of certain insects (Kaplanis *et al.*, 1968). Although the ecdysones and analogues generally are not as active as the most active JH compounds, they still are quite potent inhibitors that disrupt insect development and reproduction when applied directly to the insect or added to the diet or medium in μg or ppm quantities, respectively (Table 2).

However, a major advantage of the ecdysones and analogues is that their primary lethal effect is to disrupt development during any larval or nymphal stage, and these disruptive effects are usually associated with the processes of molting and metamorphosis (Hora *et al.*, 1966; Robbins *et al.*, 1968, 1970; Sato, 1968; Earle *et al.*, 1970; Thompson *et al.*, 1971a). The ecdysones and analogues also terminate pupal diapause in certain lepidopterans (Williams, 1968; Kaplanis *et al.*, 1969), and, as with the termination of adult diapause, this effect could also be used for insect control.

Another important lethal action of the ecdysones and analogues is their female chemosterilant activity in insects (Robbins *et al.*, 1968, 1970; Rezabova *et al.*, 1968; Kaplanis *et al.*, 1968; Earle *et al.*, 1970; Wright and Kaplanis, 1970). The effect of the MH compounds on reproduction is to inhibit ovarian development and egg production as well as to reduce both egg hatch and the viability of the progeny. Interestingly, this chemosterilant action can be irreversible. With certain of the synthetic ecdysone analogues both the lethal effect on immature insects and the female chemosterilant action may be enhanced by greater than 10-fold with certain of the insecticide synergists, such as piperonyl butoxide (Kaplanis *et al.*, 1968).

A most promising approach to the use of hormonal chemicals to disrupt the MH regulated processes of insects, however, would be the use of compounds that interfere with the biosynthesis and metabolism of the ecdysones. These compounds could either be much simpler steroids, such as the inhibitory azasteroids (Svoboda and

TABLE 2 Lethal Effects of Ecdysones and Analogues When Injected into Insects or Added to the Larval Diet or Medium

Insect	Stage	Compound	Dose per Insect or Concentration	Effect	Reference
Injection					
Cynthia silkworm (*Samia cynthia*)	Diapausing pupa	Cyasterone	0.6 μg	Abnormal adults	*b*
House fly	Adult	Ponasterone A	2.0 μg	Inhibits ovarian development	*c*
Diet or medium					
Yellow fever mosquito	Larva	22,25-Bisdeoxy-5β-hydroxyecdysone	7 ppb	No pupation[a]	*d*
Boll weevil (*Anthonomus grandis*)	Larva	22,25-Bisdeoxyecdysone	10 ppm	No adults[a]	*e*
House fly	Larva	22,25-Bisdeoxyecdysone	75 ppm	No adults[a]	*f*

[a] Disruption of molting occurs at all larval stages.
[b] Williams, 1968.
[c] Kaplanis *et al.*, 1970.
[d] Thompson *et al.*, 1971a.
[e] Earle *et al.*, 1970.
[f] Robbins *et al.*, 1970.

Robbins, 1971; Robbins *et al.*, 1971b), or perhaps even nonsteroidal chemicals. Such compounds should be less expensive to produce commercially and at the same time would retain the major advantage of MH insecticides—their lethal effect on the larval or nymphal stages.

ADVANTAGES OF HORMONAL INSECTICIDES

A number of desirable characteristics have been visualized for hormonal insecticides, and certain of these have been realized in compounds developed thus far, including (1) high biological activity, (2) specificity, and (3) low vertebrate toxicity. High biological activity could require that less of the chemical be used in insect control and this would be important both from the standpoint of economy or cost and residue. As indicated in the previous discussion (Tables 1 and 2), high biological activity is a salient feature of the hormonal compounds. Under laboratory conditions, the most active of the hormonal chemicals disrupt development when nanogram quantities are applied directly to the insect or when concentrations in the ppb range are present in the diet or medium. As we increase our understanding of the relationship of structure to biological activity, it is certain that even more active compounds will be synthesized.

Specificity could be an important feature in a pesticide in that it would permit us to control the target pest without affecting beneficial insects or upsetting the ecological balance. The high specificity that may be attained with hormonal chemicals is well documented by the research of the Czechoslovakian group with the pyrrochorids, particularly with *P. apterus.* Certain JH compounds, such as methyl *trans*-7,11-dichlorofarnesenate (VI, Figure 2), will produce sterility in *P. apterus* females at doses as low as 0.1 μg per insect (Masner *et al.*, 1968) and less than 1 ng will interfere with the nymphal–adult molt in this insect (Slama *et al.*, 1969), while microgram to milligram quantities of this same compound are ineffective against a number of other insects (Slama *et al.*, 1970). However, many of the hormonal compounds are much less specific, and thus could be used against a variety of insects. For example, the methylenedioxyaromatic-terpenoid ethers, which include some of the most active JH compounds known (Bowers, 1969), are active against a wide spectrum of insects (Bowers, 1971).

It was reasoned that insect hormone chemicals should be relatively safe for vertebrates since they selectively disrupt biochemical and physiological processes peculiar to insects, and preliminary toxicological studies with certain of these chemicals appear to confirm this premise. In a recent study, Siddall and Slade (1971) found that one of the Cecropia juvenile hormones (V, Figure 2) and the ethyl ester of compound II (Figure 2), ethyl *trans, trans*-10,11-epoxyfarnesenate, produced no signs of toxicity when given to mice in a single oral dose of 5,000 mg/kg. When the animals were observed for 21 days following treatment and then subjected to comprehensive blood and tissue analyses no adverse effects were noted. The diethyl homologue of the 6,7-epoxygeranyl ether of methylenedioxyphenol (X, Figure 2), an extremely active JH compound for a number of insects (Bowers, 1969, 1971), also has low toxicity in mice; its acute oral LD_{50} is 4,000 ± 270 mg/kg and its skin absorption LD_{50} is greater than 4,000 mg/kg (Hoffmann–LaRoche, Nutley, N.J., unpublished data). An insect ecdysone 20-hydroxyecdysone, and the phytoecdysone cyasterone (Figure 3) were without toxicity to mice when administered orally at 100 μg/mouse/day for 90 days, which in growing mice is equivalent to an approximate dosage rate of 4 to 10 mg/kg/day (Hikino *et al.*, 1969). The authors report that both steroids slightly increased growth and stimulated protein biosynthesis (anabolic effect). The stimulatory effect on protein biosynthesis had previously been reported (Otaka *et al.*, 1968, 1969a, 1969b) for a number of ecdysteroids that contain C-20 and C-22 hydroxyl groups, including 20-hydroxyecdysone, cyasterone, and ponasterone A, while α-ecdysone (Figure 3), which lacks the C-20 hydroxyl group, did not show this effect. Masuoka *et al.* (1970) found ponasterone A to be nontoxic to rats when administered orally at doses as high as 50 mg/animal/day for 5 days (equivalent to approximately 350 to 500 mg/kg/day). No abnormalities were detected in the treated rats and these workers report that neither ponasterone A nor 20-hydroxyecdysone had an anabolic effect in the rat. These preliminary oral and dermal toxicity studies indicate that certain of the insect hormones or hormonal compounds have a low toxicity in mice or rats. More extensive toxicological and pharmacological evaluation of the effects on mammals of any potential hormonal insecticide must be made prior to its practical use in insect control.

Another potential advantage that has been suggested is that insects will not readily become resistant to hormonal chemicals; however, Schneiderman *et al.* (1970) have advanced arguments that resistance is likely to develop. The belief that resistance will not occur is based on the assumption that mechanisms for resistance to essential biological regulators, such as hormones, would work to the detriment of the insect, and this might well be true if we were only considering the insect's own hormones from an endogenous source or the receptor site(s) involved in the hormone's action. However, insect control with hormonal insecticides involves the use of chemicals that enter the insect from an exogenous source and consequently the physiological and biochemical systems involved in their absorption and transport, as well as their metabolism, may be distinct and different from analogous systems for the insect's endogenous hormones. In addition, the hormonal insecticide may so differ structurally from the insect's hormones (see Figure 2) that it would be treated as a xenobiotic rather than as an essential metabolite. Thus, the insect's hormones may not even serve as substrates for those enzymes that could inactivate or degrade the hormonal pesticide. However, if the above-mentioned desirable characteristics of hormonal compounds continue to hold true, then the possibility of insect resistance to the hormonal chemicals and the eventual need to replace these chemicals in time with new hormonal compounds would not be sufficient to negate the overall positive characteristics and potential value of these compounds as hormonal insecticides.

RESEARCH NEEDS AND THE STATUS OF HORMONAL INSECTICIDES

Research Needs

As we survey our progress in the development of hormonal insecticides it becomes apparent that there are a number of research areas that require additional emphasis and effort. We have an immediate need for carefully conducted field tests with certain of our most potent hormonal compounds against insects that have been shown under laboratory conditions to be especially susceptible to these chemicals. If, when applied at high levels in the field, these compounds fail to reduce the insect population, then research should be carried out to determine why control was not achieved. If our available active compounds do not work under field conditions because of certain unfavorable features, such as volatility or the lack of stability, then only when we know of these features can we begin to remedy them either through formulation or through the design and synthesis of new compounds. Such field testing could be expedited if insects of economic importance were more extensively used in our basic research on insect hormones as well as in our synthesis and testing programs. Indeed, the high degree of specificity observed for certain of the insect hormonal chemicals emphasizes the need for using important insect pests in all of our insect hormone research. We also need to employ more extensively laboratory assay systems that measure the cumulative effects of hormonal chemicals on insect populations for a generation or more during intermittent or continuous exposure to the chemicals. Such tests, involving continuous exposure, would ensure that the insect is in contact with the hormonal compound at any susceptible or vulnerable period. These tests would also measure any delayed action effects on insect development and reproduction as well as the additive effects of disruptive activity for more than one of the developmental stages. The results of such assays would be particularly useful in assessing the potential insecticidal value of the compounds for insects in habitats where continuous exposure could be effected, such as soil insects or stored products pests.

Another aspect that deserves additional emphasis is the search for and the attempted synthesis of compounds other than those with hormone activity that might be used to disrupt the hormone-regulated processes of insects. This includes both hormone antagonists or antihormones and compounds that interfere with hormone biosynthesis or metabolism. As previously discussed, particularly interesting and useful types of hormonal compounds could result from such studies, including antijuvenile hormones that would block early larval development and simple steroids (Svoboda and Robbins, 1971; Robbins *et al.*, 1971b) or nonsteroidal compounds that interfere with the MH regulated processes by blocking ecdysone biosynthesis and metabolism. This research would call for modifications of our existing assays or the development of new ones, since most of the assays currently in use would not detect such compounds.

More effort should also be directed to the isolation and identification of additional hormones from insects. The natural insect hormones serve as molecular models for the preparation of synthetic compounds and more such models are needed, particularly with the juvenile hormones. We know only two insect juvenile hormones, both isolated and identified from two closely related species of moths, and both from adult insects (Roller *et al.*, 1967; Meyer *et al.*, 1968; Dahm and Roller, 1970). The high specificity observed for some of the compounds with JH activity certainly suggests the existence of different juvenile hormones in different species and clearly the juvenile hormones also need to be isolated and identified from immature insects, where they normally function. Although we know a total of only three insect ecdysones, these have been identified from several different insects and a crustacean, and the information on the insect ecdysones is well supplemented by our knowledge of thirty or more structurally related phytoecdysones, many of which have MH activity (Horn, 1971). However, the hormones should be isolated and identified from different developmental stages to determine whether qualitive differences exist in the ecdysones at different times during insect development.

A particularly serious deficiency in our knowledge of the insect juvenile hormones is the lack of basic information on terpenoid and juvenile hormone biosynthesis and metabolism in insects. By comparison, we have a rather extensive body of knowledge on steroid and ecdysone biochemistry in insects and already a number of the intermediates and end products in the pathways of ecdysone biosynthesis and metabolism have been verified or identified (Robbins *et al.*, 1971a; Thompson *et al.*, 1971b). Similar information on the insect juvenile hormones will be of particular value both for the design and preparation of compounds that are not readily inactivated by insects as well as the search for and synthesis of chemicals that interfere with JH biosynthesis or metabolism.

Status of Hormonal Insecticides

In summation, the current status of hormonal insecticide development is as follows: A number of natural and synthetic insect hormonal chemicals are known. These include insect hormones, hormone analogues, and a diversity of synthetic compounds that mimic the action of insect hormones and/or interfere with the hormone-regulated processes of insects. Under laboratory conditions these compounds cause diverse disruptive effects on insect development, molting, metamorphosis, and reproduction when they are either applied directly to the insect or added to the insect's diet or medium in extremely small quantities. Certain of the synthetic compounds, particularly some of the more active JH mimics, are relatively simple molecules that would make their commercial production feasible. A number of pharmaceutical and agricultural chemical companies have become actively involved in hormonal insecticide development and certain of these firms have extensive programs involving synthesis and laboratory and field testing. Finally, a few compounds with JH activity have been tested as insect control chemicals in small scale field-plot and cage tests and even though these tests were quite preliminary, some of the results were promising (Entomology Research Division, A.R.S., U.S.D.A., unpublished data). However, carefully conducted field tests are now needed to resolve two important questions that confront us in the development of hormonal insecticides: How do we translate the disruptive effects of the insect hormonal chemicals observed under laboratory conditions into terms of their use in insect control? Are the hormonal compounds that we are developing (and the biological assays used in their development) providing us the most appropriate and useful chemicals for insect control? The field tests will help us answer these questions and provide us with new or added directions in our research that should expedite the development and the application of these promising chemicals for insect control.

REFERENCES

Berkoff, C. E. 1969. The chemistry and biochemistry of insect hormones. Quart. Rev. (London) 23:372–391.

Bowers, W. S. 1968. Juvenile hormone: Activity of natural and synthetic synergists. Science 161:895–897.

Bowers, W. S. 1969. Juvenile hormone: Activity of aromatic terpenoid ethers. Science 164:323–325.

Bowers, W. S. 1971. Juvenile hormones. *In* Naturally occurring insecticides. M. Jacobson and D. G. Crosby (eds). M. Dekker, Inc., New York.

Bowers, W. S., and C. C. Blickenstaff. 1966. Hormonal termination of diapause in the alfalfa weevil. Science 154:1673–1674.

Bowers, W. S., H. M. Fales, M. J. Thompson, and E. C. Uebel. 1966. Juvenile hormone: Identification of an active compound from balsam fir. Science 154:1020–1022.

Bowers, W. S., and M. J. Thompson. 1963. Juvenile hormone activity: Effects of isoprenoid and straight-chain alcohols on insects. Science 142:1469–1470.

Bowers, W. S., M. J. Thompson, and E. C. Uebel. 1965. Juvenile and gonadotropic hormone activity of 10,11-expoxyfarnesenic acid methyl ester. Life Sci. 4:2323–2331.

Butenandt, A., and P. Karlson. 1954. Über die Isolierung eines Metamorphose-Hormons der Insekten in kristallisierter Form. Z. Naturforsch. 9B:389–391.

Charniaux-Cotton, H., and L. H. Kleinholz. 1964. Hormones in invertebrates other than insects. Hormones 4:135–198.

Cleveland, L. R., A. W. Burke, Jr., and P. Karlson. 1960. Ecdysone induced modifications in the sexual cycles of the protozoa of *Cryptocercus*. J. Protozool. 7:229–239.

Connin, R. V., O. K. Jantz, and W. S. Bowers. 1967. Termination of diapause in the cereal leaf beetle by hormones. J. Econ. Entomol. 60:1752–1753.

Dahm, K. H., and H. Roller. 1970. The juvenile hormone of the giant silk moth *Hyalophora gloveria* (Strecker). Life Sci. Pt. II 9:1397–1400.

Dahm, K. H., H. Roller, and B. M. Trost. 1968. The juvenile hormone-IV. Stereochemistry of juvenile hormone and biological activity of some of its isomers and related compounds. Life Sci. 7:129–137.

Dahm, K. H., B. M. Trost, and H. Roller. 1967. The juvenile hormone. V. Synthesis of the racemic juvenile hormone. J. Am. Chem. Soc. 89:5292–5294.

Earle, N. W., I. Padovani, M. J. Thompson, and W. E. Robbins. 1970. Inhibition of larval development and egg production in the boll weevil following ingestion of ecdysone analogues. J. Econ. Entomol. 63:1064–1069.

Hampshire, F., and D. H. S. Horn. 1966. Structure of crustecdysone, a crustacean moulting hormone. Chem. Commun. (2):37–38.

Highnam, K. C., and L. Hill. 1969. The comparative endocrinology of the invertebrates. Edward Arnold Ltd., London.

Hikino, H., S. Nabetani, K. Nomoto, T. Arai, T. Takemoto, T. Otaka, and M. Uchiyama. 1969. Effect of long term administration of insect-metamorphosing substances on higher animals. I. Yakugaku Zasshi 89:235–240.

Hocks, P., A. Jager, U. Kerb, R. Wiechert, A. Furlenmeier, A. Furst, A. Langemann, and G. Waldvogel. 1966. Synthetic steroids with moulting hormone activity. Angew. Chem. Int. Ed. Engl. 5:673.

Hocks, P., and R. Wiechert. 1966. 20-Hydroxy-ecdyson, isoliert aus insekten. Tetrahedron Letters (26):2989–2993.

Hoffmeister, H., and H. F. Grutzmacher. 1966. Zur chemie des ecdysterons. Tetrahedron Letters (33):4017–4023.

Hora, J., L. Labler, A. Kasal, V. Cerny, F. Sorm, and K. Slama. 1966. On steroids CIII. Molting deficiencies produced by some sterol derivatives in an insect (*Pyrrhocoris apterus* L.). Steroids 8:887–914.

Horn, D. H. S. 1971. The ecdysones. *In* Naturally occurring insecticides. M. Jacobson and D. G. Crosby (eds). M. Dekker, Inc., New York.

Huber, R., and W. Hoppe. 1965. Zur chemie des ecdysons, VII die kirstall- und molekulstrukturanalyse des insektenverpuppungshormons ecdyson mit der automatisierten faltmolekulmethode. Chem. Ber. 98:2403–2424.

Ilan, J., J. Ilan, and S. Ricklis. 1969. Inhibition of juvenile hormone of growth of *Crithidia fasciculata* in culture. Nature 224:179–180.

Jizba, J., V. Herout, and F. Sorm. 1967. Polypodine B—A novel ecdysone-like substance from plant material. Tetrahedron Letters (51):5139–5143.

Johnson, W. S., S. F. Campbell, A. Krishnakumaran, and A. S. Meyer. 1969. Total synthesis of the racemic form of the second juvenile hormone (methyl 12-homojuvenate) from the cecropia silk moth. Proc. Nat. Acad. Sci. U.S. 62:1005–1009.

Kaplanis, J. N., W. E. Robbins, M. J. Thompson, and A. H. Baumhover. 1969. Ecdysone analog: Conversion to alpha ecdysone and 20-hydroxyecdysone by an insect. Science 166:1540–1541.

Kaplanis, J. N., L. A. Tabor, M. J. Thompson, W. E. Robbins, and T. J. Shortino. 1966a. Assay for ecdysone (molting hormone) activity using the house fly, *Musca domestica* L. Steroids 8:625–631.

Kaplanis, J. N., M. J. Thompson, and W. E. Robbins. 1968. The effects of ecdysones and analogs on ovarian development and reproduction in the housefly *Musca domestica* (L.). Proc. Int. Congr. Entomol., 13th, Moscow, 1968.

Kaplanis, J. N., M. J. Thompson, R. T. Yamamoto, W. E. Robbins, and S. J. Louloudes. 1966b. Ecdysones from the pupa of the tobacco hornworm, *Manduca sexta* (Johannson). Steroids 8:605–623.

Karlson, P. 1956. Biochemical studies on insect hormones. Vitamins and Hormones 14:227–265.

Karlson, P., H. Hoffmeister, H. Hummel, P. Hocks, and G. Spiteller. 1965. Zur chemie des ecdysons, VI. Reaktionen des ecdysonmolekuls. Chem. Ber. 98: 2394–2402.

Kerb, U., P. Hocks, R. Wiechert, A. Furlenmeier, A. Furst, A. Langemann, and G. Waldvogel. 1966. Die synthese des ecdysons. Tetrahedron Letters (13): 1387–1391.

Koreeda, M., K. Nakanishi, and M. Goto. 1970. Ajugalactone, an insect moulting inhibitor as tested by the *Chilo* dipping method. J. Am. Chem. Soc. 92: 7512–7513.

Krishnakumaran, A., and H. A. Schneiderman. 1968. Chemical control of moulting in arthropods. Nature 220:601–603.

Krishnakumaran, A., and H. A. Schneiderman. 1969. Induction of molting in crustacea by an insect molting hormone. Gen. Comp. Endocrinol. 12:515–518.

Law, J. H., C. Yuan, and C. M. Williams. 1966. Synthesis of a material with high juvenile hormone activity. Proc. Nat. Acad. Sci. U.S. 55:576–578.

Masner, P., K. Slama, and V. Landa. 1968. Sexually spread insect sterility induced by the analogues of juvenile hormone. Nature 219:395–396.

Masuoka, M., S. Orita, A. Shino, T. Matsuzawa, and R. Nakayama. 1970. Pharmacological studies of insect metamorphosing hormone: ponasterone A, ecdysterone, and inokosterone in the rat. Jap. J. Pharmacol. 20:142–156.

Meyer, A. S., H. A. Schneiderman, E. Hanzmann, and J. H. Ko. 1968. The two juenile hormones from the cecropia silk moth. Proc. Nat. Acad. Sci. U.S. 60: 853–860.

Muftic, M. 1969. Metamorphosis of miracidia into cercariae of *Schistosoma mansoni in vitro.* Parasitology 59:365–371.

Nakanishi, K., M. Koreeda, S. Sasaki, M. L. Chang, and H. Y. Hsu. 1966. Insect hormones. The structure of ponasterone A, an insect-moulting hormone from the leaves of *Podocarpus nakaii* Hay. Chem. Commun. (24):915–917.

Okauchi, T. 1969. Recent studies on phytoecdysones. Botyu-Kagaku 34:140–156.

Otaka, T., S. Okui, and M. Uchiyama. 1969a. Stimulation of protein synthesis in mouse liver by ecdysterone. Chem. Pharm. Bull. 17:75–81.

Otaka, T., M. Uchiyama, S. Okui, T. Takemoto, H. Hikino, S. Ogawa, and N. Nishimoto. 1968. Stimulatory effect of insect-metamorphosing steroids from *Achyranthes* and *Cyathula* on protein synthesis in mouse liver. Chem. Pharm. Bull. 16:2426–2429.

Otaka, T., M. Uchiyama, T. Takemoto, and H. Hikino. 1969b. Stimulatory effect of insect-metamorphosing steroids from ferns on protein synthesis in mouse liver. Chem. Pharm. Bull. 17:1352–1355.

Rezabova, B., J. Hora, V. Landa, V. Cerny, and F. Sorm. 1968. On Steroids CXIII. Sterilizing effect of some 6-ketosteroids on housefly, *Musca domestica* L. Steroids 11:475–496.

Riddiford, L. M. 1970. Prevention of metamorphosis by exposure of insect eggs to juvenile hormone analogs. Science 167:287–288.

Riddiford, L. M., and C. M. Williams. 1967. The effects of juvenile hormone analogues on the embryonic development of silkworms. Proc. Nat. Acad. Sci. U.S. 57:595–601.

Robbins, W. E., J. N. Kaplanis, J. A. Svoboda, and M. J. Thompson. 1971a. Steroid metabolism in insects. Ann. Rev. Entomol. 16:53–72.

Robbins, W. E., J. N. Kaplanis, J. A. Svoboda, and M. J. Thompson. 1971b. Chemistry and biological activity of the ecdysones and ecdysone analogs. Proc. Int. Congr. Pesticide Chem., 2nd, Tel Aviv, 1971.

Robbins, W. E., J. N. Kaplanis, M. J. Thompson, T. J. Shortino, C. F. Cohen, and S. C. Joyner. 1968. Ecdysones and analogs: Effects on development and reproduction of insects. Science 161:1158–1160.

Robbins, W. E., J. N. Kaplanis, M. J. Thompson, T. J. Shortino, and S. C. Joyner. 1970. Ecdysones and synthetic analogs: Molting hormone activity and inhibitive effects on insect growth, metamorphosis and reproduction. Steroids 16: 105–125.

Roller, H., K. H. Dahm, C. C. Sweely, and B. M. Trost. 1967. The structure of the juvenile hormone. Angew. Chem. Int. Ed. Engl. 6:179–180.

Romanuk, M., K. Slama, and F. Sorm. 1967. Constitution of a compound with a pronounced juvenile hormone activity. Proc. Nat. Acad. Sci. U.S. 57:349–352.

Sato, Y. 1968. Insecticidal action of phytoecdysones. Appl. Entomol. Zool. 3: 155–162.

Sato, Y., M. Sakai, S. Imai, and S. Fujioka. 1968. Ecdysone activity of plant-originated molting hormones applied on the body surface of lepidopterous larvae. Appl. Entomol. Zool. 3:49–51.

Schmialek, P. 1961. Die identifizierung zweier im tenebriokot und in hefe vorkommender substanzen mit juvenilhormonwirkung. Z. Naturforsch. 16b: 461–464.

Schneiderman, H. A., and L. I. Gilbert. 1964. Control of growth and development in insects. Science 143:325–333.

Schneiderman, H. A., A. Krishnakumaran, P. J. Bryant, and F. Sehnal. 1970. Endocrinological strategies in insect control. Agr. Sci. Rev. 8:13–25.

Shanta, C. S., and E. Meerovitch. 1970. Specific inhibition of morphogenesis in *Trichinella spiralis* by insect juvenile hormone mimics. Can. J. Zool. 48: 617–620.

Siddall, J. B., A. D. Cross, and J. H. Fried. 1966. Steroids. CCXCII. Synthetic studies on insect hormones. II. The synthesis of ecdysone. J. Amer. Chem. Soc. 88:862–863.

Siddall, J. B., and M. Slade. 1971. Absence of acute oral toxicity of *Hyalophora cecropia* juvenile hormone in mice. Nature New Biol. 229:158.

Slama, K., K. Hejno, V. Jarolin, and F. Sorm. 1970. Natural and synthetic materials with insect hormone activity. 5. Specific juvenile hormone effects in aliphatic sesquiterpenes. Biol. Bull. 139:222–228.

Slama, K., M. Romanuk, and F. Sorm. 1969. Natural and synthetic materials with insect hormone activity. 2. Juvenile hormone activity of some derivatives of farnesenic acid. Biol. Bull. 136:91–95.

Slama, K., and C. M. Williams. 1966a. The juvenile hormone. V. The sensitivity of the bug, *Pyrrhocoris apterus,* to a hormonally active factor in American paper-pulp. Biol. Bull. 130:235–246.

Slama, K., and C. M. Williams. 1966b. 'Paper factor' as an inhibitor of the embryonic development of the European bug, *Pyrrhocoris apterus.* Nature 210: 329–330.

Svoboda, J. A., and W. E. Robbins. 1971. The inhibitive effects of azasterols on sterol metabolism and growth and development in insects with special reference to the tobacco hornworm. Lipids 6: 113–119.

Takemoto, T., Y. Hikino, K. Nomoto, and H. Hikino. 1967. Structure of cyasterone, a novel C_{29} insect-moulting substance from *Cyathula capitata.* Tetrahedron Letters (33):3191–3194.

Thompson, M. J., J. N. Kaplanis, W. E. Robbins, and R. T. Yamamoto. 1967. 20,26-Dihydroxyecdysone, a new steroid with moulting hormone activity from the tobacco hornworm, *Manduca sexta* (Johannson). Chem. Commun. (12): 650–653.

Thompson, M. J., W. E. Robbins, C. F. Cohen, J. N. Kaplanis, S. R. Dutky, and R. F. N. Hutchins. 1971a. Synthesis and biological activity of 5β-hydroxy analogs of α-ecdysone. Steroids 17:399–409.

Thompson, M. J., W. E. Robbins, J. N. Kaplanis, C. F. Cohen, and S. M. Lancaster. 1970. Synthesis of analogs of α-ecdysone. A simplified synthesis of 2α,3β,14α-trihydroxy-7-en-6-One-5β-steroids. Steroids 16:85–104.

Thompson, M. J., J. A. Svoboda, J. N. Kaplanis, and W. E. Robbins. 1971b. Metabolic pathways of steroids in insects. Proc. Roy. Soc. (London) Ser. B. In press.

Walker, W. F., and W. S. Bowers. 1970. Synthetic juvenile hormones as potential coleopteran ovicides. J. Econ. Entomol. 63:1231–1233.

Wigglesworth, V. B. 1964. The hormonal regulation of growth and reproduction in insects. Advances in Insect Physiol. 2:247–336.

Williams, C. M. 1956. The juvenile hormone of insects. Nature 178:212–213.

Williams, C. M. 1967. Third-generation pesticides. Sci. Am. 217:13–17.

Williams, C. M. 1968. Ecdysone and ecdysone-analogues: Their assay and action on diapausing pupae of the cynthia silkworm. Biol. Bull. 134:344–355.

Williams, C. M., and W. E. Robbins. 1968. Conference on insect–plant interactions. BioScience 18:791–792, 797–799.

Willis, J. H., and P. A. Lawrence. 1970. Deferred action of juvenile hormone. Nature 225:81–83.

Wright, J. E. 1969. Hormonal termination of larval diapause in *Dermacentor albipictus.* Science 163:390–391.

Wright, J. E., and J. N. Kaplanis. 1970. Ecdysones and ecdysone-analogues: Effects on fecundity of the stable fly, *Stomoxys calcitrans.* Ann. Entomol. Soc. Am. 63:622–623.

Yamazaki, M., and M. Kobayashi. 1969. Purification of the proteinic brain hormone of the silkworm, *Bombyx mori.* J. Insect Physiol. 15:1981–1990.

Chemical Weed Control Strategies for the Future

W. C. Shaw and L. L. Jansen

A basic goal of our research, regulatory, and educational programs is the establishment of chemical weed control practices that are compatible with high-yielding agroecosystems and with a quality environment. Fundamental to the success of this concept is the knowledge that the world's human population, as well as its domestic animals and all wildlife, exists in an ever-changing environment and cannot, in fact, exist in a natural or static condition.

Those Americans who have never lived on farms probably do not understand the vital processes involved in the production of food and fiber. Many overlook the elementary fact that plants are the sole source of energy from which to derive an abundant supply of high quality and wholesome food for man, livestock, and wildlife. The extent to which plants and animals have to be protected against damage by pests must be more effectively communicated to the general public and understood by them.

Many Americans, including a segment of the scientific community, apparently do not realize that the native or natural vegetation in most areas in the United States was originally not very efficient or economical as a source of food for livestock and humans. When native vegetation is disturbed or when cultivated fields are abandoned, the sequence of plant succession is well known. Weeds comprise an early stage of these plant successions, one that terminates with the climax vegetation characteristic of the area. Plant successions involve well-established ecological principles, and they establish the basic need for efficient chemical weed control technology (Figure 1).

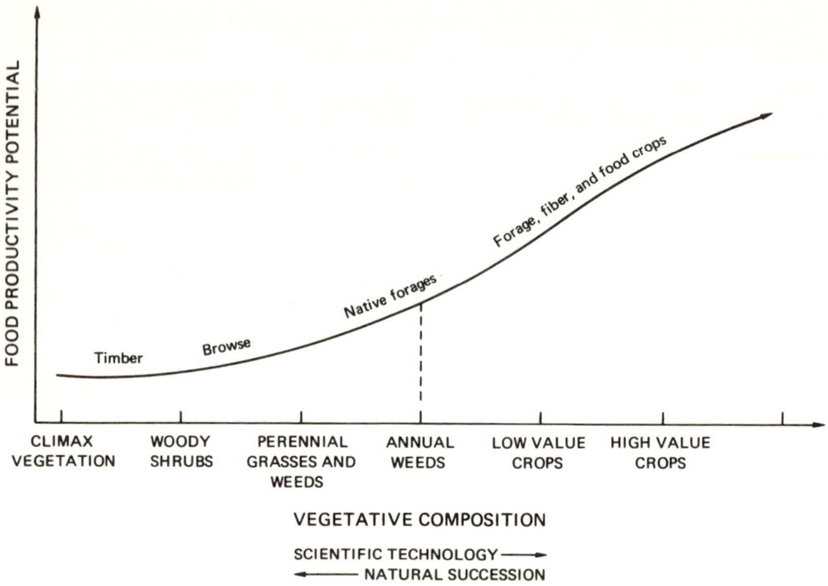

FIGURE 1 Relations of weed control to ecological principles.

ECOLOGICAL ASPECTS OF WEED CONTROL

Man's standard of living, and indeed his survival, depend on his manipulation of the environment. He must produce an ecological situation favorable to the production of food crops and livestock, to the protection of his natural resources, and to other environmental values. Weed control is a major environmental manipulation. It is necessary for the safe, efficient, and economical production of food crops and for the protection of our health and welfare.

The general public needs to understand the ecological situation in which crops and livestock are produced. These compete in a complex environment that is shared by about 30,000 species of weeds distributed throughout the world. More than 1,800 of these cause serious economic losses each year. Most cultivated crops are subject to competition from about 200 weed species. From 10 to 50 different weeds infest each major food crop, and these must be controlled each year.

Because there are many kinds of weeds, including annuals, biennials, and perennials, with varying periods of germination and with highly divergent life cycles, they obviously cannot be controlled by a single method. The control of weed populations of this type requires an integrated systems approach that includes cultural, mechanical, biological, ecological, bioenvironmental, and chemical methods. The chemical control of such diverse weed populations requires a broad spectrum of selective herbicides, mixtures of herbicides, and combination chemical treatments.

Moreover, crops and livestock are attacked by about 50,000 species of fungi that cause more than 1,500 diseases. About 15,000 species of nematodes attack crop plants and more than 1,500 of these cause damage. More than 10,000 species of pest insects add to the serious losses that occur each year. In spite of the best pest-control technology that we have been able to develop, insects, diseases, nematodes, and weeds cause losses that amount to about 33 percent of potential agricultural production each year.

Although our best weed control technology has been used effectively, weeds continue to reduce our agricultural production by 10 percent yearly. Also, farmers spend more than $2.5 billion each year for weed control. The general public benefits from this investment. Because farmers follow effective weed control practices, we have fewer weeds in our lawns, gardens, waterways, and recreation sites. Farm weed control practices also reduce human health problems that are often caused by allergenic and poisonous weeds. Cultural and other ecological and biological methods of weed control are used on more than 300 million acres of harvested crops each year. More than 100 million acres of intertilled row crops receive intensive tillage and cultural weed control. Moderate cultural weed control is practiced on more than 200 million acres of drill crops. The intertilled row crops usually receive two to five seedbed preparation treatments, and one to as many as seven cultivations in a single growing season. Cultural weed control practices are also used on more than one billion acres of hay, pasture, and rangelands, and on millions of acres of nonagricultural lands and acquatic sites.

Herbicides are now used on approximately 150 million acres of harvested cropland each year. This amounts to about 50 percent of our total harvested crop acreage. The production and use of herbicides has grown at the rate of about 26 percent a year during the past 5 years, as compared to a 9 percent growth rate for all other synthetic organic pesticides. The domestic use of herbicides increased from 125 million pounds in 1963 to more than 335 million pounds in 1969. Currently, herbicides account for about 34 percent of the total pesticides used in the United States, and they represent about 58 percent of the value of total pesticide sales.

Shifts to newer and more effective herbicides have increased average unit costs from about \$0.94 per pound in 1963 to about \$1.52 per pound in 1970. Both rate of usage and chemical costs per acre have increased, primarily because of shifts to herbicides that are applied to the soil as preplanting or preemergence treatments for weed control. Such treatments control a broader spectrum of weeds, but usually require higher dosages.

ADVANCES IN CHEMICAL WEED CONTROL TECHNOLOGY

Advances in chemical weed control technology are difficult to assess unless they are evaluated as a part of our total production and protection technology. The total technology includes genetically improved, high-yielding, and high-quality crop varieties; improved crop management practices; better plant and animal nutrition; improved farm equipment and mechanization practices; improved irrigation equipment, principles, and practices; and efficient control of diseases, insects, nematodes, parasites, and other pests. Chemical weed control practices, along with other production and protection inputs have been integrated into high-yielding agroecosystems compatible with a quality environment. These have had far-reaching benefits.

Progress in the development of chemical weed-control technology has been unparalleled during the past 20 years. The discovery of several diverse classes of highly selective herbicides helped to revolutionize modern weed control. Some of the most important discoveries include the phenoxy compounds, *s*-triazines, substituted ureas and uracils, chloroalphatic acids, carbamates, chloroacetamides, benzoic acids, substituted anilines, toluidines (such as trifluralin), and bipyridyliums (such as diquat).

Introduction of new techniques—such as versatile preplanting; preemergence, and other soil application and incorporation methods; granular formulations; low-pressure low-gallonage application techniques; and technology for improved distribution and deposit of sprays, including low drift formulations, improved the effectiveness and safe use of herbicides. Other significant advances were the development and use of cultural, ecological, bioenvironmental, and mechanical methods in an agroecosystems approach that included herbicide mixtures and crop and herbicide rotations.

Important progress has also been made in the development of efficient herbicide detection, screening, evaluation systems, and equipment. The standardization and wide use of improved evaluation techniques gave impetus to the industrial development of new herbicides. In 1950 we had about 15 basic herbicides. Today there are more than 180 and about 6,000 formulated products.

Outstanding progress has also been made in understanding the penetration, absorption, translocation, mechanisms of action, and metabolism of herbicides in plants and their behavior and fate in soils, water, air, and in other components of the environment. Our understanding of the influence of surfactants, cosolvents, and other adjuvants on the selective action of herbicides on crop plants and weeds and the effects of such additives on spray deposit, drift, persistence, and fate in the environment has also improved.

Improved analytical methods for the determination of minute quantities of herbicide residues have provided an additional margin of safety in the use of herbicides. They have also established a sound basis for understanding the fate of residues in the environment. These advances stimulated improvements in procedures for the registration of herbicides that provide greater assurance of herbicide utility and safety.

We have increased the number of weed scientists equipped to provide knowledge for safe and effective use of herbicides. Improved education of herbicide users has also been accomplished.

The Weed Science Society of America and the strong regional and national conferences have provided effective forums for discussion of weed control research developments and technology. They have played an important role by stimulating cooperation among state, federal, and industrial scientists and have provided a system for coordinating their programs.

THE IMPACT OF CONTROL ON CROP PRODUCTION PRACTICES

These unparalleled advances in chemical weed control technology have had a far-reaching impact on all phases of crop production. New chemical methods of weed control have exerted an important influence on the selection of crops and varieties, on seedbed preparation, on methods of seeding, and on seeding rates.

New chemical weed control practices have made it possible to exercise a wide choice of row spacings, plant spacings in the row, and plant populations. Chemical weed control practices also influence fertilizer practices, including time of application and placement.

Tillage and cultivation techniques have been revolutionized. For many crops, chemical weed control now makes minimum tillage possible. For a few crops on certain soil types, zero tillage now seems feasible. Chemical weed control has had a major impact on irrigation practices, harvesting, seed-cleaning operations, and erosion control.

New herbicides have tremendously increased the weed-control efficiency of fallow practices. The use of chemicals has greatly reduced tillage in some of the fallow-farming areas. The adoption of minimum tillage and chemical fallow practices greatly reduces wind erosion; improves moisture conservation; reduces sheet erosion; improves fertilizer utilization; improves soil structure and drainage; and increases yields and quality of crops.

Improved chemical weed control has also improved the effectiveness of various practices directed to the control of diseases, nematodes, and insects. Herbicides have increased the effectiveness of pasture renovation techniques. They have also improved the productivity of pastures, rangelands, and forests. The use of farm water resources for irrigation, recreation, and the maintenance of reservoirs, drainage ditches, ditch banks, irrigation canals, and farm roadsides have been significantly improved by the use of herbicides.

Crop production is on the horizon of a second-generation breakthrough in production efficiency because of the impact of chemical weed control. These new practices have already established the necessity for a new research cycle in crop production that can result in new plateaus of crop yields, crop quality, and production and harvesting efficiency.

BENEFITS OF CHEMICAL WEED CONTROL

The use of herbicides and associated chemical weed control technology has accounted for at least 10 percent of the total increase in farm outputs since 1940. Currently we have about 6 percent excess production capacity. Any combination of factors that results in a 10 percent reduction in production would cause serious and untenable economic consequences. This knowledge alone explains why herbicides are so important to the production of our current food supplies.

Current efficiency in production, which has resulted from the use of herbicides and other agricultural advances, permits each farm worker to produce enough for 45 other persons. Only 5 percent of the population of the United States is engaged in agriculture and we spend only 17 percent of our income for food. In the Soviet Union 39 percent of the population is engaged in agriculture and they spend 56 percent of their income for food.

What are some of the positive and direct benefits derived from the use of herbicides? In 1939, the average yield of wheat in the United States was 14 bushels per acre. This year the average yield is 31 bushels per acre. During the past 20 years, the use of 2,4-D to control weeds in more than 460 million acres of wheat increased the yield of wheat by about 2 billion bushels–enough for a bonus of 130 billion extra loaves of bread. The increased farm income from this single practice was $3.25 billion.

This morning, in suburban Maryland, several major supermarket chain stores are selling five 1-lb loaves of bread at the retail price of 20 cents per loaf. The farmer receives about 3 cents for the wheat used in each loaf. Even if the farmer provided the wheat free of charge, the bread would cost 17 cents per loaf. For many foods, the farmer's share of the retail price is so small that if he gave his products away, retail prices would decline very little. The use of 2,4-D, and other advanced practices, has helped to keep bread prices low and quality high. The use of such technology has enabled producers to maintain acceptable profits without price increases. The bread is not only a bargain from the standpoint of cost, but it is also wholesome and nutritious. It has a higher protein content and a better-balanced complement of amino acids, minerals, and other improved nutritional properties.

In the past 30 years, wheat, rice, and potato yields in the United States have doubled, and corn yields have tripled. Improved chemical and cultural weed control practices in these crops alone have reduced labor requirements by 30 to 50 percent. In Japan, rice yields doubled between 1951 and 1961 while labor requirements for rice production were reduced 42 percent. Improvements in weed control accounted for 34 percent of the savings in labor.

The benefits of herbicides can also be illustrated by determining the economic impact of banning herbicides. Banning the use of 2,4,5-T would increase costs to domestic users about $47 million per year, even if other chemical herbicides were not restricted. However, if other phenoxy herbicides were not available, costs could increase an estimated $167 million per year. In the short run these additional costs would be borne by farmers, governmental units, and the recreational, industrial, and timber industries. However, in the long run consumers will pay the bill.

Herbicides have greatly reduced the labor and equipment requirements in crop and livestock production. Production is also less dependent on manual labor. The yields and quality of crops have increased and losses caused by weeds have been significantly reduced. The chemical control of weeds that produce allergenic pollens greatly reduces human illness, and control of poisonous weeds prevents deaths of livestock and wildlife.

Chemical weed control has contributed to economic stability in the United States through the creation of new occupations and a wide variety of employment opportunities at increased pay levels. The use of herbicides is an effective and economical technique for the maintenance of airports, industrial sites, and rights-of-way. Effective and safe practices also contribute to aesthetic values of rural and urban living. Herbicides improve the maintenance of waterways, watersheds, and impounded waters for commercial, agricultural, and recreational uses. They also improve the safety of our highways by removing vegetation hazards, thereby improving visibility and reducing the number of accidents and deaths of humans and of wildlife.

PROBLEMS AND RISKS IN CHEMICAL WEED CONTROL

Perhaps the single greatest risk involved in chemical weed control is misuse. Inadequate training of operators as to proper safety precautions is common. Among the commonest errors are the transfer of herbicides to improper containers and the failure to establish safety measures for the benefit of children and animals. Improper handling, storage, transport, and disposal of unused herbicides, herbicide wastes, and herbicide containers rank high among the problems and risks.

Environmental contamination, or side effects not anticipated or fully comprehended, including human injury through unsuspected toxic metabolites that may be detrimental to humans, animals, or the environment, must be monitored. Risks are also greatly increased by the illegal use of herbicides that may result in residues in food or feed.

Volatility and drift of sprays are serious problems. Drift not only causes damage to beneficial plants, but it is also an important potential source of environmental contamination. We do not know what effects, if any, airborne herbicide molecules have on the photosynthetic efficiency of crops in large herbicide use areas.

We need a better understanding of the actual or potential long-range effects of continued and repeated use of herbicides. However, the evidence, after 20 years of rather intensive use, indicates that current practices will not cause serious risks in the future.

The national pesticide monitoring programs on food, water, soil, plants, domestic animals, and wildlife provide the best assessment of actual as well as potential risks from current use of herbicides. These results show that herbicides are rarely found in the food monitoring studies conducted by the Department of Health, Education and Welfare. Trace amounts that are occasionally found are below tolerances established for these chemicals in food.

The results of our national monitoring studies also show that herbicides have not been biomagnified in humans, domestic animals, wildlife, or other objects or organisms in the environment. There is no evidence that agricultural uses of herbicides cause an accumulation of residues in humans, domestic animals, wildlife, and fish, birds, or other objects in the environment. Herbicides used at agricultural rates on croplands dissipate rapidly. There is little evidence from our monitoring programs of any accumulation or buildup of herbicidal residues in soils. We do know that herbicides volatilize readily and that drift is a continuing problem. How much herbicides may contaminate the air, and cause chronic damage, is not known. This problem needs further study.

Although there is some concern that weeds may become resistant to herbicides, most scientists believe that this is not a significant problem. There is greater concern over the increase in weed populations that are resistant to control by herbicides. Although these ecological shifts may involve some risks, they will not become serious if cultural, mechanical, ecological, biological, and chemical methods are combined in an integrated systems approach.

Many scientists believe that the risk of using herbicides can also be further reduced by the development of a better knowledge of their toxicological effects. More emphasis is needed on understanding the carcinogenic, mutagenic, and teratogenic effects of herbicides during the research and development phase, and prior to widespread use.

Industrial organizations are beginning to experience difficulty in justifying the unusually high commercial costs of developing and registering new herbicides. In 1956 the overall cost of research to develop a pesticide was about $1.2 million. In 1970 it cost more than $4 million. If such costs continue to increase, a shortage of herbicides could occur. If this happens, it will weaken integrated systems of control and increase our vulnerability to weed attacks.

The amount of time and resources used by scientists, administrators, and other personnel to defend current registered uses of herbicides have become excessive and counterproductive. If industrial scientists are required to spend a disproportionate amount of their time, talents, and resources in defensive activities, the discovery and development of new herbicides and other pesticidal chemicals will be inhibited. Many already believe that work toward the discovery of improved pesticides has suffered. There is a serious need for a significant increase in trained weed research scientists and regulatory personnel, to ensure safe and effective use of herbicides.

Improved methods for control of aquatic weeds are needed for the solution of this increasingly serious problem. We need better methods for control of broadleaf weeds in soybeans. Current herbicides are not effective on many of the weeds that are rapidly becoming serious in this important crop. Herbicides that do not have excessive residual action are needed for the control of perennial weeds such as nutsedges, Canada thistle, and milkweeds on cropland.

CHEMICAL WEED CONTROL STRATEGIES FOR THE FUTURE

In developing strategic plans for chemical weed control in the future we must keep in mind that our most important objectives are to safely and effectively reduce losses in agricultural production. Renewed efforts must also be made to reduce all other risks that are associated with the use of herbicides, and to maintain productive agroecosystems that are compatible with a quality environment.

Research Needs

We need to intensify our research programs in order to develop a better understanding of the ecology, phenology, physiology, biochemistry, and botanical aspects of weeds. More knowledge is needed on their complete life cycles, so that we can determine the growth stages and the physiological and biochemical systems that are most vulnerable to control.

A much better understanding is needed of the acute and chronic toxicological effects of herbicides and their impact on the environment. Criteria and methods for quantitating their environmental impact must be developed. A fundamental pool of information is gradually providing the quantitative components and knowledge necessary for making technological assessments of their potential as well as actual environmental impact before they are widely used.

We need to emphasize control of weeds as a part of total farm and regional agroecosystems, rather than to limit research and control strategies to a single weed in a single crop. In developing an agroecosystems approach to chemical weed control, we need to rotate our crops, to rotate herbicides on crops, and to use combination herbicide treatments, sequential treatments, and mixtures of herbicides. Chemical control must be integrated with cultural practices and ecological techniques (Tables 1 and 2).

A better understanding of weed population dynamics, as influenced by chemical and nonchemical treatments, would enable us to predict ecological changes. This would facilitate an adjustment of treatments to prevent those undesirable shifts that make control more difficult.

More research is needed on the effects of contaminants, analogues,

TABLE 1 Rotation of Different Herbicides on the Same Crop[a]

Year	Crop Sequence	Chemical Weed Control Treatments		
		Preplanting	Pre-emergence	Postemergence
First	corn	EPTC	simazine	2,4-D
Second	corn	atrazine	CDAA + 2,3,6-TBA	linuron
Third	corn	EPTC	atrazine	dicamba
Fourth	soybeans	nitralin	chlorpropham + NPA	2,4-DB

[a]Hypothetical rotations used to illustrate the principle of keeping maximum pressure on the weed population by the use of a series of herbicides and mixtures that differ in their effectiveness in controlling various weed spectrums. This procedure reduces the chance of a species that is tolerant to a specific herbicide from becoming dominant. It also reduces the chance of an accumulation of herbicide residues in the soil.

TABLE 2 Rotation of Different Herbicides on All Crops[a]

Year	Crop Sequence	Chemical Weed Control Treatments		
		Preplanting	Pre-emergence	Postemergence
First	corn	EPTC	atrazine	2,4-D
Second	peanuts	vernolate	sesone	DNBP + 2,4-DEP
Third	cotton	trifluralin	diuron	MSMA
Fourth	soybeans	nitralin	amiben	chloroxuron

[a]Hypothetical rotations used to illustrate the principle of keeping maximum pressure on the weed population by the use of a series of herbicides and mixtures that differ in their effectiveness in controlling various weed spectrums. This procedure reduces the chance of a species that is tolerant to a specific herbicide from becoming dominant. It also reduces the chance of an accumulation of herbicide residues in the soil.

isomers, and other ingredients in herbicides and formulations. Cross-contamination of herbicides and other pesticides is a growing problem, one of national concern, and research and guidelines are needed to solve it.

We need to determine which herbicides and combination treatments are most effective for the implementation of minimum tillage and zero tillage strategies in crop production. These practices conserve water and soil, reduce herbicide volatility, and reduce the loss of herbicides in sediment, sheet erosion, and runoff.

The effectiveness of most herbicides is reduced because of inadequate residual activity. In order to overcome these limitations, initial applications are usually higher than needed. In our future strategies, we shall need to develop herbicides with controlled release characteristics. These would enhance the release of active moieties uniformly, and over predetermined periods of time. In fact, the successful development of this technology would revolutionize chemical weed control. Initial rates of application could be reduced. Losses from volatility, undesirable downward movement through the soil profile, sheet erosion, drift, and other environmental problems could be reduced.

Recent developments in the synthesis of polymerized herbicides with controlled-release characteristics offer promise of solving many of these problems. Intensive research is needed to exploit the potential controlled-release characteristics of polymerized herbicides, copolymers, encapsulation, and related formulation techniques.

Faulty disposal of unused pesticides, pesticide wastes, and pesticide containers are a threat to a quality environment. Disposal practices of the past will not be acceptable in the future. A significant expansion of research for the development of better disposal technology will be essential to the effective and safe use of herbicides. Unless better disposal techniques are developed, constraints on current methods will undoubtedly be imposed that will inhibit expansion in chemical weed control.

The development of weed seed germination stimulants could significantly improve weed control technology. If uniform germination for most weed species that infest a single crop could be induced through the use of germination stimulants, chemical weed control could be significantly improved. Many of the growth-regulating chemicals that are known to stimulate weed seed germination are not effective because of their rapid degradation in the soil. Techniques to enhance their residual activity would greatly improve the chances of effective use.

Future chemical control strategies must emphasize mission-oriented research for the development of safer and more effective herbicidal use patterns, formulations, and methods of treatment. We need improved methods for the detection, measurement, and elimination or minimization of herbicidal residues in plants, animals, soils, air, and water. Processes in need of more intensive research are adsorption, photodecomposition, absorption by plants, degradation by microorganisms, volatilization, drift, aerial dissipation, chemical reactions, leaching, and surface runoff.

Regulatory Needs

Preventive weed-control technology in the United States is relatively ineffective. Much consideration has been given to regulatory needs for the registration and safe use of herbicides. However, preventive weed control regulations have received too little attention.

State and Federal legislative authority, regulations, and current programs do not provide for adequate protection against the importation or interstate movement of noxious weeds, weed-propagation parts, and weed seeds, or of materials such as soil, manure, hay, straw, sod, feed grains, equipment, nursery stock, livestock, and other materials infested with noxious weeds, weed-propagation parts, or weed seeds.

The Federal Seed Act and State Seed Laws permit relatively high tolerances for weed seeds in crop seeds. The provisions in these laws are not effective in preventing the importation, interstate and intrastate movement, and dissemination of weed seeds in crop seeds. About one third of the states have no limitation on total weed seed permitted to be sold in crop seeds. Limitations that do exist in some states range from 1 to 4 percent by weight. Regulations of this type do not take into account the advances in weed control of the past 20 years. Effective preventive weed control technology cannot be developed unless the Federal Seed Act and State Seed Laws are amended and weed seed tolerances in crop seeds are established that are scientifically feasible and reasonable.

Policies and procedures for registration of herbicides need a careful review and assessment. Complaints that research is not directed toward the support of regulatory programs are often made. Criteria and protocols for the registration of herbicides need to be established and published. Only if these requirements are known and understood by the scientific community can research be effectively directed in support of registration objectives.

Guidelines and formulas need to be developed for registration of herbicides. What are the components of the registration formula? Research has identified some of the requirements for registration. They include the following: an understanding of acute and chronic toxicological properties of herbicides; absorption and translocation; mechanisms of action; sites of action; metabolism in plants and soils; behavior and fate in plants, animals, soil, water, and air; persistence; effects on target and nontarget organisms; patterns of use; and impact on the environment.

We are beginning to develop a balance-sheet approach to understanding the full spectrum of events associated with the uses of herbicides and their dissipation from the environment (Figure 2). As our knowledge of these functions increases, we should be able to identify those that are essential to the registration process. As we do so, we should also be able to arrive at a quantitative evaluation, at least to the extent of our knowledge, of each component in the registration formula (Figure 3). Quantitation of the registration process should enhance understanding by scientists in government service and by industrial scientists and should permit a better direc-

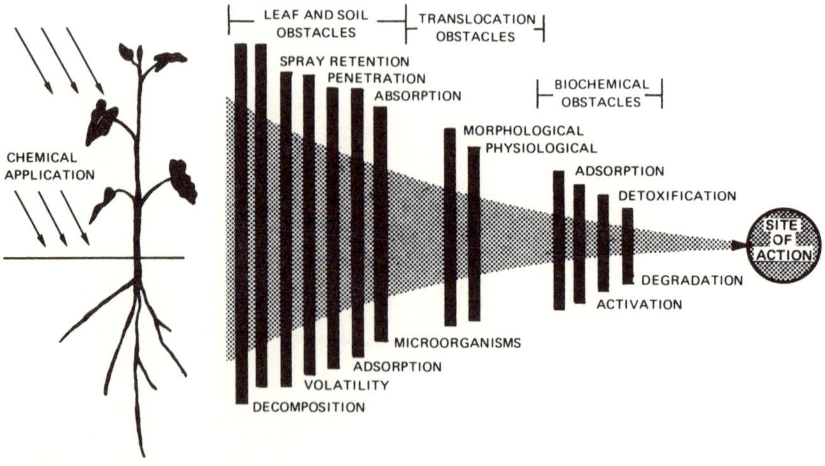

FIGURE 2 Obstacles that determine the concentration of toxic material at the site of action.

SUGGESTED HERBICIDE REGISTRATION REQUIREMENTS

PROTOCOLS		KNOWLEDGE INDEX
Chemical Characteristics		9
+		
Acute and Chronic Toxicity		8
+		
Absorption and Translocation		9
+		
Mechanisms of Action		9
+		
Sites of Action		8
+		
Metabolism and Metabolites		8
+		
Persistence and Residual Activity		4
+		
Behavior and Fate in the Environment		5
+		
Environmental Exposure Index		5
+		
Effectiveness on Target and Nontarget Org.		5
Registration Score	=	70

FIGURE 3 Suggested herbicide registration requirements.

tion of the latter's research to serve basic needs and registration requirements.

We need to establish classes of herbicides on the basis of their immediate as well as their long-range impacts on the environment. Only after we have developed and published such information can we expect intelligent use by consumers and greater acceptability by those concerned about the effects of chemicals on environmental values. Everyone agrees that scientific assessments should replace emotional concerns and reactionary responses. How can this be done? As our knowledge of the behavior and fate of pesticides grows, we should be able to develop a relative environmental exposure index for each pesticide. Such an index would require assessment of toxicity, dosage, total use, and persistence, among other factors.

At the present time our knowledge of the behavior and fate in plants and soils of some of the important pesticides is sufficiently complete to permit an approximation of their environmental impact. A number of these pesticides are applied directly to soil. Major quantities of others eventually reach the soil either during application to other targets or afterwards. The soil, then, is a major source of potential to human exposure to pesticide residues, either jointly or individually, through food crops and the food chains. From the known rates of application, which remain relatively constant, and the time required for dissipation of 75 to 100 percent of the quantity applied in a typical application, we can calculate the annual residues.

We have used a maximum persistence value of 25 percent and the number of persistence periods per year as an exponential function of this value. The exposure index is the ratio of average daily human exposure in mg/kg to a reference mammalian toxicity value, the acute oral LD_{50} to rats. In Figures 4, 5 and 6, we have shown comparative indexes and other parameters for selected herbicides, fungicides, and insecticides.* Indexes for 1968 were calculated from per capita do-

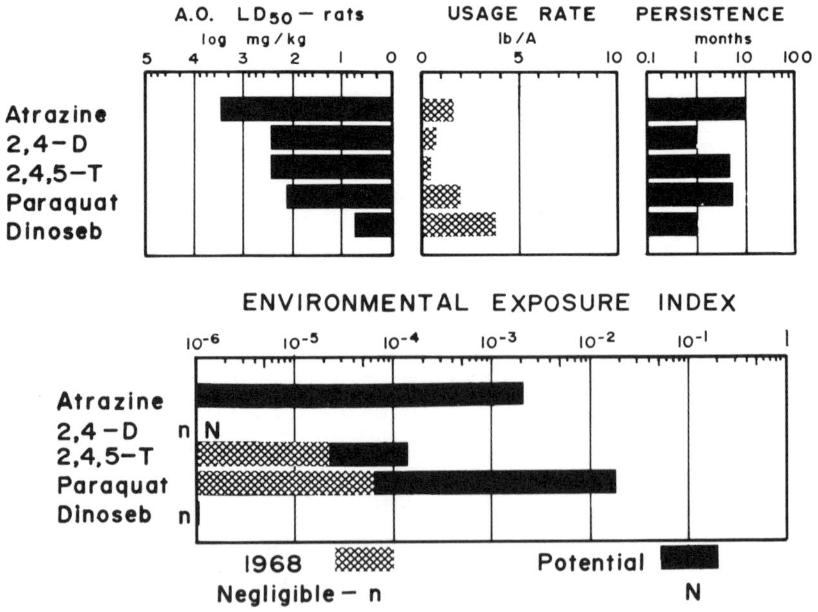

FIGURE 4 Herbicide-environmental interactions.

*Sources of data:

U.S. Department of Agriculture, Economic Research Service. 1970. Quantities of pesticides used by farmers in 1966. Agr. Econ. Rep. 179.

U.S. Department of Agriculture, Agricultural Stabilization and Conservation Service. 1971. The pesticide review, 1970.

U.S. Department of Agriculture. 1969. Agricultural statistics–1969.

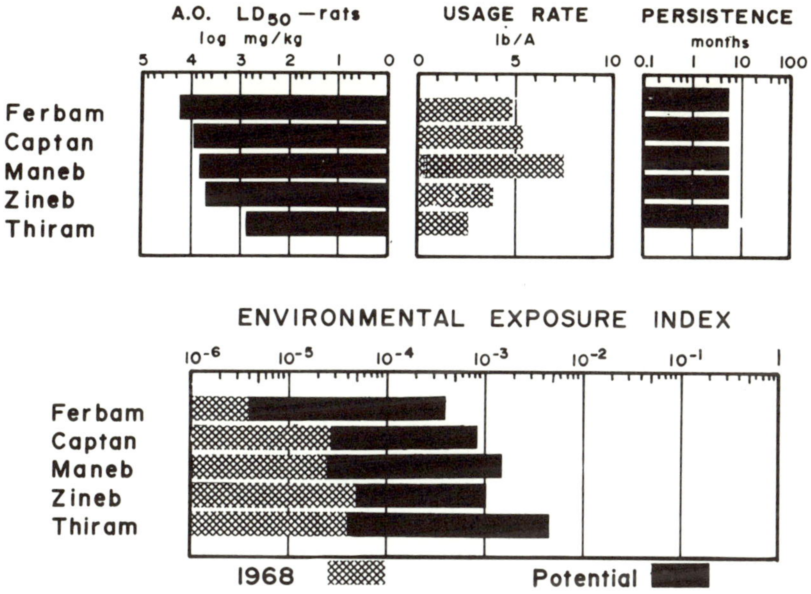

FIGURE 5 Fungicide–environmental interactions.

mestic use of the pesticides. Potential indexes reflect the impact from an unrealistic assumed use of each pesticide on all of the 1.22 acres per capita required for the production of all food and fiber.

Needless to say, the environmental exposure indexes for pesticides in the soil environment leave much to be desired in total quantitative assessment of environmental impact. They do not take into account exposures from other parts of the biosphere or the hazards from accidental or excessive exposures in handling and applying pesticides. However, we believe that such values represent a significant step in quantitative assessment.

Educational Needs

Our educational needs for the efficient and safe use of herbicides are extensive and growing. We are entering an era of intensive chemical use in agricultural production. Education toward the understanding and appreciation of chemicals and their use should begin at the elementary school level and should be intensified in our high schools and universities. Many of our universities are granting Bachelor of

Science degrees in the agricultural sciences without providing adequate training in the use of agricultural chemicals. A revision and upgrading of college courses on ecology, pesticide chemistry, physiology, biochemistry, and toxicology are urgently needed. In addition, intensive educational programs are needed to inform consumers on the proper handling and use of pesticides.

SUMMARY

Agricultural technology has assured us an abundant and wholesome food supply, even on a continuously reduced cropland acreage. It has permitted us to reserve more than 300 million acres of land for natural beauty, hunting, fishing, parks, wildlife preserves, and other forms of recreation. This contribution to an improved environment for all Americans must also be included in the benefit/risk evaluations when agricultural practices are being judged.

We must communicate the benefits of agricultural research and technology to a concerned general public. Scientists must play a more

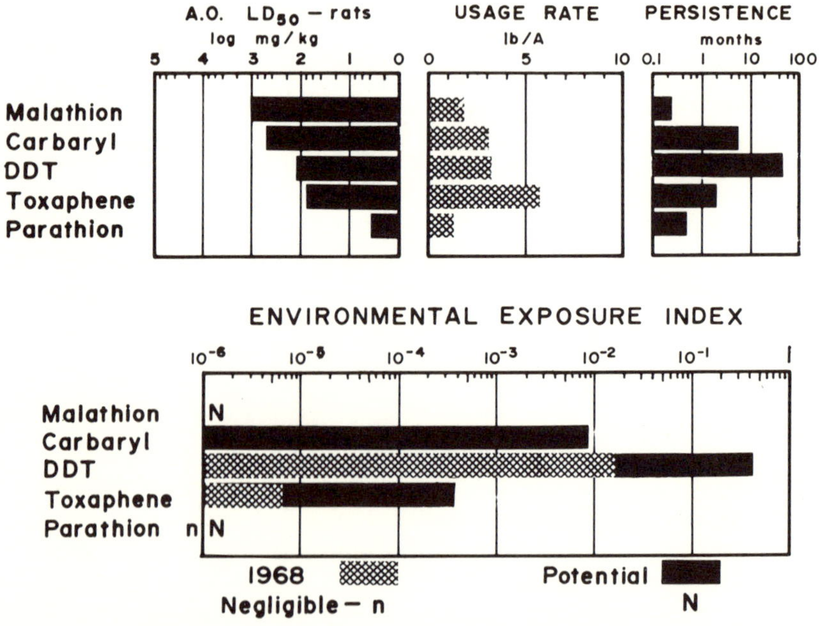

FIGURE 6 Insecticide-environmental interactions.

significant role. We all know that serious or potentially serious problems have been caused by the misuse of agricultural chemicals. We are also confident that many of the fears about agricultural chemicals are unjustified.

As agricultural scientists and administrators, we must be responsive to the problems and concerns of all citizens. The need for agricultural chemicals and all other agricultural technology must be related to the needs of society as a whole. We cannot ignore agricultural principles and practices that impair the environment. We must continue to identify, define, and assess actual and potential environmental problems. We must devote more effort to advanced assessments of technology. Moreover, our criteria must evaluate social, aesthetic, and environmental parameters.

We must examine carefully those herbicide, fertilizer, and other agricultural chemical uses that have risks that outweigh their benefits. When this is done, using acceptable scientific criteria, and the risks remain questionable, such chemical uses should be discontinued and replaced with safer alternatives. If such alternatives are not available, researchers should develop them.

We must achieve and maintain a fine balance. On the one hand, we must protect our health and comfort and our capacity for producing food and fiber from the hazards of weeds. On the other hand, we must understand the ecological significance of weeds and the methods of control. We must protect all of the important values in our environment against the potential hazards. Benefits and risks must be carefully weighed before decisions can be made that are clearly in the public interest.

Selective Chemicals for Plant Disease Control

James G. Horsfall

In my paper I shall limit myself to fungi that cause plant disease. The same principles apply to viruses, nematodes, and bacteria, although our detailed ignorance of these organisms is much greater than for fungi. In other words, more can be said about compounds selective for fungi than for the other pathogens of plants.

In plant disease, as in human disease, we have the classical interaction of environment, pathogen, and host. Even though the effect of each is difficult to separate from the others, we shall attempt to do so in the interests of clarity.

SELECTIVITY RELATED TO THE ENVIRONMENT

It is fashionable these days to talk about the environment. So let us talk about the selectivity of antifungal compounds that is related to the environment. The old classic in this field, of course, is the case of sulfur painted on greenhouse steam pipes. The heat sublimes the sulfur and it settles on the plants in the greenhouse and controls the powdery mildew diseases.

Vapor Phase Activity We have a recent example in our laboratory. We have often wondered why zineb controls *Alternaria solani* on tomato foliage better than the copper in the old standby, Bordeaux mixture. Within the last year (Lukens and Horsfall, 1971) we have established that *A. solani* spores can germinate and infect the leaf in high humidity without free water. Under these conditions a zineb deposit will kill the spores. Bordeaux mixture will not. Zineb or probably some product of it has enough vapor pressure to kill in the vapor phase. Bordeaux mixture does not.

Bent (1967) has shown that several new organic fungicides can control powdery mildew on leaves in the vapor phase. He lists in this category drazoxolon, oxythioquinox, binapacryl, dinitrooctylphenyl crotonate, and phthalimido-phosphonothionate. Hislop (1967) listed maneb and dichlofluanid as effective on *Botrytis* in the vapor phase.

Redistribution by Rain Another striking example of the effect of the environment on selectivity is that of redistribution. As Rich (1954) has shown, particles of copper fungicides carry an electrokinetic positive charge. This means that the particles cling less readily to each other than to the leaf. Accordingly the rain dislodges particles from the clumps in the original deposit, moves them to uncovered tissue below, and redeposits them there. Thus, lower leaves, not easily covered by the original spraying, are covered by redistribution in rain. Zineb particles are not so charged and, hence, are not redistributed by rain. They must be applied to the lower leaves by nozzles that trail through the foliage.

Counteraction by Environment Of course, the environment is unfavorable for most antifungal compounds. Most candidate protective fungicides are beaten to death by the rain or decomposed by the intense rays of the sun. Thus the commercially successful protectants can be said to be selective against environmental degradation.

SELECTIVITY RELATED TO THE HOST

Compounds That Make the Host Resistant

The most exciting new idea is to develop antifungal compounds whose selectivity is related to the host. They make the host resistant to disease.

Compounds That Alter the Host Physiology Mineral oil is the host-altering substance that finds the largest commercial use. It is used on bananas for the dreaded Sigatoka disease caused by *Cercospora musicola*. Calpouzos (1966) has discussed this interrelation. Mineral oil is not fungitoxic. It can protect leaves from successful invasion and it can also exert a chemotherapeutic effect on established infections. Calpouzos concludes that the oil changes the physiology of the leaf. The disease, he says, is a high-sugar disease in the sense of Horsfall and Dimond (1957). This means that the fungus is most destructive on young foliage high in sugar. The oil reduces the sugar level and reduces disease.

This leads us to the interesting problem of low- and high-sugar diseases and the relation to selectivity. Target spot disease of tomatoes, caused by *Alternaria solani*, is a low-sugar disease of foliage, and root rot of barley, caused by *Helminthosporium sativum*, is a low-sugar disease of the roots.

Maleic hydrazide damages phloem and reduces sugar transport, thereby acting to increase sugar in foliage and reduce it in the roots. It is not surprising then that maleic hydrazide reduces the low-sugar disease (*Alternaria*) on the foliage (O'Brien and Bornmann, 1952, and von Ramm *et al.*, 1962) and increases the root disease of barley (Richardson, 1957).

The data of Sijpesteijn and Pluijers (1962) show that cucumber scab, caused by *Cladosporium cucumerinum*, is a high-sugar disease. A non-fungitoxic substance, S-carboxymethyl-N,N-dimethyldithiocarbamate, reduces the sugar content of the foliage and makes the plant resistant to infection.

Similarly, leaf rust of wheat is a high-sugar disease. Bromfield and Emge (1958) have shown that maleic hydrazide, and DDT as well, increase the sugar content of foliage and increase infection with rust.

The bean rust fungus enters the leaf through the stomates. α-Hydroxydecanesulfonic acid closes the stomates, and Dimond (1963) cites data to show that the compound can prevent penetration by the fungus through the stomates.

Destruction of the pectins of the middle lamella between host cells is often involved in the disease process, as in fusarial wilt of tomato. Several growth hormones inhibit the methylation of pectic acid. If calcium is present in sufficient supply, calcium bridges are formed; the pectin is less available to the fungus and the fungus produces less disease (Edgington *et al.*, 1961). This story itself is tied to the sugar story outlined above. The *Fusarium* fungus obtains glucose for energy from the pectins such as galacturonic acid that it hydrolyzes.

The polygalacturonase of the tomato *Fusarium* is produced only if needed. Patil and Dimond (1968) considered then that galactose should inhibit the formation of galacturonase in the diseased plant and should inhibit pathogenesis. This hypothesis was confirmed by experiment. Galactose, then, can be considered as a therapeutant to make the tomato plant more resistant to fusarial wilt. And this experiment gives a biochemical base for some of the sugar story described above.

We must mention at this point a paper by Grossman (1962) who showed that rufianic acid acts as a synthetic chemotherapeutant for fusarial wilt of tomato by inhibiting the action of polygalacturonase.

No Helping Phagocytes in Plants Efforts to alter the host to make it more resistant have shown remarkable strides and surely should be pursued much further. These efforts, however, have a very serious built-in weakness. They do not deal directly with the organism. They don't kill it, and it hangs around ready to pounce on the host at any moment. The folklore statement for this situation is that the medicine was a success but the patient died.

Human and animal pathologists have a great advantage that plant pathologists do not have: Animals have phagocytes to clean up the stragglers; plants do not. If plants did, then the compounds that alter host metabolism would have a better chance of success.

Systemic Fungicides That leads us to systemic fungicides, and we are back to the killers. Much gold is being discovered in those hills. Several compounds have appeared almost simultaneously on the experimental market.

The compound that first achieved publication was thiabendazole, which is 2-(4-thiazolyl)benzimidazole (Staron and Allard, 1964; Staron *et al.*, 1966). The compound enters the root, is translocated to all parts of the plant, and kills organisms in the tissue. In 1968 a related compound called benomyl was published by Delp and Klopping (1968). This compound is methyl-1-(butylcarbamoyl)-2-benzimidazole carbamate. It, too, is systemic and kills invading fungi. A look at the structural formulas shows that these two fungicides differ chiefly in the substituent at the 2-position on the benzimidazole nucleus. One has a thiazolyl group at that point, the other a methyl carbamate group. The butyl carbamoyl group in the 1-position of benomyl presumably is a fat-solubilizing group that aids permeation but not fungitoxicity, according to Clemons and Sisler (1969).

It should be said that these benzimidazole compounds had their antecedents, too. Samborski *et al.* (1958) had shown several years earlier that benzimidazole reduced rust on wheat very significantly. As far as I know, no one has shown exactly what the 2-substitution of thiabendazole and benlate does. Wang *et al.* (1961) had investigated a few 2-substituted benzimidazoles on wheat rust. Although it was phytotoxic, 5-chloro-2-methylbenzimidazole seemed more effective on rust at lower concentrations than benzimidazole itself. In 1952 Dr. David Davis of our laboratory (unpublished) had tested five benzimidazoles as chemotherapeutants on fusarial wilt of tomato. 2-Phenylbenzimidazole was the best of the five but three others showed some effect.

A third new series of systemic fungicides is comprised of the 1,4-oxathiins as originally reported by von Schmeling and Kulka (1966). Edgington *et al.* (1966) showed that this series is especially selective for basidiomycetes. The oxathiins are able to penetrate the seed coat of barley and wheat seeds and eliminate loose smut. This is a dramatic effect on an extremely recalcitrant disease.

Another systemic fungicide is Milstem (Bent, 1969). It is highly selective for powdery mildew diseases. Lastly, there is phenylthiourea and 1-phenylthiosemicarbazide. Both are systemic. The former is not fungitoxic *in vitro;* the last one is (see Sijpesteijn and Pluijers, 1962). They are very active therapeutants against cucumber scab.

Iatrogenic Compounds If a medicine induces a new disease in the host, we call the new disease an iatrogenic disease—a disease caused by the doctor. Plants have iatrogenic diseases, too. Perhaps the classic case of iatrogenic disease is the well-known occurrence of powdery mildew on grapes sprayed with zineb in Europe or of powdery mildew on apple sprayed with captan in the United States. In another case, zineb controls leaf diseases on tomato but increases *Botrytis* rot on the fruit, according to Cox and Hayslip (1957). Gibson (1956) reported that ethylmercury phosphate increased damping-off of pine seedlings, Erwin *et al.* (1961) that phorate increased *Pythium* on cotton seedlings, Farley and Lockwood (1968) that pentachloronitrobenzene increased *Fusarium* root rot. All authors concluded that the effect is to eliminate the sensitive microbial inhibitors of the pathogens without eliminating the pathogens. Smith *et al.* (1970) ran up a small red flag on benomyl. They say it encourages a basidiomycete on lawn grass while controlling fusarial rot.

SELECTIVITY RELATED TO THE PATHOGEN

Lethal Synthesis

If the pathogen creates its own death potion from a nontoxic compound, we call it lethal synthesis. In other words, the pathogen synthesizes a toxic compound from a nontoxic one. This is an exciting source of selectivity but one that has been little explored.

Karathane is a highly selective spray for powdery mildews. Yet it has no effect on *Alternaria solani* or on rusts (Doling and Hepple, 1959). We have investigated its effect on *A. solani*. Karathane is a crotonic acid ester of a dinitrophenol. The phenol nucleus of Karathane is toxic to *A. solani* and so is the acetic acid ester. Presumably *A. solani* can de-esterify the acetic acid ester, not the crotonic acid ester. Possibly the conjugated double bond of the latter is involved. If the side chain is lengthened from 3 to 5 carbons and the double bond moved to the end carbon, conjugation is lost and the ester is toxic. We conclude that *A. solani* can de-esterify that ester (Horsfall and Lukens, 1969) but not the crotonic acid ester. It seems probable, then, that powdery mildews can also de-esterify the crotonic acid ester and the host cannot. If so, this accounts for the selectivity.

Activity Decreased by Pathogen Detoxication

We have more data by far on detoxication than on lethal synthesis. Woodcock (1967) has partially reviewed this field as has van der Kerk (1969) and Horsfall and Lukens (1966). Since this is only peripherally related to selectivity we shall drop it there.

Compounds That Alter the Biology of the Pathogen

Most fungitoxic substances are not very selective, but we must remember that fungi differ rather drastically from their hosts and from man who might happen to ingest the compound. We shall explore the differences and see how selectivity can be related to it.

Compounds That Affect Cell Walls The cell walls of the majority of fungi contain large amounts of chitin, a substance that does not occur in higher plants or animals. Chitin is a polymer of N-acetylglucosamine. Chitinase is an enzyme that can be obtained from the gut of a snail. It inhibits germination of spores of *Aspergillus niger* (Hilborn and Farr, 1959). Griseofulvin was originally suggested (Brian, 1949) to act selectively on chitin. Evans and White (1967) have modified this concept somewhat, but griseofulvin still acts selectively on the fungus wall. Since man does not synthesize chitin, griseofulvin can be taken safely by mouth to control the fungus infections between the toes.

Griseofulvin causes bloated cells of fungi, as does pentachloronitrobenzene. Macris and Georgopoulos (1969) say that this compound actually reduces the glucosamine content of fungal cell walls. Horsfall (1956) lists several bloat-inducing compounds. It may be that some of these inhibit glucosamine deposition in cell walls.

Compounds That Affect Sporulation Fungi produce spores called conidia, but higher plants and man do not. Can we selectively inhibit sporulation? In our laboratory, we have tried for several years to make a practical, effective compound without noteworthy success. Compounds must be screened empirically. Glycolic acid analogues are successful in the laboratory but not in the field, because they are too sensitive to degradation on the leaf (Horsfall and Lukens, 1968). Nickel compounds are potent antisporulants for rust (Jensen and Daly, 1960), but they too have not reached commercial adoption because the nickel shows up in the cereal grain.

Compounds That Disrupt Integrity of Membranes Much has been done to disrupt the cell membrane. It is probable that dodine, a commercial fungicide, owes its effectiveness to the leakage of metabolites from treated fungal cells (Brown and Sisler, 1960). The polyene antibiotics like filipin and nystatin probably affect the sterol components of membranes and cause critical metabolites to leak out (see Gottlieb and Shaw, 1970).

Pathogens That Develop Resistance Legend tells us that Mithradates was the first "pathogen" to develop resistance to a pesticide. He was said to have fed himself low doses of arsenic and he became so resistant that he could imbibe poisonous potions with his enemies at parties. They would die; he would live.

Insects and bacteria develop resistance to pesticides so rapidly as to create an important practical problem. In the field, fungi have rarely developed economically important resistance in farmers' crops. In the laboratory, they often do. The general theory for this is that most fungicides have a broad spectrum of activity. Some fungicides seem to have a limited spectrum of activity and resistance has developed to these. Georgopoulos (1969) has reviewed this subject. Perhaps the first is the report of Farkas and Aman (1940) that the *Penicillium* that spoils oranges in transit had become resistant to diphenyl.

Georgopoulos lists two other cases of field resistance, bunt of wheat to hexachlorobenzene and *Helminthosporium* on oats to organic mercury. Schroeder and Provvidenti (1969) reported that the cucumber powdery mildew fungus has become resistant to benomyl. Szkolnik and Gilpatrick (1969) say that *Venturia inaequalis*, which causes apple scab, is becoming resistant to dodine in New York State. Webster *et al.* (1970) reported that *Botrytis cinerea* on stone fruits is becoming resistant to 2,6-dichloronitroaniline in California.

These compounds probably affect only one or a very few sites in the fungus cell. These, then, are the practical hazards of selectivity. Selective compounds are surely to be desired ecologically, but they may not last long in fighting the fungi for which they are selective.

REFERENCES

Bent, K. J. 1969. Fungicides in perspective. Endeavour 28:129–134.

Bent, K. J. 1967. Vapor action of fungicides against powdery mildews on cucumbers. Ann. Appl. Biol. 60:251–263.

Brian, P. W. 1949. Studies on the biological activity of griseofulvin. Ann. Bot. N.S. 13:60–77.

Bromfield, K. R., and R. G. Emge. 1958. The effect of DDT on the rust reaction of several wheat varieties to wheat stem rust. Plant Dis. Rep. 42:354–359.

Brown, I. F., and H. D. Sisler. 1960. Mechanisms of fungitoxic action of *n*-dodecylguanidine acetate. Phytopathology 50:830–839.

Calpouzos, Lucas. 1966. Action of oil in the control of plant disease. Ann. Rev. Phytopathol. 4:369–390.

Clemons, G. P., and H. D. Sisler. 1969. Formation of a fungitoxic derivative from Benlate. Phytopathology 59:705–706.

Cox, R. S., and N. C. Hayslip. 1957. Recent developments on the control of foliar diseases of tomato in South Florida. Plant Dis. Rep. 41:878–883.

Delp, C. J., and H. L. Klopping. 1968. Performance attributes of a new fungicide and mite ovicide candidate. Plant Dis. Rep. 52:95–99.

Dimond, A. E. 1963. The selective control of plant pathogens (with fungicides). World Rev. Pest Contr. 2:7–17.

Doling, D. A., and Shirley Hepple. 1959. "Karathane": the value of its differential fungicidal effect in relation to mildew and yellow rust of wheat. Nature (London). 183:622.

Edgington, L. V., M. E. Corden, and A. E. Dimond. 1961. The role of pectic substances in chemically induced resistance to *Fusarium* wilt of tomato. Phytopathology 51:179–182.

Edgington, L. V., G. S. Walton, and P. M. Miller. 1966. Fungicide selective for basidiomycetes. Science 153:307–308.

Erwin, D. C., H. T. Reynolds, and M. J. Garber. 1961. Predisposition to *Pythium* seedling disease and an activated charcoal-fungicide interaction as factors influencing emergence of cotton seed treated with phorate. J. Econ. Entomol. 54:855–858.

Evans, G., and N. H. White. 1967. Effect of the antibiotics, radicicolin and griseofulvin, on the fine structure of fungi. J. Exp. Bot. 18:465–470.

Farkas, A., and J. Amman. 1940. The action of diphenyl on *Penicillium* and Diplodia moulds. Palestine J. Bot. Jerusalem Ser 2:38–45.

Farley, J. D., and J. L. Lockwood. 1968. Effect of pentachloronitrobenzene (PCNB) on the soil microflora. Phytopathology 58:1050 (Abstract).

Georgopoulos, S. G. 1969. The problem of fungicide resistance. BioScience 19: 971–973.

Gibson, I. A. S. 1956. An anomalous effect of soil treatment with ethyl-mercury phosphate on the incidence of damping-off in pine seedlings. Phytopathology 46:181–182.

Gottlieb, D., and P. Shaw. 1970. Mechanisms of action of antifungal antibiotics. Annu. Rev. Phytopathol. 8:371–402.

Grossman, F. 1962. Untersuchungen uber die Hemmung pektolytischer Enzyme von *Fusarium oxysporum* f. *lycopersici* III. Wirkung einiger Hemmstoffe in vivo. Phytopathol. Zeitschr. 45:139–159.

Hilborn, M. T., and Wanda K. Farr. 1959. A biologically produced fungicide. Phytopathology 49:541.

Hislop, E. C. 1967. Observations on the vapor phase activity of some foliage fungicides. Ann. Appl. Biol. 60:265–279.

Horsfall, James G. 1956. Principles of Fungicidal Action. Chronica Botanica Co., Waltham, Mass.

Horsfall, J. G., and A. E. Dimond. 1957. Interactions of tissue sugar, growth substances, and disease susceptibility. Z. Pfl. Krankh. 64:415–421.

Horsfall, J. G., and R. J. Lukens. 1969. The strategy of finding fungicides. *In* Residue reviews 25:81–91. F. A. Gunther (ed.), Springer-Verlag, N. Y.

Horsfall, J. G., and R. J. Lukens. 1968. Aldehyde traps as antisporulants for fungi. Conn. Agr. Exp. Stn. Bull. 694:1–27.

Horsfall, J. G., and R. J. Lukens. 1966. Selectivity of fungicides. Conn. Agr. Exp. Stn. Bull. 676:1–24.

Jensen, S. G., and J. M. Daly. 1960. Some effects of the chemotherapeutant $NiCl_2$ on the physiological and morphological development of rust-infected wheat plants. Phytopathology 50:640–41.

Kerk, G. J. M. van der. 1969. The development of synthetic fungicides; trends and prospects. Neth. J. Pl. Path. 75 (1969 Suppl. 1):5–20.

Lukens, R. J., and J. G. Horsfall. 1971. Spore germination and appressorial formation, a new assay for fungicides. Phytopathology 61:130 (Abstract).

Macris, B., and S. G. Georgopoulos. 1969. Reduced hexosamine content of fungal cell wall due to the fungicide pentachloronitrobenzene. Phytopathology. 59:879–880.

O'Brien, George, and Adelaide Bornmann. 1952. Fungicidal properties of maleic hydrazide (unpublished report). *In* Literature summary on maleic hydrazide, J. W. Zukel (ed.), U.S. Rubber Co. MHIS No. 6, p. 19.

Patil, S. S., and A. E. Dimond. 1968. Repression of polygalacturonase in *Fusarium oxysporum* f. sp. *lycopersici* by sugars and its effect upon symptom reduction in infected tomato plants. Phytopathology 58:676–682.

Ramm, C. von, G. B. Lucas, and H. V. Marshall, Jr. 1962. The effect of maleic hydrazide on brown spot of flue-cured tobacco. Tobacco 154:30–33.

Rich, S. 1954. Dynamics of deposition and tenacity of fungicides. Phytopathology 44:203–213.

Richardson, L. T. 1957. Effect of insecticides and herbicides applied to soil on the development of plant diseases. I. The seedling disease of barley caused by *Helminthosporium sativum* P, K, and B. Can. J. Plant Sci. 37:196–204.

Samborski, D. J., F. R. Forsythe, and C. Person. 1958. Metabolic changes in detached wheat leaves floated on benzimidazole and the effect of these changes on rust reaction. Can. J. Bot. 36:591–601.

Schmeling, B. von, and Marshall Kulka. 1966. Systemic fungicidal activity of 1,4-oxathiin derivatives. Science 152:659–660.

Schroeder, W. T., and R. Provvidenti. 1969. Resistance to benomyl in powdery mildew of cucurbits. Plant Dis. Rep. 53:271–275.

Sijpesteijn, A. K., and C. W. Pluijers. 1962. Phenylthioureas as systemic fungicides. Meded. Landbouw hogesch. Ghent 27:1199–1203.

Smith, A. M., B. A. Stynes, and K. J. Moore. 1970. Benomyl stimulates growth of a Basidiomycete on turf. Plant Dis. Rep. 54:774–775.

Staron, T., C. Allard, H. Darpoux, H. Grabowski, and A. Kollman. 1966. Phytoatrie-Phytopharmacie 15:129–134.

Staron, T., and Allard, C. 1964. Phytoatrie-Phytopharmacie 13:163–168.

Szkolnik, Michael, and J. D. Gilpatrick. 1969. Apparent resistance of *Venturia inaequalis* to dodine in New York apple orchards. Plant Dis. Rep. 53:861–864.

Wang, D., M. S. H. Hao, and E. R. Waygood. 1961. Effect of benzimidazole analogues on stem rust and chlorophyll metabolism. Can. J. Bot. 39:1029–1036.

Webster, R. K., J. M. Ogawa, and Elaine Bose. 1970. Tolerance of *Botrytis cinerea* to 2,6-dichloro-4-nitroaniline. Phytopathology 60:1489–1492.

Woodcock, D. 1967. Microbiological detoxication and other transformations. p. 613–642. *In* Fungicides, D. C. Torgeson (ed.), Academic Press, New York.

Attractants and Repellents for Insect Pest Control

Morton Beroza

Rachel Carson in her book *Silent Spring* (Carson, 1962) referred to the use of attractants and repellents for the control of insect pests as one of the "new, imaginative, and creative approaches to the problem of sharing our earth with other creatures." While many concerned with environmental quality applaud this approach, its practicability, except in isolated instances, remains to be demonstrated.

My purpose here is to review the extent to which attractants and repellents are now used—about 9 years after the appearance of *Silent Spring*—and then to suggest roles these materials might play in future pest-control operations. Emphasis has been placed on objectives that are believed to be practical and attainable.

ATTRACTANTS

Let us consider attractants first. Many insects have developed highly efficient olfactory organs as a result of evolutionary pressures, and they rely heavily on this sense of perception to secure vital needs. We want to use this trait to cause the destruction of harmful species.

Interest in attractants was sporadic until the problem of insect resistance to insecticides arose in the 1950's, and the prospect of escalating insecticide dosages for crop protection triggered the search for alternative means of pest control—or, at least, the search for means of increasing the efficiency and selectivity of insecticide applications.

Impetus in this direction has mounted as additional drawbacks to the use of insecticides were recognized—the appearance of secondary pests, such as mites, as target insects were controlled; the finding that certain pesticides were persisting as residues in our food chain and in our body fat, especially through biological magnifications of nonbiodegradable pesticides; and finally the exceptional vulnerability of certain nontarget species to the nonbiodegradable pesticide residues in the environment. These difficulties have provided ample reason to explore further the use of attractants in pest control.

Since the brevity of this presentation forces my coverage to be selective rather than comprehensive, I shall stress concepts rather than specifics. For additional information, the reader may wish to consult recent reviews on insect attractants (Beroza, 1970; Butler, 1967; Eiter, 1970; Jacobson, 1965, 1966; Karlson, 1969; Muto, 1968; National Academy of Sciences, 1969a; Otto, 1968; Pyatnova *et al.*, 1969; Regnier and Law, 1968; Shorey and Gaston, 1967; Shorey *et al.*, 1968; Tomida, 1968; Warren, 1969; Wood *et al.*, 1970).

Synthetic Attractants

Insect attractants may arbitrarily be divided into two groups—those of natural and those of synthetic origin. The synthetic attractants were the first to become available on a large scale, so they have been used much more than those of natural origin. A listing of the synthetics currently in use (or proposed for such use) is shown in Table 1. These materials have been invaluable, particularly in survey applications. Traps are baited with the attractant; responding insects upon being caught signal the presence of their species in the vicinity. This information is used to direct control or eradication operations and thereby prevent the establishment or spread of insect infestations.

The number of attractants in Table 1, only nine for the seven species, is hardly impressive. We would like to have more, but effective attractants are not easily found. Those in Table 1 were obtained by testing a large number of chemicals (Beroza and Green, 1963) and then synthesizing compounds related to the most attractive candidates to find a more potent lure. We do not know why insects find these compounds so attractive (e.g., we are not aware that a chemical similar to trimedlure exists in nature), nor do we know why the first three compounds attract only males, the next four both sexes, and the last two yellow jacket workers, which are females.

TABLE 1 Synthetic Insect Attractants

Formula	Common or Chemical Name	Species Attracted	References
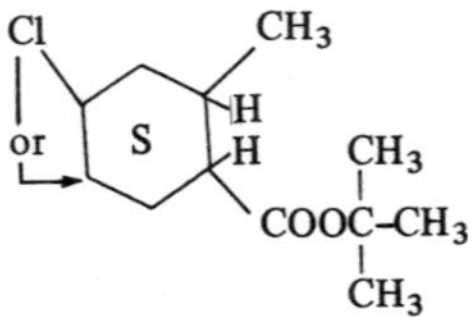	Trimedlure	Mediterranean fruit fly, *Ceratitis capitata*	Beroza *et al.*, 1961
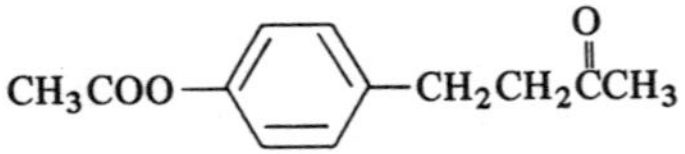	Cue-lure	Melon fly, *Dacus cucurbitae*	Alexander *et al.*, 1962; Beroza *et al.*, 1960
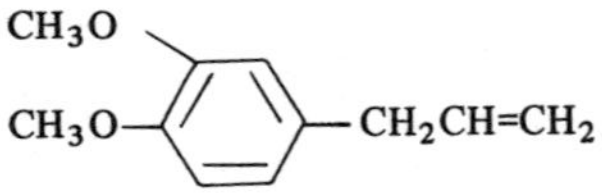	Methyl eugenol	Oriental fruit fly, *Dacus dorsalis*	Steiner, 1952
$CH_3CH{=}CH{-}CH{=}CHCOOC_4H_9$	Butyl sorbate	European chafer, *Amphimallon majalis*	Tashiro *et al.*, 1964

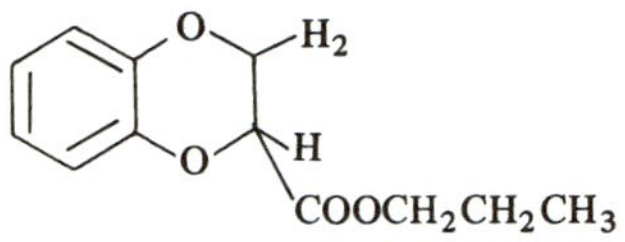	Amlure	European chafer	McGovern *et al.*, 1970c
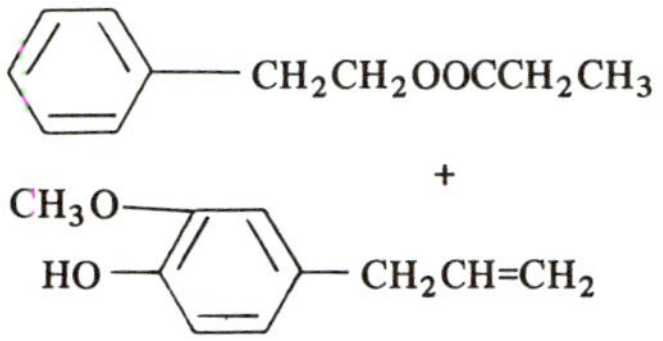	Phenethyl propionate and Eugenol	Japanese beetle, *Popillia japonica*	McGovern *et al.*, 1970a
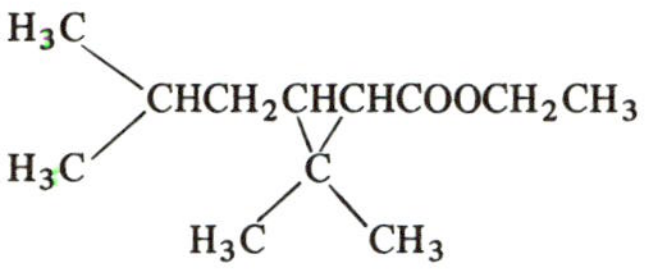	Ethyl dihydrochrysanthemumate (ethyl 3-isobutyl-2,2-dimethylcyclopropanecarboxylate)	Coconut rhinoceros beetle, *Oryctes rhinoceros*	Barber *et al.*, 1971
$CH_3CH{=}CH{-}CH{=}CHCH_2OOCCH_2CH_2CH_3$	2,4-Hexadienyl butyrate	Yellow jacket wasps, *Vespula* spp.	Davis *et al.*, 1967, 1968
$CH_3(CH_2)_5CH_2OOCCH_2CH_2CH_3$	Heptyl butyrate	Yellow jacket wasps	Davis *et al.*, 1969

Some of the synthetic attractants have already proved their worth. For example, the first three compounds of Table 1 are attractants for three of the world's worst insect pests, and these pests have been kept out of this country by ringing our ports of entry with traps, each baited with the three lures; infestations of any of the three species are quickly detected and eradicated before they can become established. These early detections have saved the government many millions of dollars in potential eradication costs. Should an infestation become established, the traps facilitate eradication by pinpointing the locations of the target insect and indicating where insecticides should be applied and when their application may be terminated; insecticides are therefore not spread needlessly in the environment. The power of this approach may be judged from the fact that the Mediterranean fruit fly, found in 1956 on about 1 million acres in Florida, was eradicated before the end of the next year. The synthetic attractant siglure (Gertler *et al.*, 1958) (subsequently replaced by the more potent trimedlure) was used in this campaign, and the insecticide malathion plus a protein hydrolyzate food lure were used to eliminate the insect wherever found (Steiner *et al.*, 1961).

While the first three compounds of Table 1 have been eminently successful in helping to keep nonindigenous insect pests from becoming established in the United States, the remaining compounds have not yet achieved such wide usage.

The European chafer is found in several localities in the northeastern part of the country, and scientists of our Plant Protection Division are using butyl sorbate to detect infestations as part of their efforts to try to prevent the spread of this insect. Amlure (McGovern *et al.*, 1970c) which is longer lasting and traps about three times as many chafers in field tests, is not yet in use.

This year the Plant Protection Division of USDA adopted a mixture of phenethyl propionate plus eugenol (McGovern *et al.*, 1970a) as the latest of a series of lures (McGovern *et al.*, 1970b) used during the past 30 years to monitor the location of the Japanese beetle, which has now spread over most of eastern United States. The combined action of the two chemicals is superior to the additive effect of the two chemicals separately. A means of dispensing the mixtures in optimum proportion for attraction has been reported (McGovern and Beroza, 1970).

Ethyl dihydrochrysanthemumate attracts a rhinoceros beetle (Barber *et al.*, 1971) that causes severe damage to coconut palms. The beetle has seriously affected the economy of many Pacific islands largely dependent on copra production.

The last two lures were developed to help beekeepers, orchardists, and others plagued by yellow jacket wasps. One firm is planning to use heptyl butyrate to lure yellow jackets to a trap containing a source of food spiked with a slow-acting insecticide; the yellow jackets take the poisoned food back to their colonies, causing their destruction. This technique, which avoids indiscriminate use of insecticide in the environment by placing it exactly where it is needed, has been most effective against *Vespula* spp. on the West Coast.

Except for the fact that most are esters, these synthetic attractants appear to be characterized by their diversity rather than by any similarity in structure. Since good attractants are highly specific for a given species, such differences are to be expected. However, even very similar compounds can differ markedly in attraction. Thus, one of the four major isomers of trimedlure is not attractive; it differs from the three attractive isomers only in the position of the chlorine atom (McGovern and Beroza, 1966; McGovern *et al.*, 1966).

Although the attractants are invaluable for detection, their use in direct control or eradication remains to be explored. The unique means of controlling yellow jacket wasps, i.e., getting the insect to spread the insecticide, has already been noted, but this approach is suitable only for the social insects, those that collect food for their colonies. In only one instance was eradication achieved directly using an attractant. The oriental fruit fly was eliminated from a 33-square-mile Pacific island (Steiner *et al.*, 1965) by dropping by airplane every few weeks small cane-fiber squares containing a combination of methyl eugenol, a powerful lure for the male fly, and naled, a fast-acting insecticide. The males sought out the attractive squares and in a few months annihilated themselves; with no males, the females failed to reproduce and the species on the island perished. Use of synthetic attractants in this manner is ecologically acceptable because action of the lures is highly specific, and only the responding insects are eliminated. In principle then, attractants (plus an insect control agent) can be used for control and possibly eradication. However, methyl eugenol happens to be an unusually powerful lure and eradication with a less potent attractant is less than certain. Should we be satisfied with control rather than eradication, this approach may be successful even if it only results in a reduction in the amount of pesticide needed and diminishes side effects in the environment.

If, in principle, synthetic attractants can be used for improving our pest control efforts, why are they not more widely sought after and used? There are many reasons. Considerable patience and expense are required to find a truly effective attractant. To my knowledge this approach has been explored extensively only by federal and state governments. Private companies have avoided this work because they probably could not recoup their research costs. Bear in mind that not only do thousands of compounds have to be procured or synthesized, but a satisfactory bioassay has to be developed, one that caters to the idiosyncrasies of each species. Some insects respond in free flight, some crawl. Vapor concentration can be important, e.g., some attractants become repellent at high concentrations and can be missed in screening. Field testing is the ultimate criterion of effectiveness, but it is time-consuming, expensive, and limited to times when the insect (usually the adult) is available in the field. A good laboratory bioassay, by making testing possible throughout the year, can speed progress, but its development entails considerable study. A means is needed for rearing healthy insects. This involves special problems because insects are often vulnerable to disease and some to cannibalism, and they often require special diets. In addition to normal biological variations, there are differences in response due to day or season, age of insect, trap position, and light effects. Means of measuring the response of the insect to candidate chemicals must also be devised. Testing for an attractant is much more difficult and less certain than testing for an insecticide.

If an attractant is found in laboratory screening, it must still prove itself in the field under actual conditions of use. Trap design is important for maximum catches. Optimum type and size of wick, amount of chemical, size of opening, color of trap, and trap height must be determined. Persistence of the chemical and the trapping means (sticky material, insecticide) must be studied. Traps must be positioned to allow their chemical vapors to travel downwind freely.

Inasmuch as each species has to be handled differently, considerable work is involved in developing attractants, even if efforts are simply trials or modifications of procedures currently in use. Conversely, when an attractant is found, it is generally useful for only one species in a given area, although several species may require control. One must also recognize that the larvae usually do the damage, while most attractants draw only the adults, and usually one sex.

Assuming an attractant and method of using it are devised, the attractant in most cases can only be used effectively over a large area. An individual trying to use an attractant on or near his crop may be worse off than without it because he will draw the offending insects from surrounding areas to his plot. If only for this reason, it makes sense for governments or large producer organizations rather than individuals to explore the use of attractants for control. The same is true in the use of attractants for survey and detection. An exception is the use of a few traps for timing insecticide sprays to coincide with the appearance of the insect.

Perhaps the major reason attractants are not being used for control is that few properly organized attempts to do so have been carried out. Such experiments are necessarily large and therefore expensive. Expensive new projects are not popular these days.

Despite the many problems, there are compelling reasons to develop attractants. Attractants are very much needed to cut down on environmental effects of pesticides. Their value in detection, especially in eradication programs, is well established. The time expended in developing an attractant for use can be justified if the insect is economically important. An attractant once found will be useful, hopefully, for all time. Synthetic attractants have already saved our government many millions of dollars, especially by helping keep out certain nonindigenous insect pests.

Natural Attractants

About 10 years ago I stated (Beroza, 1960b) that attractants were easier to find by the volume screening of synthetic chemicals of many different types than by isolation of attractants known to be present in insects, such as the sex attractants. This statement is no longer true. Chemical instrumentation and techniques—infrared, ultraviolet, NMR (nuclear magnetic resonance spectroscopy), mass spectrometry, gas and thin-layer chromatography, reaction chromatography—have improved to such a degree that identifications of compounds at the level of 0.1 milligram and less are now being reported. However, identifications at these low levels are most difficult and require exceptional care and skill.

These developments have exciting implications for insect pest control; they are making possible the isolation and identification of odorous chemicals that are vital to insects, chemicals that are usually available only in very minute amount. Some of these are the pheromones (Karlson and Luescher, 1959), compounds emitted by one member of a species to influence the behavior or development of others of the same species. (Only behavioral aspects will be discussed; developmental aspects are outside the scope of this presentation.) Pheromones can be sex attractants, aphrodisiacs, assembling scents, masking scents, or chemicals inducing alarm, foraging, recruitment, trail following, or swarming. Through the use of pheromones, social insects organize and coordinate the complex activities of their nests.

Sex Attractants

Since it will not be possible to discuss pheromones in detail, I would like to direct attention to one type of pheromone, the sex attractant, because it probably has the greatest potential for pest control. The sex attractant is usually emitted by one sex to call the opposite sex (and sometimes both sexes) for mating. Many insect species have persisted in great numbers because of their high reproductive potential, which depends upon their ability to find each other, often by means of a sex attractant, and mate. Hundreds of species, many of which are economically important pests, have now been shown to utilize sex attractants. There is no dearth of data to demonstrate the fantastic potency of these attractants. Amounts needed are often incredibly small (e.g., 10^{-14} gram), and thousands of insects have been drawn to a single virgin female in the field (Coppel *et al.*, 1960). As with the synthetic attractants, we know we can use the sex attractants for survey purposes, but we believe they are also potentially valuable either for direct pest control or as an aid in the execution of an integrated control program.

The great difficulty in identifying sex attractants is the large number of insects needed to get even a tiny amount of material. Furthermore, the material may be unstable or it may be accompanied by a masking agent, i.e., a chemical that in some way negates the action of the sex attractant.

A decade ago the structure of only one sex pheromone had been deciphered, that of the silkworm moth (Butenandt *et al.*, 1959), and this after 20 years of effort. Nature appears to give up its secrets grudgingly. Fortunately, our knowledge in this area has increased considerably. Table 2 is a listing of the insect sex pheromones that are now known to attract in the field. The structures of about 90 percent of these were determined in the past 5 years, coinciding with the availability of the new sophisticated instrumentation and techniques.

In some instances the sex attractant is a multicomponent mixture, and the ingredients do not attract individually. Needless to state, this situation multiplies the difficulties of bioassay, isolation, identification, and synthesis of the attractant. The determination of the sex lure of the boll weevil was particularly refractory because it has four exotic constituents (Tumlinson *et al.*, 1969). The elucidation of the sex lures (or assembling scent) of certain bark beetles has been most difficult, not only because they are multicomponent, but also because odors of the host plant appear to be required for optimum attraction (Bedard *et al.*, 1969; Vité, 1970).

One of the unexpected findings of the past few years is that several species utilize the same chemical as a sex lure, yet we know of no cross-mating occurring in nature. Reproductive isolation of the various species could be maintained in the same locality if insects mated at different times of the day or season or under different conditions (Roelofs and Comeau, 1969). However, in many instances one or more additional chemicals that are themselves inert appear to be secreted by insects to maintain the integrity and exclusiveness of the communication system of each species. We now realize that these secondary chemicals can either greatly enhance or counteract the action of the sex attractant (Roelofs and Comeau, 1971a); they have been referred to as synergists, activators, antipheromones, and inhibitors. Thus, with species utilizing the same sex attractant, it is speculated that the secondary chemical may cause one species to be attracted and another to be repelled. Of course, the admixture of a third chemical can further restrict the exclusiveness of chemical communication lines.

As with the synthetic lures, a laboratory bioassay for sex attraction, preferably one available throughout the year, is needed to test fractions isolated from the insect if reasonable progress is to be made. Difficulties in devising a bioassay have already been mentioned. Unfortunately some of the sex pheromones identified with laboratory bioassays excite the opposite sex but do not attract in the field. These excitants or aphrodisiacs are probably emitted during courtship to increase the receptivity of the opposite sex to mating. Although these excitants may also play a role in maintaining the reproductive isolation of a species, they do not appear to have any practical value in pest control at present and therefore are not listed in Table 2. Clearly, in bioassays, it is essential to distinguish between excitants and attractants.

The screening approach has also been used to find sex attractants. When it became apparent that the sex lures of several lepidopterous species were long-chain unsaturated C_{12}–C_{16} alcohols or their acetates, many of these compounds were synthesized and exposed in the field. Hexalure, an attractant for the pink bollworm (Green *et al.*,

TABLE 2 Insect Sex Attractants Active in Field Tests

Formula[a]	Chemical Name	Insect	References
$CH_3(CH_2)_2(CH{=}CH)_2(CH_2)_8CH_2OH$	(*E,Z*)-10,12-Hexadecadien-1-ol	Silkworm, *Bombyx mori*	Butenandt *et al.*, 1959
$CH_3(CH_2)_3CH{=}CH(CH_2)_6OAc$	(*Z*)-7-Dodecenyl acetate	Cabbage looper, *Trichoplusia ni*	Berger, 1966
$CH_3(CH_2)_2CH{=}CH(CH_2)_7OAc$	(*Z*)-8-Dodecenyl acetate	Oriental fruit moth, *Grapholitha molesta*	Roelofs *et al.*, 1969
$CH_3(CH_2)_3CH{=}CH(CH_2)_8OAc$	(*Z*)-9-Tetradecenyl acetate[b]	Southern armyworm, *Spodoptera eridania* (*Prodenia eridania*)	Jacobson *et al.*, 1970; Redfern *et al.*, 1971
		A gelechiid moth (B1), *Bryotopha similis*	Roelofs and Comeau, 1969
$CH_3(CH_2)_3CH{=}CH(CH_2)_8OAc$	(*E*)-9-Tetradecenyl acetate	A gelechiid moth (B2)	Roelofs and Comeau, 1969
$CH_3CH_2CH{=}CH(CH_2)_{10}OAc$	(*Z*)-11-Tetradecenyl acetate	Redbanded leafroller, *Argyrotaenia velutinana*	Roelofs and Arn, 1968
		European corn borer, *Ostrinia nubilalis*	Klun and Brindley, 1970
		Oblique banded leafroller, *Choristoneura rosaceana*	Roelofs and Tette, 1970
$CH_3CH{=}CHCH_2CH{=}CH(CH_2)_8OAc$	(*Z,E*)-9,12-Tetradecadienyl acetate[b]	Southern armyworm	Jacobson *et al.*, 1970;
		Indian meal moth, *Plodia interpunctella*	Brady *et al.*, 1971; Kuwahara *et al.*, 1971
		Almond moth, *Cadra cautella*	Brady *et al.*, 1971; Kuwahara *et al.*, 1971

$CH_3(CH_2)_9\underset{\diagdown O \diagup}{CH{-}CH}(CH_2)_4CH(CH_3)_2$	*cis*-7,8-Epoxy-2-methyloctadecane	Gypsy moth, *Porthetria dispar*	Bierl *et al.*, 1970
$CH_3(CH_2)_9CHO$	Undecanal	Greater wax moth, *Galleria mellonella*	Roeller *et al.*, 1968
$CH_3(CH_2)_{14}CH(CH_3)_2$	2-Methylheptadecane	Tiger moths, *Holomelina aurantiaca* complex	Roelofs and Carde, 1971
$CH_3\overset{O}{\overset{\|}{C}}(CH_2)_5CH{=}CHCOOH$	(*E*)-9-Oxo-2-decenoic acid	Honeybee, *Apis mellifera*	Gary, 1962
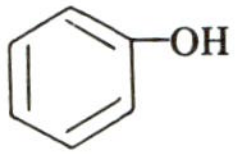	Phenol	Common grass grub beetle, *Costelytra zealandica*	Henzell and Lowe, 1970
$CH_3(CH_2)_3COOH$	Valeric acid	Sugar beet wireworm, *Limonius californicus*	Jacobson *et al.*, 1968
$CH_3(CH_2)_7(CH{=}CH)_2CH_2COOH$	(*E,Z*)-3,5-Tetradecadienoic acid	Black carpet beetle, *Attagenus megatoma*	Silverstein *et al.*, 1967
$CH_3CH_2\overset{CH_3}{\overset{\vert}{C}}H(CH_2)_4CH{=}CH(CH_2)_6COOCH_3$	(–)-Methyl (*Z*)-14-methyl-8-hexadecenoate	A grain beetle, *Trogoderma inclusum*	Rodin *et al.*, 1969
$CH_3CH_2\overset{CH_3}{\overset{\vert}{C}}H(CH_2)_4CH{=}CH(CH_2)_6CH_2OH$	(*Z*)-(–)-14-Methyl-8-hexadecen-1-ol	*Trogoderma inclusum*	Rodin *et al.*, 1969
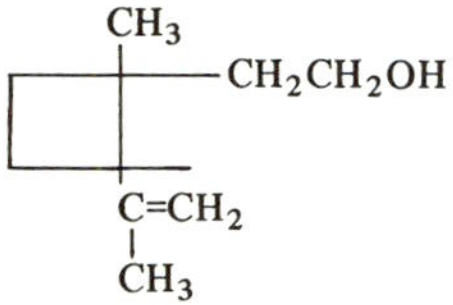	*cis*-2-Isopropenyl-1-methylcyclobutaneethanol[c]	Boll weevil, *Anthonomus grandis*	Tumlinson *et al.*, 1969
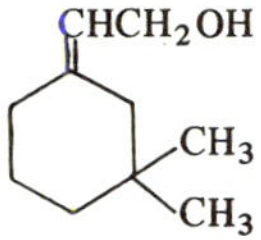	(*Z*)-3,3-Dimethyl-$\Delta^{1,\beta}$-cyclohexaneethanol[c]	Boll weevil	Tumlinson *et al.*, 1969

TABLE 2 Insect Sex Attractants Active in Field Tests (Continued)

Formula[a]	Chemical Name	Insect	References
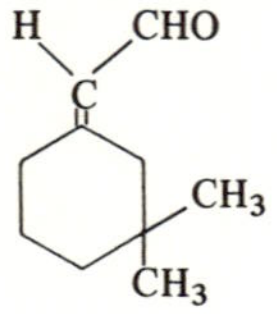	(*Z*)-3,3-Dimethyl-$\Delta^{1,\alpha}$-cyclohexaneacetaldehyde[c]	Boll weevil	Tumlinson *et al.*, 1969
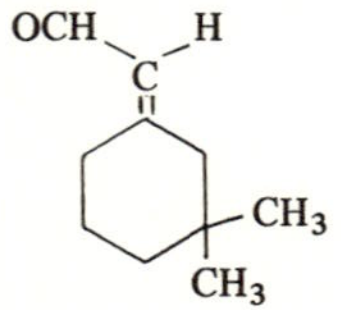	(*E*)-3,3-Dimethyl-$\Delta^{1,\alpha}$-cyclohexaneacetaldehyde[c]	Boll weevil	Tumlinson *et al.*, 1969
$(CH_3)_2CHCH_2CH(OH)\text{–}CH_2C(=CH_2)\text{–}CH=CH_2$	(–)-2-Methyl-6-methylene-7-octen-4-ol[d]	California fivespined ips, *Ips confusus*	Silverstein *et al.*, 1966
$(CH_3)_2C=CH\text{–}CH(OH)\text{–}CH_2\text{–}C(=CH_2)\text{–}CH=CH_2$	(+)-2-Methyl-6-methylene-2,7-octadien-4-ol[d]	California fivespined ips,	Silverstein *et al.*, 1966

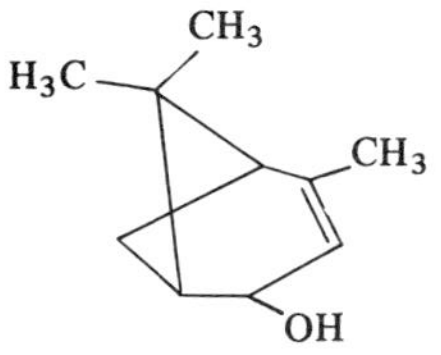	*cis*-(+)-Verbenol[d]	California fivespined ips	Silverstein *et al*., 1966
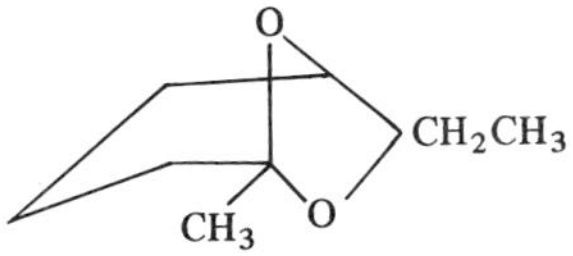	*exo*-7-Ethyl-5-methyl-6,8-dioxabicyclo [3.2.1] octane[e]	Western pine beetle, *Dendroctonus brevicomis*	Silverstein *et al*., 1968
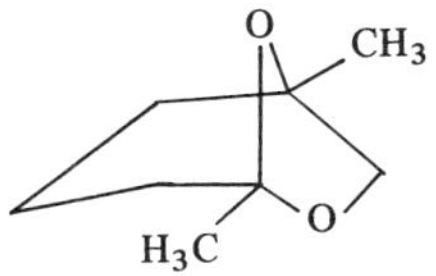	1,5-Dimethyl-6,8-dioxabicyclo-[3.2.1] octane[e]	Southern pine beetle, *Dendroctonus frontalis*	Kinzer *et al*., 1969

[a]Ac = acetate.
[b]Two compounds needed for field attraction of southern armyworm; see other compound.
[c]Four compounds needed for maximum response.
[d]Produced by male, attracts both sexes. Mixture of three compounds needed for maximum response.
[e]Produced by female, attracts both sexes.

1969), and a number of other lures were found in this way. For pest control, these materials are useful only if they attract economically important species.

Another procedure that has been exceptionally fruitful for finding sex attractants of the long-chain acetate or alcohol variety, devised by W. L. Roelofs (Roelofs and Comeau, 1971b), utilizes the electroantennogram (EAG) technique of D. Schneider (Schneider, 1957; 1969) to measure the electrical response of an excised insect antenna to odors. Roelofs first determines the gas chromatographic retention times of the sex attractant extracted from a few insects of a given species, using a polar and nonpolar column. From these data, together with results from several chemical tests that help designate the functional group(s) in the molecule, he can decide on the probable size of the molecule and the number of unsaturated sites. He then checks EAG's of model compounds, and the one giving the greatest response is tested in the field. Dr. Roelofs has found sex lures for at least 50 species in this manner and in field testing (Roelofs and Comeau, 1971b), some being of considerable economic importance. The method has the limitation that model compounds are required, but many are now available as a result of much synthesis along these lines.

It is possible that a sex attractant found by field screening or EAG is very similar rather than identical to the natural attractant. This conclusion is based on a recently completed study in which we determined the attraction of the gypsy moth to 55 chemicals structurally related to disparlure (Sarmiento *et al.*, 1972), the true sex attractant of this insect. The simplest variation of disparlure's structure resulted in at least a 100-fold drop in activity, which gives an indication of the remarkable specificity of this sex attractant. However, this compound, 100 times less active than disparlure, is still a very potent attractant; and it does give a strong EAG response. Therefore, though the EAG and field-screening procedures are invaluable for finding attractants, proof that a compound is the true attractant of a species will require more adequate characterization, e.g., by classical isolation and spectral analyses. From a practical standpoint, the fact that a sex attractant is not the natural sex attractant of a species is immaterial if it is truly effective, although we expect that the natural attractant will be the most attractive compound for each species.

The sex attractants available today are splendid tools for detecting infestations, which in itself justifies the effort made to find these materials. Practicality of their use in direct control or eradication remains to be established.

One procedure advanced by E. F. Knipling (Knipling and McGuire, 1966) of the USDA is suggested for the eradication of light infestations of insect pests, and, in at least one limited field trial, results were promising (Roelofs *et al.*, 1970). Presently we have a unique opportunity to use this procedure on the gypsy moth which has recently spread from New England to parts of Virginia. The general idea is to drop by airplane large numbers of inexpensive mass-produced traps (small tubes with an adhesive inside) containing attractant to capture the males before they can find the nonflying females and mate. If the ratio of traps to females is high enough, the population of the lightly infested areas could be eradicated. This overflooding approach is somewhat similar to the one devised by Dr. Knipling to eradicate the screwworm (*Cochliomyia hominivorax*) from Florida with sterile males (Knipling, 1960). Should this approach be successful, thought should be given to applying insecticides (or another insect control agent) where gypsy moth populations are too high to apply this technique and then overflooding with traps to eliminate the few survivors.

In a second approach to control, called the confusion method, the atmosphere is permeated with the sex attractant, e.g., disparlure (Bierl *et al.*, 1970) for the gypsy moth. It is postulated that with attractant everywhere, the males should not be able to locate the female scent, or their odor receptors may simply become fatigued or saturated, thereby preventing them from responding. Although the idea was suggested 11 years ago (Beroza, 1960a) and showed promise in small trials with several insect species (Gaston *et al.*, 1967; Klun and Robinson, 1970; Shorey *et al.*, 1967), the confusion method has never been adequately tested. It holds promise of suppressing both heavy and light infestations.

In disparlure, we have a powerful and readily available lure; when suitably formulated, 1 nanogram of it caught insects 3 months after it was placed in the field (Beroza *et al.*, 1971a). With a lure of this kind, it would be most unfortunate if the trapping and confusion methods were not tried on the gypsy moth quickly, i.e., before the present lightly infested areas become heavily infested. The gypsy moth is causing extensive damage to our forests. If we are serious about protecting our environment, the use of attractants should be fully explored in the control of the gypsy moth, especially since no satisfactory alternatives are available.

Other Behavior-Releasing Chemicals

The secondary chemicals that enhance or negate attraction when combined with a sex attractant are potentially useful in pest control. For example, we recently found that tetradecyl acetate inhibited hexalure's attraction of pink bollworm moths *Pectinophora gossypiella* (Beroza *et al.*, 1971b). This finding suggests the possibility that permeating the air with this readily available and inexpensive chemical will inhibit mating of the moths. It is anticipated that this approach, which could be considered a modification of the confusion method, will be tried this summer. Similar inhibitors have been found for the red-banded leaf roller and the European corn borer (Roelofs and Comeau, 1968; Klun and Robinson, 1971).

Many ants lay trails to guide others of their colony to a source of food (Blum, 1970; Gabba, 1967). Stingless bees and bumble bees use marking scents, i.e., a chemical spotted periodically, to designate a path to be followed (Blum, 1970; Kullenberg *et al.*, 1970; Lindauer and Kerr, 1958). To my knowledge, these and other intraspecies communicative chemicals (excluding sex lures) have not yet been applied in practical pest control, although some may have this potential.

Odors are frequently a powerful force in directing one species to another. We have recently isolated and identified what we believe is the first known chemical that directs a parasite to its host, the corn earworm in this case (Jones *et al.*, 1971). It turned out to be a most unusual hydrocarbon, one with 32 carbon atoms. The chemical may help improve biological control by keeping parasites in areas where needed, especially when host populations are low.

Chemicals attracting an insect species may also attract a predator of that species (Bedard *et al.*, 1969; Vité, 1970; Vité and Williamson, 1970; Wood *et al.*, 1968). Since elimination of the predator would be detrimental, such interactions must be taken into account.

We would like to use attractants to manipulate bees more effectively. For example, bees might be induced to work on a given field if it were sprayed with an attractive chemical (Woodrow *et al.*, 1965).

The complexity of the attraction phenomenon in insects can often be underestimated. Lactic acid plus CO_2 was shown to attract yellow fever mosquitoes even though individually they were inactive (Acree *et al.*, 1968). Moisture and physical agents, such as heat, have also been implicated in the attraction of this insect (Daykin *et al.*, 1965; Smith *et al.*, 1970).

Mixing of insecticides with a food, which is usually a weak

short-range lure, was tried for crop protection as long ago as the late 1800's (Dethier, 1947), at which time it did not succeed. However, the procedure can be successful with the proper combination of food and insecticide. The protein hydrolyzate-malathion mixture used to combat the Mediterranean fruit fly (Steiner *et al.*, 1961) is one such instance. With this combination, much less toxicant was required than without the food, and complete coverage was not necessary because the insect came to the food lure. At least with fruit flies, this technique is being used. In another instance, soya bean oil, a food lure, was mixed with a slow-acting insecticide to combat the imported fire ant (*Solenopsis saevissima richteri*) (Lofgren *et al.*, 1963). Control was almost complete with a few grams of insecticide per acre. These short-range food attractants complement the more specific long-range chemical lures (e.g., those in Table 1) that help detect the insect. Food lures are generally inexpensive and therefore suitable for direct control measures requiring large amounts of attractant.

In an attempt to refine the food-insecticide approach, food stimulants (gustatory stimulants) have been sought (Gueldner *et al.*, 1970; Hedin *et al.*, 1968). They are usually multicomponent and partly volatile. These materials should be worth exploring.

REPELLENTS

Repellents are chemicals whose vapors keep insects away, presumably by action on the insect olfactory senses. Like the attractants they may be divided into two groups—those of synthetic and of natural origin. Reviews describing both types of repellents should be consulted for more information than can be presented here (Garson and Winnike, 1968; Jacobson, 1966; National Academy of Sciences, 1969b; Painter, 1967; Shambaugh *et al.*, 1957; Smith, 1970).

From the standpoint of ecology repellents are desirable control chemicals. They are generally not toxic; they offer protection and comfort to individuals while leaving ecosystems relatively undistrubed.

Synthetic Repellents

During World War II the Armed Forces of the United States had to operate in areas infested with disease-carrying arthropods, such as malaria-carrying mosquitoes and typhus-carrying lice. Since eliminating these pestiferous arthropods from extensive areas was impractical, a concerted effort was made to find effective repellents to protect military personnel. More than 20,000 chemicals have now been tested against a variety of arthropods under the auspices of the U.S. Department of Agriculture and the Defense Department (Entomology Research Division, 1967; King, 1954). Although many of the compounds were effective, very few were safe and comfortable enough for use on skin or clothing and at the same time had the many other requisites of a suitable repellent. The repellent had to be nontoxic, long-lasting, inoffensive in odor, nonallergenic, harmless to clothing, stable to sunlight, and, if possible, effective against a broad range of insect pests. Skin repellents had to persist during sweating and to resist loss by wiping, while clothing repellents had to resist loss during rain and laundering.

The chemical that best met these requirements for use on skin is deet (*N,N*-diethyl-*m*-toluamide) (Gilbert *et al.*, 1957; McCabe *et al.*, 1954), now the repellent most widely used by civilians and the military. In tests conducted throughout North America and elsewhere, deet protected subjects for 3 to 8 hr against at least a dozen pests including mosquitoes, fleas, chiggers, ticks, deer flies, sand flies, biting gnats, and land leeches. Other chemicals with good repellency for certain insects are ethyl hexanediol, dimethyl phthalate, and dimethyl carbate.

Repellents have made it possible for people to work and enjoy our environment when they could not do so otherwise. However, there are pestiferous species for which we have no effective repellents, such as tabanids and tsetse flies (*Glossina* spp.) (National Academy of Sciences, 1969b). Finding an effective repellent generally involves the development of a bioassay, which is subject to many of the same difficulties encountered with a bioassay for an attractant.

Because deet on clothing has poor resistance to leaching by rain or laundering, several more persistent clothing repellents are used. Formulation M-1960, one of the best, is a mixture of three chemicals and an emulsifier: 30% benzyl benzoate (repels mites), 30% *N*-butylacetanilide (repels ticks), 30% 2-butyl-2-ethyl-1,3-propanediol (repels mosquitoes), 10% Tween 80* (emsulsifer) (Hall *et al.*, 1957). Applied at the rate of 2 g/sq ft, it protects for several weeks. The three chemicals repel more species than is possible with any of the chemicals alone.

*Mention of a proprietary product in this paper is for information only and does not constitute endorsement by the USDA.

Adequate protection from insects requires complete coverage with the repellent, so care in its application is important.

Attempts to relate repellency to specific chemical or physical properties of compounds have not enabled the prediction of good repellent structures for synthesis (Garson and Winnike, 1968). Structures found repellent are amides, imides, alcohols, esters, ethers, and such polyfunctional compounds as 1,3-diols, amide-esters, hydroxy-esters, and diesters (Quintana *et al.*, 1970a, b).

The newest type of synthetic repellent is the space repellent (Gouck *et al.*, 1971; Gouck *et al.*, 1967; McGovern *et al.*, 1967; Schreck *et al.*, 1970), i.e., one whose vapor prevents the approach of an insect to within a small distance (perhaps 1 cm) from the site of application. When applied to nets with ¼-in. holes, effective space repellents can prevent insects from reaching persons behind the nets for months. In hot climates, the individual is more comfortable with a treated net than if he were to use a skin repellent (which usually lasts less than a day), or if he were behind the usual netting with tiny holes, because air circulates freely through the large holes of the treated net; the nets also provide protection with comfort when placed over open windows or when used as a cover for sleeping out of doors. Their use over clothing rather than the use of a clothing repellent has been suggested (Catts, 1968). Since space repellents on nets generally contact little (if any) of an individual's skin, they should not require rigorous pharmacological clearance, a major stumbling block in finding skin repellents.

Like humans, domestic animals suffer from arthropod attack, but repellents for human use are not sufficiently long-lasting for this application. A good repellent to keep biting flies off livestock is sorely needed to reduce annoyance and blood loss, which lead to decreased milk and meat production and sometimes to transmission of disease. Despite a considerable effort over many years, a satisfactory repellent (one lasting at least several days) has not yet been found. Several toxicants, such as malathion and methoxychlor, have been applied, generally by spraying, dipping, dusting, or back rubber, but they do not protect adequately even though insects are knocked down or killed. No safe systemic repellent (one taken internally) has yet been developed.

The possibility of using a repellent on agricultural crops is not considered practical. A volatile repellent will evaporate and require replenishing frequently; for a long growing season such treatment will be too expensive. A nonvolatile "repellent" is really a feeding deterrent or antifeedant (Wright, 1967); although such materials have been effective where deposited, the plant continues to grow, and new unprotected tissue is quickly consumed.

Bees have been repelled from honeycombs at harvest time with phenol, benzaldehyde, propionic anhydride, and propionic acid (Woodrow *et al.*, 1965). A repellent is needed to keep bees off insecticide-treated fields; to be effective the repellent should persist until the insecticide concentration falls to a safe level.

Natural Repellents

The defensive secretions of arthropods (Weatherston and Percy, 1970), particularly those involved in predator–prey relationships, would appear to be a good source of repellent compounds. Although more than 100 compounds have been identified in these secretions, including benzoquinones, aromatic compounds, terpenoids, acids, phenols, esters, carbonyl compounds, and steroids, these materials have not found their way into pest-control use. Interestingly, hydrogen cyanide, which has been used in pest control, is a defensive secretion.

Plants contain both attractants and repellents for arthropods; they also contain feeding deterrents (not really repellents) that discourage insects from feeding on them (Wright, 1967). Pyrethrum is a plant-derived material that is both repellent and insecticidal (Sylvester and Weaving, 1967; Wright and Burton, 1969). It has been used with synergists on livestock and on food-containing packages for protection against insects; it has also been burned to clear an area of insect pests. Some plant extracts such as creosote and oil of cedarwood find use occasionally in the protection of wood against insect attack, but these applications are very limited.

SUMMARIZING COMMENTS

The use of insect attractants, both of synthetic and of natural origin, has proven extremely valuable in detecting infestations of nonindigenous species and in guiding measures to prevent their spread. Their use in direct control, i.e., with an insect control agent, although demonstrably feasible, cannot be considered generally applicable. Much depends on the potency of the attractant, and each proposed use will have to be studied carefully. Some potent attractants, sex attractants in particular, are available now and these should be tried for direct control. With a potent attractant consideration might also be given to eradication programs. At the very minimum, attractants can be adapted into integrated control programs, e.g., to monitor insect populations, time sprays, combine with pesticides, and make insecticide applications more efficient and more selective in their attack.

If attractants are to be used successfully for pest control, much research will be needed to find them, to learn more about olfaction and chemoreception generally, to study the actions of secondary chemicals in the olfactory process, e.g., masking agents, synergists, inhibitors, and antipheromones, and to develop the technology needed to take full advantage of attractants and their inhibitors. This research will require support from scientists of many disciplines–entomologists, chemists, biochemists, physiologists, and engineers, to name a few.

Other attractants worth studying are those leading predators to their prey and those used by social insects.

As for repellents, they have great importance in protecting man from disease-transmitting and pestiferous insects since they allow him to work in and enjoy areas that he normally could not enter without considerable risk or discomfort. Thanks to the great amount of work on synthetic repellents, we have excellent ones for use on skin and clothing. Repellents should be sought for important pests for which effective ones are not now available. Work on space repellents should be speeded by more vigorous support, and efforts to find an arthropod repellent to protect animals should be continued. Repellents of natural origin might also be examined for possible use in pest control.

There are some excellent opportunities for the use of attractants and repellents in pest control. Each successful use will decrease the need for pesticides and, overall, will benefit our environment, ecology, and ourselves.

REFERENCES

Acree, F., Jr., R. B. Turner, H. K. Gouck, M. Beroza, and N. Smith. 1968. *L*-Lactic acid: A mosquito attractant isolated from humans. Science 161: 1346–1347.

Alexander, B. H., M. Beroza, T. A. Oda, L. F. Steiner, D. H. Miyashita, and W. C. Mitchell. 1962. The development of male melon fly attractants. J. Agr. Food Chem. 10:270–276.

Barber, I. A., T. P. McGovern, M. Beroza, C. P. Hoyt, and A. Walker. 1971. Attractant for the coconut rhinoceros beetle. J. Econ. Entomol. 64:1041–1044.

Bedard, W. D., P. E. Tilden, D. L. Wood, R. M. Silverstein, R. G. Brownlee, and J. O. Rodin. 1969. Western pine beetle: Field response to its sex pheromone and a synergistic host terpene, myrcene. Science 164:1284–1285.

Berger, R. S. 1966. Isolation, identification, and synthesis of the sex attractant of the cabbage looper, *Trichoplusia ni.* Ann. Entomol. Soc. Am. 59:767–771.

Beroza, M. 1960a. Insect attractants are taking hold. Agr. Chem. 15 (7):37–40.

Beroza, M. 1960b. Insect attractants. Soap Chem. Specialties 36 (2):74, 76, 101, 103.

Beroza, M. (ed.). 1970. Chemicals controlling insect behavior. Academic Press, New York.

Beroza, M., B. H. Alexander, L. F. Steiner, W. C. Mitchell, and D. H. Miyashita. 1960. New synthetic lures for the male melon fly. Science 131:1044–45.

Beroza, M., B. A. Bierl, J. G. R. Tardif, D. A. Cook, and E. C. Paszek. 1971a. Activity and persistence of synthetic and natural sex attractants of the gypsy moth in laboratory and field trials. J. Econ. Entomol. 64:1499–1508.

Beroza, M., and N. Green. 1963. Materials tested as insect attractants. Agriculture Handbook No. 239. Agricultural Research Service, U.S. Department of Agriculture, Washington, D.C.

Beroza, M., N. Green, S. I. Gertler, L. F. Steiner, and D. H. Miyashita. 1961. New attractants for the Mediterranean fruit fly. J. Agr. Food Chem. 9:361–365.

Beroza, M., R. T. Staten, and B. A. Bierl. 1971b. Tetradecyl acetate and related compounds as inhibitors of attraction of the pink bollworm moth to the sex lure hexalure. J. Econ. Entomol. 64:580–582.

Bierl, B. A., M. Beroza, and C. W. Collier. 1970. Potent sex attractant of the gypsy moth: Its isolation, identification, and synthesis. Science 170:87–89.

Blum, M. S. 1970. The chemical basis of insect sociality, p. 61–94. *In*: Chemicals controlling insect behavior. M. Beroza (ed.). Academic Press, New York.

Brady, U. E., J. H. Tumlinson III, R. G. Brownlee, and R. M. Silverstein. 1971. Sex stimulant and attractant in the Indian meal moth and in the almond moth. Science 171:802–804.

Butenandt, A., R. Beckmann, D. Stamm, and E. Hecker. 1959. Ueber den Sexual-Lockstoff des Seidenspinners *Bombyx mori.* Reindarstellung und Konstitution. *Z. Naturforsch.*, B 14:283–284.

Butler, C. G. 1967. Insect pheromones. Biol. Rev. 42:42–87.

Carson, R. 1962. Silent spring. Houghton Mifflin Co., Boston.

Catts, E. P. 1968. Deet-impregnated net shirt repels biting flies. J. Econ. Entomol. 61:1765.

Coppel, H. C., J. E. Casida, and W. C. Dauterman. 1960. Evidence for a potent sex attractant in the introduced pine sawfly, *Diprion similis* (Hymenoptera: Diprionidae). Ann. Entomol. Soc. Am. 53:510–512.

Davis, H. G., G. W. Eddy, T. P. McGovern, and M. Beroza. 1967. 2,4-Hexadienyl butyrate and related compounds highly attractive to yellow jackets (*Vespula* spp.). J. Med. Entomol. 4:275–280.

Davis, H. G., G. W. Eddy, T. P. McGovern, and M. Beroza. 1969. Heptyl butyrate, a new synthetic attractant for yellow jackets. J. Econ. Entomol. 62:1245.

Davis, H. G., T. P. McGovern, G. W. Eddy, T. E. Nelson, K. M. R. Bertun, M. Beroza, and J. C. Ingangi. 1968. New chemical attractants for yellow jackets (*Vespula* spp.). J. Econ. Entomol. 61:459–462.

Daykin, P. N., F. E. Kellog, and R. H. Wright. 1965. Host-finding and repulsion of *Aedes aegypti.* Can. Entomol. 97:239–263.
Dethier, V. G. 1947. Chemical insect attractants and repellents. Blakiston, Philadelphia, p. 2.
Eiter, K. 1970. Sexual attractants for insects. Chem. Pflanzenschutz-Schaedlingsbekaempfungsmittel 1970:497–522.
Entomology Research Division. 1967. Materials evaluated as insecticides, repellents, and chemosterilants at Orlando and Gainesville, Fla., 1952–1964. Agr. Handbook No. 340. Agricultural Research Service, U. S. Department of Agriculture, Washington, D.C.
Gabba, A. 1967. Aspetti dell' organizzazione degli insetti sociali. 2. La sostanza della traccia nei Formicidae. Natura (Milan) 58:150–172.
Garson, L. R., and M. E. Winnike. 1968. Relationships between insect repellency and chemical and physical parameters–a review. J. Med. Entomol. 5:339–352.
Gary, N. E. 1962. Chemical mating attractants in the queen honeybee. Science 136:773–774.
Gaston, L. K., H. H. Shorey, and C. A. Saario. 1967. Insect population control by the use of sex pheromones to inhibit orientation between the sexes. Nature 213:1155.
Gertler, S. I., L. F. Steiner, W. C. Mitchell, and W. F. Barthel. 1958. Esters of 6-methyl-3-cyclohexene-1-carboxylic acid as attractants for the Mediterranean fruit fly. J. Agr. Food Chem. 6:592–594.
Gilbert, I. H., H. K. Gouck, and C. N. Smith. 1957. New insect repellent. Soap Chem. Spec., Part I 33(5):115–117, 129–133. Part II 33(6): 95–99, 109.
Gouck, H. K., D. R. Godwin, K. Posey, C. E. Schreck, and D. E. Weidhaas. 1971. Resistance to aging and rain of repellent-treated netting used against salt-marsh mosquitoes in the field. Mosquito News 31:95–99.
Gouck, H. K., T. P. McGovern, and M. Beroza. 1967. Chemicals tested as space repellents against yellow-fever mosquitoes. I. Esters. J. Econ. Entomol. 60: 1587–1590.
Green, N., M. Jacobson, and J. C. Keller. 1969. Hexalure, an insect sex attractant discovered by empirical screening. Experientia 25:682–683.
Gueldner, R. C., A. C. Thompson, D. D. Hardee, and P. A. Hedin. 1970. Constituents of the cotton bud. XIX. Attractancy to the boll weevil of the terpenoids and related plant constituents. J. Econ. Entomol. 63:1819–1821.
Hall, S. A., N. Green, and M. Beroza. 1957. Insect repellents and attractants. J. Agr. Food Chem. 5:663–669.
Hedin, P. A., L. R. Miles, A. C. Thompson, and J. P. Minyard. 1968. Constituents of a cotton bud. Formulation of a boll weevil feeding stimulant mixture. J. Agr. Food Chem. 16:505–513.
Henzell, R. F., and M. D. Lowe. 1970. Sex attractant of the grass grub beetle. Science 168:1005–1006.
Jacobson, M. 1965. Insect sex attractants. Wiley (Interscience), New York.
Jacobson, M. 1966. Chemical insect attractants and repellents. Ann. Rev. Entomol. 11:403–422.
Jacobson, M., C. E. Lilly, and C. Harding. 1968. Sex attractant of sugar beet wireworm: Identification and biological activity. Science 159:208–210.
Jacobson, M., R. E. Redfern, W. A. Jones, and M. H. Aldridge. 1970. Sex pher-

omones of the southern armyworm moth: Isolation, identification, and synthesis. Science 170:542–544.

Jones, R. L., W. J. Lewis, M. C. Bowman, M. Beroza, and B. A. Bierl. 1971. A host-seeking stimulant for a parasite of corn earworm: Isolation, identification, and synthesis. Science 173:842–843.

Karlson, P. 1969. Insect hormones and insect pheromones as possible agents for insect control: A critical review. J. S. Afr. Chem. Inst. 22 (spec. issue), 41S–55S.

Karlson, P., and M. Luescher. 1959. Pheromones: A new term for a class of biologically active substances. Nature 183:55–56.

King, W. V. 1954. Chemicals evaluated as insecticides and repellents at Orlando, Fla. Agriculture Handbook No. 69. Agricultural Research Service, U.S. Department of Agriculture, Washington, D.C.

Kinzer, G. W., A. F. Fentiman, Jr., T. F. Page, Jr., R. L. Foltz, J. P. Vité, and G. B. Pitman. 1969. Bark beetle attractants: Identification, synthesis and field bioassay of a new compound isolated from *Dendroctonus.* Nature 221:477–478.

Klun, J. A., and T. A. Brindley. 1970. *cis*-11-Tetradecenyl acetate, a sex stimulant of the European corn borer. J. Econ. Entomol. 63:779–880.

Klun, J. A., and J. F. Robinson. 1970. Inhibition of European corn borer mating by *cis*-11-tetradecenyl acetate, a borer sex stimulant. J. Econ. Entomol. 63: 1281–1283.

Klun, J. A., and J. F. Robinson. 1971. European corn borer moth: sex attractant and sex attraction inhibitors. Ann. Entomol. Soc. Am. 64:1083–1086.

Knipling, E. F. 1960. The eradication of the screw-worm fly. Sci. Am. 203 (4): 54–61.

Knipling, E. F., and J. U. McGuire, Jr. 1966. Population models to test theoretical effects of sex attractants used for insect control. Agriculture Inform. Bull. 308. Agricultural Research Service, U.S. Department of Agriculture, Washington,D.C.

Kullenberg, B., G. Gergstrom, and S. Stallberg-Stenhagen, 1970. Volatile components of the cephalic marking secretion of male bumble bees. Acta Chem. Scand. 24:1481–1483.

Kuwahara, Y., C. Kitamura, S. Takahashi, H. Hara, S. Ishii, and H. Fukami. 1971. Sex pheromone of the almond moth and the Indian meal moth: *cis*-9,*trans*-12-Tetradecadienyl acetate. Science 171:801–802.

Lindauer, M., and W. E. Kerr. 1958. Die gegenseitige Verstandigung bei den stachellosen Bienen. Z. Vergleich. Physiol. 41:405–434.

Lofgren, C. S., F. J. Bartlett, and C. E. Stringer. 1963. Imported fire ant toxic bait studies: Evaluation of carriers for oil baits. J. Econ. Entomol. 56:62–66.

McCabe, E. T., W. F. Barthel, S. I. Gertler, and S. A. Hall. 1954. Insect repellents. III. *N,N*-Diethylamides. J. Org. Chem. 19:493–498.

McGovern, T. P., and M. Beroza. 1966. Structure of the four isomers of the insect attractant trimedlure. J. Org. Chem. 31:1472–1477.

McGovern, T. P., and M. Beroza. 1970. Volatility and compositional changes of Japanese beetle attractant mixtures and means of dispensing sufficient vapor having a constant composition. J. Econ. Entomol. 63:1475–1479.

McGovern, T. P., M. Beroza, and H. K. Gouck. 1967. Chemicals tested as space repellents for yellow-fever mosquitoes. II. Carbanilates, benzamides, aliphatic amides, and imides. J. Econ. Entomol. 60:1591–1594.

McGovern, T. P., M. Beroza, T. L. Ladd, Jr., J. C. Ingangi, and J. P. Jurimas. 1970a. Phenethyl propionate, a potent new attractant for Japanese beetles. J. Econ. Entomol. 63:1727–1729.

McGovern, T. P., M. Beroza, K. Ohinata, D. Miyashita, and L. F. Steiner. 1966. Volatility and attractiveness to the Mediterranean fruit fly of trimedlure and its isomers, and a comparison of its volatility with that of seven other insect attractants. J. Econ. Entomol. 59:1450–1455.

McGovern, T. P., M. Beroza, P. H. Schwartz, D. W. Hamilton, J. C. Ingangi, and T. L. Ladd. 1970b. Methyl cyclohexanepropionate and related chemicals as attractants for the Japanese beetle. J. Econ. Entomol. 63:276–280.

McGovern, T. P., B. Fiori, M. Beroza, and J. C. Ingangi. 1970c. Propyl 1,4-benzodioxan-2-carboxylate, a new attractant for the European chafer. J. Econ. Entomol. 63:168–171.

Muto, F. 1968. Insect attractants. Noyaku Seisan Gijutsu 19:1–9.

National Academy of Sciences. 1969a. Chemical attractants and insect behavior. P. 290–309. *In* Insect-pest management and control. Publ. 1695. Washington, D.C.

National Academy of Sciences. 1969b. Chemical repellents. P. 310–323. *In* Insect-pest management and control. Publ. 1695. Washington, D.C.

Otto, D. 1968. Attraktivstoff-Reaktionen bei Forstinsekten und ihre Bedeutung fuer den Forstschutz. Biol. Rundsch. 6 (1):22–32.

Painter, R. R. 1967. Repellents, p. 267–285. *In* Pest control: Biological, physical, and selected chemical methods. W. W. Kilgore and R. L. Doutt (eds.). Academic Press, New York and London.

Pyatnova, Y. B., L. L. Ivanov, and A. S. Kyskina. 1969. Insect sex attractants. Usp. Khim. 38 (2):248–275; Russ. Chem. Rev. 38 (2):126–139.

Quintana, R. P., L. R. Garson, A. Lasslo, S. I. Sanders, J. H. Buckner, H. K. Gouck, I. H. Gilbert, D. E. Weidhaas, and C. E. Schreck. 1970a. Mosquito-repellent dihydroxyacetone monoesters. J. Econ. Entomol. 63:1128–1131.

Quintana, R. P., A. Lasslo, L. R. Garson, C. N. Smith, and I. H. Gilbert. 1970b. Dermophilic insect repellents with perdurable efficacy. J. Pharm. Sci. 59:1503–1505.

Redfern, R. E., E. Cantu, W. A. Jones, and M. Jacobson. 1971. Response of the male southern armyworm, *Prodenia eridania* (Cramer), in a field cage to Prodenialure "A" and Prodenialure "B." J. Econ. Entomol. In press.

Regnier, F. E., and J. H. Law. 1968. Insect pheromones. J. Lipid Res. 9:541–551.

Rodin, J. O., R. M. Silverstein, W. E. Burkholder, and J. E. Gorman. 1969. Sex attractant of female dermestid beetle *Trogoderma inclusum* Le Conte. Science 165:904–906.

Roeller, H., K. Biemann, J. S. Bjerke, D. W. Norgard, and W. H. McShan. 1968. Sex pheromones of pyralid moths. I. Isolation and identification of the sex attractant of *Galleria mellonella* (greater wax moth). Acta Entomol. Bohemoslov. 65:208–211.

Roelofs, W. L., and H. Arn. 1968. Sex attractant of the red-banded leaf roller moth. Nature 219:513.

Roelofs, W. L., and R. T. Cardé. 1971. Hydrocarbon sex pheromone in tiger moths (Arctiidae). Science 171:684–686.

Roelofs, W. L., and A. Comeau. 1968. Sex pheromone perception. Nature 220:600–601.

Roelofs, W. L., and A. Comeau. 1969. Sex pheromone specificity; Taxonomic and evolutionary aspects in Lepidoptera. Science 165:398–400.

Roelofs, W. L., and A. Comeau. 1971a. Sex pheromone perception: Synergists and inhibitors for the red-banded leaf roller attractant. J. Insect Physiol. 17: 435–448.

Roelofs, W. L., and A. Comeau. 1971b. Sex attractants in Lepidoptera. Proc. 2nd Int. Congr. Pest. Chem., Tel Aviv, Israel, 1971. In press.

Roelofs, W. L., A. Comeau, and R. Selle. 1969. Sex pheromone of the oriental fruit moth. Nature 224:723.

Roelofs, W. L., E. H. Glass, J. Tette, and A. Comeau. 1970. Sex pheromone trapping for red-banded leaf roller control: Theoretical and actual. J. Econ. Entomol. 63:1162–1167.

Roelofs, W. L., and J. P. Tette. 1970. Sex pheromone of the oblique-banded leaf roller moth. Nature 226:1172.

Sarmiento, R., M. Beroza, B. A. Bierl, and J. G. R. Tardif. 1972. Activity of compounds related to disparlure, the sex attractant of the gypsy moth. J. Econ. Entomol. In press.

Schneider, D. 1957. Elektrophysiologische Untersuchungen von Chemo und Mechano-rezeptoren der Antenne des Seidenspinners *Bombyx mori* L. Z. Vergl. Physiol. 40:8–41.

Schneider, D. 1969. Insect olfaction: deciphering system for chemical messages. Science 163:1031–1037.

Schreck, C. E., I. H. Gilbert, D. E. Weidhaas, and K. H. Posey. 1970. Spatial action of mosquito repellents. J. Econ. Entomol. 63:1576–1578.

Shambaugh, G. F., R. F. Brown, and J. J. Pratt, Jr. 1957. Repellents for biting arthropods. Adv. Pest Control Res. 1:277–303.

Shorey, H. H., and L. K. Gaston. 1967. Pheromones, p. 241–265. *In* Pest control: Biological, physical, and selected chemical methods. W. W. Kilgore and R. L. Doutt (eds.). Academic Press, New York.

Shorey, H. H., L. K. Gaston, and R. N. Jefferson. 1968. Insect sex pheromones. Adv. Pest Control Res. 8:57–126.

Shorey, H. H., L. K. Gaston, and C. A. Saario. 1967. Sex pheromones of noctuid moths. XIV. Feasibility of behavioral control by disrupting pheromone communication in cabbage loopers. J. Econ. Entomol. 60:1541–1545.

Silverstein, R. M., R. G. Brownlee, T. E. Bellas, D. L. Wood, and L. E. Browne. 1968. Brevicomin: Principal sex attractant in the frass of the female western pine beetle. Science 159:889–891.

Silverstein, R. M., J. O. Rodin, W. E. Burkholder, and J. E. Gorman. 1967. Sex attractant of the black carpet beetle. Science 157:85–87.

Silverstein, R. M., J. O. Rodin, and D. L. Wood. 1966. Sex attractants in frass produced by male *Ips confusus* in ponderosa pine. Science 154:509–510.

Smith, C. N. 1970. Repellents for anopheline mosquitoes. Misc. Publ. Entomol. Soc. Amer. 7 (1):99–117.

Smith, C. N., N. Smith, H. K. Gouck, D. E. Weidhaas, I. H. Gilbert, M. S. Mayer, B. J. Smittle, and A. Hofbauer. 1970. *L*-Lactic acid as a factor in the attraction of *Aedes aegypti* (Diptera: Culicidae) to human hosts. Ann. Entomol. Soc. Am. 63:760–770.

Steiner, L. F. 1952. Methyl eugenol as an attractant for oriental fruit fly. J. Econ. Entomol. 45:241–248.

Steiner, L. F., W. C. Mitchell, E. J. Harris, T. T. Kozuma, and M. S. Fujimoto. 1965. Oriental fruit fly eradication by male annihilation. J. Econ. Entomol. 58:961–964.

Steiner, L. F., G. G. Rohwer, E. L. Ayers, and L. D. Christenson. 1961. The role of attractants in the recent Mediterranean fruit fly eradication program in Florida. J. Econ. Entomol. 54:30–35.

Sylvester, N. K., and A. J. S. Weaving. 1967. Pyrethrum as an insect repellent. III. The persistence of pyrethrin films, when applied in three types of formulation, as insect repellents to human skin. Pyrethrum Post 9 (2):8–14.

Tashiro, H., S. I. Gertler, M. Beroza, and N. Green. 1964. Butyl sorbate as an attractant for the European chafer. J. Econ. Entomol. 57:230–233.

Tomida, I. 1968. Pheromone and its application to agricultural chemicals. Jap. Agr. Res. Quart. 3 (3):14–19.

Tumlinson, J. H., D. D. Hardee, R. C. Gueldner, A. C. Thompson, P. A. Hedin, and J. P. Minyard. 1969. Sex pheromones produced by male boll weevil: Isolation, identification, and synthesis. Science 166:1010–1012.

Vité, J. P. 1970. Pest management systems using synthetic pheromones. Contrib. Boyce Thompson Inst. Plant Res. 24 (13):343–350.

Vité, J. P., and D. L. Williamson. 1970. *Thanasimus dubius*: Prey perception. J. Insect Physiol. 16:233–239.

Warren, F. L. 1969. Sex pheromones of insects. J. S. Afr. Chem. Inst. 22 (spec. issue), 150S–154S.

Weatherston, J., and J. E. Percy. 1970. Arthropod defensive secretions, p. 95–144. *In* Chemicals controlling insect behavior. M. Beroza (ed.). Academic Press, New York.

Wood, D. L., L. E. Browne, W. D. Bedard, P. E. Tilden, R. M. Silverstein, and J. O. Rodin. 1968. Response of *Ips confusus* to synthetic sex pheromones in nature. Science 159:1373–1374.

Wood, D. L., R. M. Silverstein, and M. Nakajima (eds.). 1970. Control of insect behavior by natural products. Academic Press, New York.

Woodrow, A. W., N. Green, H. Tucker, M. H. Schonhorst, and K. C. Hamilton. 1965. Evaluation of chemicals as honey bee attractants and repellents. J. Econ. Entomol. 58:1094–1102.

Wright, D. P., Jr. 1967. Antifeedants, p. 287–293. *In* Pest control: Biological, physical, and selected chemical methods. W. W. Kilgore and R. L. Doutt (eds.). Academic Press, New York.

Wright, R. H., and D. J. Burton. 1969. Pyrethrum as a repellent. Pyrethrum Post 10 (2):14–16, 21.

IV
Alternative Means to Pest Control

Crop Resistance to Pests and Pathogens

P. R. Day

The promise of resistance to pests and diseases has attracted plant breeders ever since the demonstration near the turn of the century that resistance could be selected and that it is often simply inherited. The "built-in" protection of resistance, effective throughout a crop plant's life, offers a compelling alternative to fungicides and insecticides. In the early years, minimizing the use of protective chemicals was desirable because it saved money, but today there is the added advantage of reducing their hazard to the environment. Resistance really came into its own when used against those pests and diseases such as the cereal rusts, for which no other control was practicable. I shall not summarize the achievements of breeding for resistance to pests and diseases nor will I try to document how much our crops depend on it today. My emphasis will be on unresolved problems rather than on recent advances.

If we look back dispassionately, as others have done (Stakman, 1954; Caldwell, 1966), we find failures along with the successes. In 1971 plant breeders are still introducing genes for resistance to stem rust of wheat in much the same fashion as they did in 1918. Borlaug (1965) remarked that the average useful life in some parts of the world of a rust-resistant variety was as short as 5 years before it succumbed to a new race of rust. We can point to other examples of continued breeding effort: crown rust of oats, Hessian fly on wheat, powdery mildew on barley, *Cladosporium* leaf mold and tobacco mosaic virus on tomato, and golden nematode and late blight on potato, to name but a few. In each case the life of resistant varieties has been short. Why did these efforts fail? Each was the subject of a long-continued series of efforts.

Other epidemics seem to come out of the blue. Why did the 1968 sorghum crop experience a greenbug (*Schizaphis graminum*) infestation affecting 7.34 million acres in nine states and causing in Kansas alone an estimated loss exceeding $12 million (Hackerott and Harvey, 1969)? Why in 1969–1970 was there one of the most devastating epidemics in the recent history of plant pathology—the southern corn leaf blight (*Helminthosporium maydis*)—a crisis that reverberates about us even now? This article examines these questions first by summarizing our present knowledge of pest and disease resistance and how we have deployed it, by reviewing some of the mistakes made and how we might profit by them, and finally by considering the areas in which we must invest future efforts.

By pests and diseases I mean the parasites: nematodes, insects, viruses, bacteria, and fungi whose depredations reduce the yield and quality of our crops. The subject is very large and its literature formidable and I think the occasion calls for some bold, and therefore controversial, generalizations.

THE NATURE OF RESISTANCE—SOME GENERALIZATIONS

It is well established that all plants show defense reactions after physical or chemical injury. We need not be concerned with the details of these reactions here (see, for example, Kuc, 1966; Day, 1968) except to note that they usually involve synthesis of polyphenols that may be toxic to any organisms present and to the plant itself. One important type of resistance to parasites results from the initiation of the defense mechanisms at the time of host-plant invasion. Susceptible and resistant reactions differ principally in the speed and scale of host response. In resistant reactions the same things happen as in susceptible reactions but they happen more rapidly. As a consequence the development of the invading parasite is restricted. A good example is the hypersensitive reaction where the host's response results in localized death of its own tissue as well as of the pathogen. Hypersensitive resistance to viruses, bacteria, and fungi is well known, and there are also examples of such resistance to insects (Stevenson, 1970).

If the speed and scale of host response are reduced, then a parasite will be capable of greater development so that symptoms of injury caused by its growth and reproduction are more obvious. At this point its development may be additionally slowed in other ways unrelated to the defense mechanism. For example, a deficiency or imbalance in nutrient factors may limit its growth (Hsiao, 1969). Even when parasite development may seem to proceed without hindrance from resistance, there may still be useful differences among varieties. Thus, varieties that are tolerant produce greater agronomic yields than other susceptible varieties (Bawden, 1964; Simons, 1966).

In the search for resistance, dramatic effects catch the eye. Breeders naturally selected the best resistance they could. The snag is that the more dramatic resistance is frequently very specific and not effective at all against some races. The reason for this is as follows: If a parasite, as a result of a mutation, can avoid triggering the defense mechanism, its development will be unchecked. Any resistance the host plant may still have will depend on other mechanisms. This is one reason why new mutant forms of pest and disease organisms appear which are no longer controlled by existing resistant varieties. The new races or biotypes are usually not resistant to the host's defenses but can steal in by the backdoor undetected. We can, of course, introduce another gene whose effect is to restore the tripping mechanism that sets off the alarm. All too often it suffers the same fate. In apples, for example, there are at least 19 different ways of initiating what seems to be the same basic mechanism to resist the scab fungus (*Venturia inaequalis*) and 19 corresponding ways for the fungus to avoid tripping the alarm (Bagga and Boone, 1968). Van der Plank (1963) called resistance with this drawback *vertical resistance*. He also coined the term *horizontal resistance* to describe resistance that is equally effective against all pathogen races. It is also known as uniform resistance, mature plant resistance, field resistance, and generalized or nonspecific resistance. Horizontal disease resistance often lacks the dramatic effect of vertical resistance. It may, for example, reduce the probability of infection by interfering with host penetration, slow the development of infection, and reduce the number of spores produced. In this way more time is required for a disease to spread to other plants and reach epidemic proportions. In artificial inoculations the large numbers of spores applied and the optimum conditions for infection may obscure the value of this type of resistance. To correct for this in a breeding program, epidemics can be created by inoculating spreader rows of a susceptible variety so that reactions can be observed under more natural conditions of infection.

The mechanisms of horizontal resistance are not well understood and are likely to be varied. Their most valuable feature is their stability and continued effectiveness against all forms of the pathogen.

If we examine resistance to insects we find the counterparts of vertical and horizontal resistance. There are grape phylloxera strains that produce galls on the leaves of resistant grapes. The larvae of some biotypes of Hessian fly flourish on the stems of several different resistant wheats because they apparently neutralize the defense reaction. This resistance is vertical and the insect biotypes are equivalent to races of disease organisms. Resistance to the golden nematode (*Heterodera rostochiensis*) in potato is also vertical (Kehr, 1966). At the same time a range of other effects may determine differences in the attractiveness and suitability of different cultivars as food plants or sites for oviposition. Many of these function in the same fashion as horizontal resistance to disease.

HOW RESISTANCE IS INHERITED

Resistance may be oligogenic, polygenic, or cytoplasmic.

Oligogenic Resistance often called major gene resistance, is determined by one or several genes whose individual effects are easy to detect. Most commonly a single allele for resistance is fully dominant in a heterozygote, but examples of incomplete dominance and recessive resistance are also known. Some resistance genes are either very closely linked or allelic (Hooker and Saxena, 1971). It is not always easy to recognize that two resistance genes are different. If they are closely linked or allelic, for instance, they may only differ in their reactions to certain vertical pathotypes. If these are not available the genes may be indistinguishable. Where they are not allelic, or closely linked, a breeding test will show that they are inherited independently. Single chromosome substitution is also a useful method of mapping resistance genes and has been used in wheat breeding (Riley and Macer, 1966).

Although most vertical resistance is also oligogenic, this is not universally true. Caldwell (1966) has reported an example of polygenic rust resistance in red winter wheat that is vertical. It is also by no means true that all oligogenic resistance is vertical. There are several examples of cultivars with single genes for resistance that have remained resistant for many years.

Polygenic Resistance is determined by many genes of individually small effect. As a result its inheritance is complex. The quantitative characters yield, quality, and date of maturity are inherited in a similar way. But such characters are amenable to breeding and selection.

Two examples will serve to illustrate the important points. Many corn inbreds from the U.S. corn belt show a mature plant resistance to common rust (*Puccinia sorghi*). In 1953 a natural rust epidemic in a breeding nursery enabled Hooker (1969) to rate 405 established inbreds for rust resistance. Some 200, or nearly 50 percent, had less than 1 percent of the leaf area infected (Table 1). Hooker took inbreds that were rust-susceptible as seedlings but showed mature plant resistance and crossed them with susceptible inbreds. The F_1's were uniform and showed intermediate resistance while the F_2's varied greatly, showing a continuous or normal distribution about a mid-parental mean. Selected F_2's from the extremes gave F_3 progenies that were more resistant or more susceptible than either parent. Evidently resistance was conditioned by several genes that show partial dominance. In Britain, Lupton and Johnson (1970) scored a much smaller sample of 41 wheat varieties for yellow rust (*P. striiformis*) resistance and also found nearly 50 percent with about 1 percent or less of the leaf area infected. They crossed Little Joss, an older wheat variety with mature plant resistance, by Nord Desprez, a modern wheat resistant to some races but highly susceptible to several others. The F_1 was as susceptible as Nord Desprez and only some 5 percent of the F_2 showed any resistance. However, the response to selection for resistance in the F_3, F_4, and F_5 generations was quite marked. In F_5, horizontally resistant lines as good as Little

TABLE 1 Distribution of Resistance among 405 Corn Belt Inbreds and 41 European Wheats to Rust (*P. sorghi* and *P. striiformis*, respectively).

Rating[a]	Corn, 1953	%	Wheat, 1968-69	%
0	9	2.2	3	7.3
1 or less	191	47.2	16	39.0
5	113	27.9	6	14.6
10	69	17.0	5	12.2
20	19	4.7	3	7.3
30	3	.7	5	12.2
40 or more	1	.3	3	7.3

Data from Hooker (1969) and Lupton and Johnson (1970).
[a]Percentage of leaf area infected.

Joss, but with the short straw and other desirable agronomic characters of Nord Desprez, were selected. Again resistance was clearly polygenic but, in this case, recessive.

Both examples included forms with vertical resistance. There are at least five such varieties among the wheats. While vertical resistance has been used to try to control the wheat rust in Europe it has not been used against the corn rust in the United States. In fact breeding for rust resistance is not an important objective in U.S. corn breeding programs. Even so, relatively few popular inbred lines are highly susceptible. Since resistance to corn rust tends to be incompletely dominant, susceptible inbreds would only be at a selective disadvantage if they were intercrossed. The wheat variety Nord Desprez illustrates what Van der Plank (1963) has called the Vertifolia effect. Selection for vertical resistance during its development entailed the loss of any horizontal resistance that may have been carried by its ancestors.

Cytoplasmic Resistance was until recently almost a curiosity, since only some five or six examples were known (Day, 1972). It became crucially important as a result of the southern corn leaf blight epidemic. The technology of hybrid seed production led in 1970 to a situation in which over 70 percent of the field corn in the U.S. was made up in a uniform cytoplasm. The Texas form of cytoplasmic male sterility (Tms) is easily introduced by backcrossing. The converted sterile inbreds are used as female or seed parents for hybrids, saving the cost and inconvenience of detasselling. Fertility is restored either genetically or by blending hybrid seed with normal (N) cytoplasm. The normal plants in a blend produce pollen enough for themselves and their neighboring sterile plants. The susceptibility of Tms to *Helminthosporium maydis* was first reported in the Philippines by Mercado and Lantican (1961). In 1969 and 1970 a new race appeared in the United States with consequences that are only too familiar. Tms is also susceptible to yellow leaf spot (*Phyllosticta* sp). Corn with normal cytoplasm is adequately resistant to the new race T of *H. maydis* and many, but not all, normal lines are resistant to *Phyllosticta.*

From the time Tms was introduced until 1961, the resistance it afforded against *H. maydis* race O was both adequate and taken for granted. But it was vertical resistance, since it proved to be ineffective against the toxin of race T. Whether N cytoplasm determines horizontal resistance remains to be seen. It is uniformly resistant to races O and T. Several other male sterile cytoplasms have been released that are also resistant to races O and T. Hooker *et al.* (1970) have shown that polygenic horizontal resistance to *H. maydis* occurs in N cytoplasm and is even expressed, to a lesser degree, in *Tms* cytoplasm to race T. Certainly we would be foolish to neglect it.

SOURCES OF RESISTANCE

The most important sources of resistant plants are the centers of diversity of our crops (Frankel and Bennett, 1970). These areas, first described by Vavilov, contain a profusion of wild relatives or forms used in primitive agriculture. The plant introduction service with its regional centers has organized the systematic collection and storage of germ plasm from these sources and arranged its release to breeders for screening. We are well supplied with materials, but are they used wisely and are the sources inexhaustible?

Until quite recently most breeders, given the choice, would elect to work with clear-cut monogenic resistance. They tended to neglect the sources of field resistance often at hand, primitive varieties from other cultures, or the older varieties our fathers and grandfathers had worked with. Undoubtedly in screening related wild species and anticipating up to six or more generations of backcrossing before obtaining agronomically useful varieties, breeders chose single dominant genes for resistance. How else could they show progress? But what horizontal resistance was lost in the accelerated evolution in their breeding plots? As we shall see presently, the best ways to put vertical resistance to work are becoming clearer. But where it has repeatedly failed, horizontal resistance is one of the most attractive answers, even though it may mean tolerating minor losses.

We must also be concerned about the continued availability of these sources of resistance. Germ plasm collections change during maintenance and storage, since some genotypes fail to survive. Only fresh collections provide insurance against such losses. But local floras change as progress invades the centers of diversity introducing roads, drainage schemes, pipelines, and other insults to ecology. The floras also change as introduced cultivars invade the original niches. It is ironic that a plant explorer looking for germ plasm should find the cultivars produced by the green revolution predominating in the centers of origin.

Some will say that we can always induce variation in the future. Indeed we can, and some disease-resistant mutants of crop plants have been produced (Favret, 1965; Sigurbjörnsson and Micke, 1969). But we have no evidence that the spectrum of useful variation will be as wide or any more useful than that from nature.

HOW RESISTANCE HAS BEEN USED

The fate of many crop varieties with vertical resistance has been described by Van der Plank (1968). Their early success in many cases led to their eventual failure. As the acreage of vertically resistant varieties increased, the selection pressure for new races increased. Because of the failure to select for any other kind of resistance, the end point was sometimes worse than the point of departure. The disease losses that the next new resistant variety would prevent were in large measure attributable to the step before. Where it was found useless, vertical resistance was abandoned in favor of horizontal resistance. Such is the present situation in breeding potatoes for resistance to late blight (*Phytophthora infestans*). On the other hand some examples of monogenic resistance have had continued and important successes in spite of new races occurring and we must examine them to see why. But first let us look at what the breeder is up against.

THE APPEARANCE OF BIOTYPES

The genetic versatility of pathogenic microorganisms needs no emphasis to biologists. Genetic studies in plant pathogenic fungi have shown that pathogenicity, like resistance, has oligogenic, polygenic, and cytoplasmic controls. At the oligogenic level the two genetic systems of host and parasite complement each other, giving rise to what Flor (1956) has called the gene-for-gene relationship. The evidence for this has been reviewed elsewhere (Day, 1972). It would be surprising if phytophagous insects were not the same. That they are was clearly demonstrated in recent work by Hatchett and Gallun (1970) who showed that the ability of Hessian fly (*Mayetiola destructor*) larvae of several biotypes to flourish on different monogenic resistant wheats was controlled by a gene-for-gene relationship. There seems little reason to doubt that biotypes of other examples of phytophagous insects are controlled in a similar way. Some examples are shown in Table 2.

Pests and diseases also have polygenic and cytoplasmic systems

TABLE 2 Some Examples of Insect Biotypes

Insect	Host	Reference
Biotypes defined by resistance genes		
Hessian fly (*Mayetiola destructor*)	Wheat	Hatchett & Gallun, 1970
Grape Phylloxera (*P. vitifoliae*)	Grape	Stevenson, 1970
Corn leaf aphid (*Rhopalosiphum maidis*)	Sorghum	Cartier and Painter, 1956
Raspberry aphid (*Amphorophora rubi*)	*Rubus*	Briggs, 1965
Biotypes not yet clearly defined by resistance genes		
Greenbug (*Schizaphis graminum*)	Wheat, sorghum	Harvey & Hackerott, 1969; Wood, 1971
Pea aphid (*Acyrthosiphon pisum*)	Peas	Herrington, 1945; Cartier *et al.*, 1965
Spotted alfalfa aphid (*Therioaphis maculata*)	Alfalfa	Pesho and Lieberman, 1960

which adjust their fitness and are responsible for the continuous variation observed in many characters associated with pathogenicity. Do these systems erode horizontal resistance? There are some indications that they may, but practical experience suggests that the risk is not high. The reason is that genetic homeostasis maintains a balance. During evolution the balance has ensured that neither partner in the host–parasite relationship has eliminated the other. Vertical resistance is a coarse control and horizontal resistance a fine control in maintaining the equilibrium. Like most fine controls horizontal resistance can only be turned up so far and the response the parasite can make is similarly limited.

VERTICAL VERSUS HORIZONTAL—ROBINSONS' RULES

The plant breeder who is introducing disease resistance may have a variety of raw materials available. These could include vertical resistance already released, which has molded the parasite genotypes he has to contend with; new unreleased monogenic resistance; and polygenic resistance. His first problem is to classify his new untried resistance as either horizontal or vertical. Prior to release both are very likely to be uniformly resistant to all the races he has at hand. He might, in theory, test the stability of a given gene for resistance by determining the induced mutation rate at the corresponding locus in the pathogen. Although mutations to new races can be induced in several fungi (Flor, 1958; Day, 1957; Day *et al.* 1971), this approach introduces the risk of mutants escaping and the problem of equating induced mutation rates with spontaneous rates. It also ignores the possibility that the race in question may already exist in the pathogen population. For the time being, we have only the rule of thumb that most vertical resistance is monogenic and is expressed in seedlings and mature plants, while most horizontal resistance is polygenic and is not expressed until after the seedling stage. However, there are exceptions to these generalizations.

Recently Robinson (1971) addressed the problem of whether to breed for vertical or horizontal resistance and how to assess the value of vertical resistance in a given situation. His analysis is that of an epidemiologist; it considers the two populations of host and pathogen and the nature of the factors that govern changes within each of them. The factors include the relative ease of breeding the host and its longevity as a crop, the size and uniformity of the host population and whether it has discontinuities in space or time, whether the pathogen spreads slowly or rapidly and if it is transmitted when the host is propagated, its rate of mutation to new races, and so on. The result is a set of 14 rules that are illuminating because they provide a logical explanation of why vertical resistance sometimes fails and sometimes succeeds. Some of the premises have been misinterpreted. In particular, the theory of stabilizing selection has been questioned (see, for example, Watson, 1970). This is the notion that when pathogens become pathogenic on certain resistant cultivars they lose fitness and cannot survive in competition with simpler races or other organisms on substrates that can support them both. There are a number of examples that show that pathogenicity and fitness are by no means always negatively correlated. In proposing the idea, Van der Plank (1968) was careful to point out that only pathogenicity to what he called "strong" resistance genes carried the penalty of lowered fitness.

An example of the application of the rules is as follows. *Fusarium* wilts spread slowly since they are for the most part soil-borne and new races thus tend to remain localized. If we accept the hypothesis that some races lack fitness in competing with other soil organisms, then crop rotation will help to reduce their frequency. If the host is an annual and easy to breed, like tomato, cabbage, or peas complete control is possible with vertical resistance as long as rotation is practiced. It is unlikely to succeed in banana with Panama disease because the crop is perennial, it is difficult to breed, single clones occupy very large acreages, and the disease is easily spread during propagation.

Every host–parasite association must be considered as a special case and each factor must be weighed in relation to all the others.

STRATEGIES FOR THE FUTURE

While it may seem absurd, because it appears to be self-evident, we should briefly examine the reason why it is necessary to breed for resistance to pests and diseases at all. After all, native wild floras have survived without our intervention. The agricultural developments made possible by plant breeding all led to the situation we call monoculture. Our crops became uniform, they produced the highest possible yields, they ripened at the same time, they flourished in stands of uniform height and density, and, to our cost, we discovered they were uniformly susceptible to devastating epidemics of pests and diseases. Those southern farmers who could harvest ears of corn from N cytoplasm plants in corn fields sown with blends in 1970 appreciated the virtues of genetic diversity. Phenotypic uniformity within a cultivar is a keystone of current levels of productivity, and the role of resistance in protecting it from loss has become vital. However, geneticists and breeders have questioned the wisdom of genotypic uniformity (Simmonds, 1962).

Most of us are aware of the dangers of monoculture but we tend to think only that it means single varieties grown on a very large scale. The fact that there are other wheat or corn or soybean varieties coming along in our breeding programs that we can turn to obscures the fact that they may be little different from those that are current now. The widespread use of Tms cytoplasm in corn showed in a very dramatic way how important the trend to uniformity can be. We need to be aware that many modern cultivars within each of the major crops differ in name more than in genotype. The history of disease epidemics shows that a narrow genetic base severely limits the capacity of a crop to respond to threats from pests and diseases. While we continue to strive for higher yield and better quality in our crops, must this necessarily lead to the erosion of their genetic base?

The answer seems to be "not necessarily," and potatoes provide an interesting case in point. The varieties grown in North America and Europe are based on a small sample of tetraploid South American clones supplemented by occasional introductions of new material. In the 1840's blight reduced variability further in that few clones survived. Breeding since that time has largely consisted of introducing dominant genes for disease resistance and, until quite recently, there has been no major introduction of variability (Simmonds, 1962). Although breeders raise millions of seedlings annually, very few survive to replace the leading varieties, many of which are more than 30 years old. There is surely a good case for making an entirely fresh approach to breeding new varieties by starting from scratch. The success of new high-yielding rice and wheat varieties shows that the example of high yield will quickly create a demand for revolutionary new cultivars. Several breeders in the United States and elsewhere have approached this problem in potatoes by selecting for increased tuber size and field resistance to late blight in populations of wild material whose germ plasm has been largely untapped by conventional breeding programs (Simmonds, 1968). Although no new cultivars have yet been released or tested commercially, the idea has merit if the introduction of genetic variability is to be a major objective.

Multilines are another approach to the introduction of variability and were first suggested by Borlaug. The theories behind their use were recently reviewed by Browning and Frey (1969). A multiline cultivar consists of a collection of lines (isolines) that are nearly isogenic except for their possession of different genes for vertical resistance. In an inbreeder like oats each line is produced by introducing resistance into a parent variety, common to all the isolines, by backcrossing. Each isoline must be at least as productive as its recurrent parent under rust-free conditions. In theory, the composition of multiline varieties can be changed as rapidly as the availability of component lines will allow. Their success depends on the correctness of the assumption that a super race will not develop that is equally pathogenic on all components. Again, our experience of this approach is too limited to judge its worth. A promising start with multiline varieties of oats with composite resistance to crown rust (*P. coronata avenae*) has been made by Frey *et al.* (in press), and several varieties with from 7 to 11 component isolines have been released in Iowa. A significant weakness of multilines is their conservatism. Much will depend on the wise choice of the recurrent backcross parent common to the component lines. The breeding effort invested in producing and testing a set of component lines will slow down change to other backgrounds when it is required. Another drawback with present multilines is that because their variability is synthetic it is polarized. It has been developed on only one plane. In the case of oats, this is resistance only to crown rust. What of resistance to stem rust and to other parasites?

Undoubtedly as we explore the multiline approach much attention will be paid to the necessity for genotypic uniformity among the components. Hopefully "heterolines" that are sufficiently uniform phenotypically may prove even more profitable than the isolines we are starting out with now. We must ask our economists if our standards of freedom from blemishes and tolerable damage, for example in fruit and vegetable crops, are not unrealistically high. Is the uniformity of quality demanded of some small grains so great that it seriously jeopardizes our ability to produce the increasing quantities required because of inherent susceptibility to epidemics? I cannot answer these questions but they must be asked since they are at the root of our dilemma of how far we can reintroduce genetic diversity and yet obtain acceptable yield and quality.

The future undoubtedly holds still other approaches. The current interest in systemic fungicides has prompted suggestions that materials may be found that potentiate innate disease resistance mechanisms. Such discoveries will be greatly helped by a better understanding of how a resistant host recognizes a parasite. We must also not neglect the mechanisms of horizontal resistance. Don't let us make the mistake of producing highly sophisticated fungicides, insecticides, and nematocides that fail because they too are vertical.

We can anticipate a continuing effort to maintain resistance in our crops. The lesson of southern corn leaf blight shows clearly a dangerous trend in our present agricultural technology. Bold new methods are needed and, indeed, are appearing. We must be prepared to let well-established principles in population genetics and epidemiology work for us, and, if some of our goals are unrealistic, we must work toward new ones.

REFERENCES

Bagga, H. S., and D. M. Boone. 1968. Genes in *Venturia inaequalis* controlling pathogenicity to crabapples. Phytopathology 58:1176–1183.

Bawden, F. C. 1964. Plant viruses and virus diseases (4th ed.). Ronald Press Co., New York.

Borlaug, N. E. 1965. Wheat, rust, and people. Phytopathology 55:1088–1098.

Briggs, J. B. 1965. The distribution, abundance and genetic relationships of four strains of the rubus aphid (*Amphorophora* rubi, Kalt) in relation to raspberry breeding. J. Hort. Sci. 40:109–117.

Browning, J. A., and K. J. Frey. 1969. Multiline cultivars as a means of disease control. Ann. Rev. Phytopathol. 7:355–382.

Caldwell, R. M. 1966. Advances and challenges in the control of plant diseases through breeding. *In* Pest control by chemical, biological, genetic and physical means. E. F. Knipling (ed.). U.S. Department of Agriculture 33–110:117–126.

Cartier, J. J., A. Isaak, R. H. Painter, and E. L. Sorensen. 1965. Biotypes of pea aphid *Acyrthosiphon pisum* (Harris) in relation to alfalfa clones. Can. Entomol. 97(7):754–760.

Cartier, J. J., and R. H. Painter. 1956. Differential reactions of two biotypes of the corn leaf aphid to resistant and susceptible varieties, hybrids, and selections of sorghums. J. Econ. Entomol. 49:498–508.

Day, P. R. 1957. Mutation to virulence in *Cladosporium fulvum*. Nature 179: 1141–1142.

Day, P. R. 1968. Plant disease resistance. Sci. Prog. 56:357–370.

Day, P. R. 1972. Genetics of host-parasite interaction. Freeman, San Francisco. In preparation.

Day, P. R., S. L. Anagnostakis, and J. E. Puhalla. 1971. Pathogenicity resulting from mutation at the b locus of *Ustilago maydis*. Proc. Nat. Acad. Sci. 68: 533–535.

Favret, E. A. 1965. Induced mutations in breeding for disease resistance. The use of induced mutation in plant breeding. Supplement to Radiation Botany 5:521–536.

Flor, H. H. 1956. The complementary genic systems in flax and flax rust. Adv. Genet. 8:29–54.

Flor, H. H. 1958. Mutation to wider virulence in *Melampsora* lini. Phytopathology 48:297.

Frankel, O. H., and E. Bennett (eds). 1970. Genetic resources in plants–their exploration and conservation. F. A. Davis Co., Philadelphia.

Frey, K. J., J. A. Browning, and R. L. Grindeland. 1971. Implementation of oat multiline cultivar breeding, symposium. Mutation Breeding for Disease Resistance IAEA, Vienna, p. 159–169.

Hackerott, H. L., and T. L. Harvey. 1969. Greenbug resistance in sorghums. Crop Sci. 9(5):656–658.

Harvey, T. L., and H. L. Hackerott. 1969. Recognition of a greenbug biotype injurious to sorghum. Econom. Entomol. 62:776–779.

Hatchett, J. H., and R. L. Gallun. 1970. Genetics of the ability of the Hessian fly, *Mayetiola destructor*, to survive on wheats having different genes for resistance. Annals Entomol. Soc. Am. 63:1400–1407.

Herrington, C. D. 1945. Biological races of the pea aphid. J. Econ. Entomol. 38:12–22.

Hooker, A. L. 1969. *In* Disease consequences of intensive and extensive culture of field crops. J. A. Browning (ed.). Iowa Agr. Exp. Sta. Spec. Rep. 64:28–34.

Hooker, A. L., and K. M. S. Saxena. 1971. Genetics of diseases resistance in plants. Ann. Rev. Genet. 5:407–424.

Hooker, A. L., D. R. Smith, S. M. Lim, and M. D. Messon. 1970. Physiological races of *Helminthosporium maydis* and disease resistance. Plant Dis. Rep. 54:1109–1110.

Hsiao, T. H. 1969. Chemical basis of host selection and plant resistance in oligophagous insects. Entomol. Exp. Appl. 12:777–788.

Kehr, A. E. 1966. Current status and opportunities for the control of nematodes by plant breeding. *In* Pest control by chemical, biological, genetic and physical means. E. F. Knipling (ed.). U.S. Department of Agriculture 33–110:126–138.

Kuc, J. 1966. Resistance of plants to infectious agents. Ann. Rev. Microbiol. 20:337–370.

Lupton, F. G. H., and R. Johnson. 1970. Breeding for mature-plant resistance to yellow rust in wheat. Ann. Appl. Biol. 66:137–143.

Mercado, A. C., Jr., and R. M. Lantican. 1961. The susceptibility of cytoplasmic male-sterile lines of corn to *Helminthosporium maydis*, Nisikado and Miyake. Philipp. Agric. 45:235–243.

Pesho, G. R., and F. V. Lieberman. 1960. A biotype of the spotted alfalfa aphid on alfalfa. J. Econ. Entomol. 53:146–150.

Riley, Ralph, and R. C. F. Macer. 1966. The chromosomal distribution of the genetic resistance of rye to wheat pathogens. Can. J. Genet. Cytol. 8:640–653.

Robinson, R. A. 1971. Vertical resistance. Rev. Plant Pathol. 50:233–239.

Sigurbjörnsson, B., and A. Micke. 1969. Progress in mutation breeding. Induced mutations in plants. IAEA, Vienna. p. 673–698.

Simmonds, N. W. 1962. Variability in crop plants, its use and conservation. Biol. Rev. 37:422–465.

Simmonds, N. W. 1968. Potato investigations. Scottish Plant Breeding Station 47th Annual Report, p 11.

Simons, M. D. 1966. Relative tolerance of oat varieties to the crown rust fungus. Phytopathology 56:36–40.

Stakman, E. C. 1954. Recent studies of wheat stem rust in relation to breeding resistant varieties. Phytopathology 44:346–351.

Stevenson, A. B. 1970. Strains of the grape Phylloxera in Ontario with different effects on the foliage of certain grape cultivars. J. Econ. Ent. 63:135–138.

Van der Plank, J. E. 1963. Plant diseases: Epidemics and control. Academic Press, New York.

Van der Plank, J. E. 1968. Disease resistance in plants. Academic Press, New York.

Watson, I. A. 1970. Changes in virulence and population shifts in plant pathogens. Ann. Rev. Phytopathol. 8:209–230.

Wood, E. A., Jr. 1971. Designation and reaction of three biotypes of the greenbug cultured on resistant and susceptible species of sorghum. J. Econ. Ent. 64:183–185.

Sterilization and Other Genetic Techniques

E. F. Knipling

It has been two decades since the first publication was issued on the feasibility of employing the sterilization technique as a means of controlling insect populations (Bushland and Hopkins, 1951). It required almost a decade of research on a number of vital basic and applied aspects of this new approach to insect control before the technique was applied for the eradication of the screwworm on the island of Curaçao and later from the southeastern United States (Baumhover *et al.*, 1955; Baumhover, 1966). The technique is now being employed in the southwestern United States and northern Mexico to manage the screwworm in an area of about 300,000 sq mi from the Gulf of Mexico to the Pacific Ocean. Infestations in livestock by this major livestock pest are reduced annually by more than 99 percent at a great savings to the livestock industry.

Two points of significance to pest-control strategies for the future can be gained from the screwworm programs. First, the principle of using insects themselves as an autocidal biological control organism to suppress pest populations has been demonstrated. Second, the programs have demonstrated that it is feasible to manage populations of a widely distributed and highly mobile insect on a regional and even on an international basis.

We hear a great deal about insect management these days. However, I wonder if most pest control strategists fully appreciate what is involved in managing an insect that is widely distributed and highly mobile, with a wide range of host plants or animals.

Aside from the direct contribution that the sterilization and other genetic techniques have already made toward more effective and more acceptable means of insect control, an important additional benefit of research on this new technique has been the stimulation of thought and investigations on the biology, behavior, ecology, and dynamics of insect populations. It has encouraged applied entomologists to think quantitatively instead of qualitatively in gaining information on insect pests. What scientists have and will learn from in-depth research in such areas could make major contributions to the improvement of insect control whether the methods involve chemical insecticides, genetic techniques, insect attractants, the use of natural biological agents, or the integration of two or more techniques. In the past we have tended to look at insect problems on a crop-to-crop and farm-to-farm basis. The availability of a wide selection of highly effective insect-control chemicals made it possible to cope with most insect pests in this manner. However, this system is breaking down because of the effect on the natural biological control organisms, the restrictions now being placed on the use of broad-spectrum insecticides, the development of insecticide-resistant insects, and the continuing high costs of control as well as continuing losses due to insect damage. Most of the old insect problems are still with us, and some are more acute after more than a quarter-century of intensive use of insecticides. New insect problems are emerging because our agroecosystem is becoming more and more favorable for insect pests. I am convinced that, for the future, we must look at major insect pest problems from a community, regional, or even from an international standpoint and explore new strategies for controlling the entire population by methods that seem feasible, practical, effective, and ecologically acceptable. Our research efforts should then be directed toward the solution of major pest problems by whatever method or combination of methods that offers the best chance of success.

The genetic technique is just one of several methods that could be useful in providing more effective and more acceptable solutions to key insect problems. It most certainly will not prove to be a panacea for the solution of all insect problems. At the same time, I think the technique offers far greater possibilities against the most important pests than most entomologists recognize today. Many entomologists still do not fully understand the principles involved in the suppression of insect populations by genetic means. It is a highly sophisticated technique, and well-trained, competent, and imaginative scientists in a number of scientific disciplines will be required to develop the technique for practical application against the pests that are good candidates. After its development from a technical standpoint, it will require good organization and precision in its application. The technique will make its greatest contribution as an adjunct to other suppression methods. However, once the system is developed and put into operation, there is probably no other method of insect suppression that can be comparable in efficiency, effectiveness, and safety in the continuous suppression and management of insect populations.

What progress has been made in the development of the sterilization technique, and what does the future hold for this and other genetic methods? I will briefly review progress in this field and then discuss the requirements to further develop the sterility technique so that it can play a more prominent role in insect pest-control strategies for the future.

CURRENT STATUS OF THE STERILITY TECHNIQUE

The success of the sterilization technique for the eradication and continuous suppression of the screwworm fly has been mentioned. The technique is being applied against two other insect pests, the Mexican fruit fly and the pink bollworm. However, research is under way on a number of other important species. Before discussing the status of research and control on various insect pests, it may be desirable to outline some of the basic requirements for successful development and application of the technique and list the type of insect problems for which the genetic technique could be useful.

The basic requirements for the practical application of the genetic technique are as follows:

1. Procedures must be developed for rearing large numbers of healthy and vigorous insects at reasonable cost.
2. Procedures must be developed for sterilizing or instilling appropriate genetic deficiencies in insects to be released without causing serious adverse effects on the vigor and mating competitiveness of the insects.
3. Suitable methods for handling and releasing such insects must be developed.
4. The insects to be released in numbers required must not be destructive or unduly annoying.

Sterile or genetically altered insects might be used in the following ways:

1. To prevent the establishment of insect pests in new areas
2. To eliminate incipient infestations
3. To eliminate or suppress established insect populations that reach low levels of abundance and have restricted distribution due to natural causes
4. To eliminate or continuously manage insect populations after they have been suppressed by other means

STATUS OF RESEARCH ON VARIOUS PESTS

It is not feasible in this paper to review and cite all of the advances that have been made in basic and applied research by many investigators. Instead, it seems more desirable to name some of the more important insects under investigation, indicating how sterile insects might be used and the stage of development of the research (Table 1). Additional comments will be made on a few of the insects named.

STATUS OF RESEARCH AND DEVELOPMENT

Screwworm

Screwworm flies are being mass produced, sterilized, and released at the rate of about 150 million per week in the southwestern states and northern Mexico. Although the program is highly successful, the maintenance of a suppression program along an 1,800-mile front from the Gulf of Mexico to the Pacific is not the best long-range solution to this problem. Discussions have been held by officials of the U.S. Department of Agriculture and of the government of Mexico in the planning of a joint effort to eliminate the pest throughout the southwestern United States and in Mexico. It should then be possible to prevent re-establishment of the insect by maintaining a sterile fly barrier at the Isthmus of Tehuantepec in southern Mexico. This would narrow the front to about 150 miles and should permit management of the population at substantially less cost than the approximately $6 million now spent each year. The plans for this joint program have

TABLE 1 Insects under Investigation, How Sterile Insects Will Be Used, and Status of Research or Practical Use

Insect	Proposed Manner of Use	Status of Research and Development
Screwworm	For suppressing populations on regional basis	In practical use
Mexican fruit fly	For preventing establishment of incipient infestations	In practical use
Pink bollworm	For preventing establishment of incipient infestations and to eliminate low level established populations	In practical use; additional improvements and pilot testing required
Oriental, Mediterranean, and melon fly	To eliminate low level populations and to prevent establishment of incipient populations	Effectiveness demonstrated in small island tests. Large pilot tests required
Codling moth	To maintain suppression after prior suppression of populations by cultural and chemical means	Effectiveness demonstrated in small orchard tests. Small pilot test under way
Boll weevil	To eliminate low level populations after prior suppression by chemical, cultural, and other means	Pilot tests planned
Bollworm and budworm	For area suppression of low level populations	Pilot tests required
Cabbage looper	For area suppression of low level populations	Pilot tests required
Fall armyworm	For area suppression of low level populations	Pilot tests required
Tobacco hornworm	For area suppression after prior suppression by cultural means	Pilot test required
Gypsy moth	For preventing spread and to eliminate incipient infestations	Pilot test required
Mosquitoes (important vector species)	To maintain suppression after prior suppression by sanitary and chemical means	One pilot test underway–others required
Tsetse flies	To eliminate low populations after prior suppression by chemicals and brush clearing	Pilot test planned
Horn fly	To eliminate low populations after animal spraying	Pilot test required

not been finalized, but such a program would be highly desirable from the standpoint of efficiency and economy.

Mexican Fruit Fly

This major pest is a continuing threat to tropical fruit in southern California because of the presence of flies emerging from infested commercial fruits shipped to marketplaces in Tijuana and other border cities. For many years insecticide sprays were applied to local fruit hosts to prevent the establishment and spread of incipient populations. The routine release of sterilized Mexican fruit flies has now replaced chemical spraying as a means of protecting fruit industries in California. The technique has been highly successful and is continuing. The release of only a few million sterilized flies in the threatened area each week ensures an overflooding ratio of sterile to native wild flies of several thousand to one. This is an excellent example of the merits of the genetic method when natural populations of the pest are extremely low.

Pink Bollworm

The spread of the pink bollworm to the irrigated cotton growing areas in Arizona, New Mexico, and southern California has created a serious cotton pest problem in those areas. The insect causes high losses in yield and quality of cotton. Costly insecticide spraying is necessary to prevent excessive losses. Moreover, the use of the insecticides has created secondary insect problems because of the destruction of natural biological agents. High populations of the pest in the southern California area and in northern Mexico represent a continuing threat of its spread into the San Joaquin Valley in California. If the insect should become firmly established in this area where more than 600,000 acres of cotton are grown, the need for control of the insect by chemical means would not only prove costly but the possibilities of maintaining effective integrated biological–chemical control programs for bollworms and other cotton pests would certainly be lost. In view of the urgency of the situation, the U.S. Department of Agriculture and the California Department of Agriculture initiated a sterile pink bollworm release program, based on laboratory and field cage studies of the technique (Graham *et al.*, in press). No other acceptable methods of eliminating incipient pink bollworm infestations are now available. The alternative is to risk establishment of the pest in the Valley or to make intensive applications of insecticides where infestations are found. Insecticides not only failed to prevent the spread of the pest into Arizona and southern California, but their use also caused the usual environmental hazard problems. Thus, the genetic approach, which is completely selective in action, offers the only hope of preventing further spread of the pest at the present time. After expected difficulty in advancing from laboratory rearing to mass insect production and handling, the U.S. Department of Agriculture succeeded in rearing more than 100 million sterilized moths for release in the lower San Joaquin Valley in 1970. The program is continuing while research is still under way on improved methods of sterilizing the insects so that they will retain maximum vigor and competitiveness. There is still general lack of confidence in the technique for pink bollworm containment, but if the program is successful in preventing the spread of the pest it should greatly advance the sterilization concept for dealing with other agricultural pests. Only further practical use and experimentation will provide the information needed to fully appraise the effectiveness of the California pink bollworm suppression program, but substantial progress has already been made in the development of the technique. Thus far, pink bollworms have not become firmly established in the San Joaquin Valley.

Codling Moth

The codling moth is the most important insect pest of apples and pears. It is responsible for most of the insecticides required in the production of these fruits. The intensive use of insecticides as a means of control for the codling moth has created secondary pest problems, especially mite problems. Laboratory and small field experiments carried out by scientists with the U.S. Department of Agriculture and with the Canada Department of Agriculture have shown substantial promise for the genetic method for codling moth control. Methods of rearing the insect at reasonable cost have been developed, and the competitiveness of the sterilized moths has been demonstrated in small-scale field tests. A small pilot test program is under way in the Yakima Valley in the State of Washington (Butt *et al.*, in press). The test is being conducted in a 25-sq-mile area in which about 500 acres of commercial fruit is produced. For years, however, a relatively few untreated noncommercial host trees have constituted a constant source of repopulation of commercial orchards by the codling moth. The test program now under way involves continuation of insecticide spraying of commercial fruit trees and the elimination or temporary spraying of the noncommercial host trees. This spring and summer, sterile moths will be released throughout the small valley and in commercial orchards at the rate of about 5,000 per acre. The natural population is now at a very low level. The release of several thousand sterile moths per acre of host trees during the season should achieve and maintain a ratio of about 50 sterile to 1 native fertile moth. If the natural population can be eliminated or greatly suppressed the first year, it may be possible to maintain suppression by the use of relatively few sterile moths alone. In such event, the need for insecticides to control the codling moth would be eliminated. If successful, the experiment should encourage the further development of the technique in controlling this major pest of fruit throughout the western region by a technique that is completely selective and that should cost much less than current chemical control methods.

Boll Weevil

The importance of the boll weevil in terms of economic losses to the cotton industry and in terms of the extensive use of broad-spectrum insecticides is well known. Losses caused by the pest are estimated to average about $200 million each year. About one third of all insecticides employed for agricultural purposes is used for controlling this insect. The U.S. Department of Agriculture and cooperating state experiment stations have undertaken a major research effort to develop methods that might be used to eliminate this costly pest from the United States or to develop procedures that will replace chemical insecticides. Progress has been made in techniques of applying chemical insecticides in the fall to suppress populations to very low levels (Brazzel *et al.*, 1961; Adkisson and Sterling, 1965; Knipling, 1968). Also, a new highly effective systemic insecticide has been developed (Coppedge *et al.*, 1969). The recent identification and synthesis of the boll weevil sex and aggregating pheromones (Tumlinson *et al.*, 1969) has provided a new tool for surveys and the attractant may also be effective in control. Methods of mass rearing the boll weevil have been developed (Gast, 1968). Promising chemicals for sterilizing the males are under investigation (Klassen *et al.*, 1968; Klassen and Earle, 1970). Plans are under way through joint efforts by the U.S. Department of Agriculture, state agricultural experiment stations, and Cotton Incorporated to undertake a large pilot test integrating all known and feasible suppression methods to determine if eradication of the boll weevil is technically and operationally feasible. The sterile males are intended to ensure the elimination of very low level populations remaining after other suppression measures have been applied. If populations can be reduced to a few boll weevils per acre, it would be practical to outnumber those that remain by a factor of about 100 to 1. There is reason to believe that such a high ratio of reasonably competitive sterile insects to fertile insects will accomplish eradication. However, adequate tests against such low natural populations in an isolated area are necessary to determine if the method is effective.

Suitable pilot tests to demonstrate elimination or suppression of highly mobile insects are difficult to carry out and are costly. Yet, there is no other way to advance the concept of insect population management regardless of the techniques employed. Results of laboratory and small-scale field tests are insufficient to appraise the effectiveness and practicability of using area-wide insect suppressive measures.

Tsetse Flies

Tsetse flies are regarded as the most important obstacle to the agricultural development of vast areas in Africa. A productive livestock economy cannot be maintained in regions where these insects occur because of the trypanosomdiasis disease organism they transit to livestock from wild game reservoirs. The insects are also a continuing threat to the health of people in many areas in Africa. Insecticides, brush clearing, and game destruction are measures now taken to eliminate the pest from areas to be developed agriculturally. We know that the eradication of any insect by the use of insecticides is difficult. Typically, it may require more insecticide applications to eliminate the last 0.1 to 1 percent of a population than is required to destroy 99 to 99.9 percent of the initial population. This has been the experience in most tsetse fly eradication programs carried out in Africa. In view of this problem, the tsetse fly was early proposed as an excellent candidate for development of an integrated chemical-sterile insect release system (Knipling, 1964a, 1964b), despite the anticipated difficulty in rearing this insect. A number of research institutions have been investigating the use of the genetic method as an adjunct to other suppressive measures, including the U.S. Department of Agriculture (under auspices of the Agency for International Development); various institutions in African countries; the University of Bristol, England; and the International Atomic Energy Agency. Research results reported by Dame and Schmidt (1970) have shown that tsetse flies can be successfully sterilized without serious adverse effects on mating competitiveness. Suppression of natural populations in a small area has been demonstrated by the release of sterilized males. Nash *et al.* (1968) have made excellent progress in developing large-scale rearing methods for tsetse flies. A pilot test is planned in Tanzania under joint sponsorship of AID and the Tanzanian government. The pilot test is to be directed by scientists with the U.S. Department of Agriculture. If this program gets under way and is successful, the genetic method may play a prominent role in dealing with one of the world's most important insect pests. Based on current progress, there is hope that the use of sterile males will help achieve eradication and reduce costs by more than one half. Perhaps of equal significance, the technique promises to reduce the amount of insecticides required for eradication to one third or less of the amount now required.

Heliothis spp. and Other Lepidopterous Pests

Research conducted by several laboratories of the Entomology Research Division on mass rearing of *Heliothis zea* (Boddie), *H. Virescens* (F.), the cabbage looper *Trichoplusia ni* (Hubner), fall armyworm *Spodoptera frugiperda* (J. E. Smith), and several other important Lepidoptera species has progressed to a degree that it seems entirely feasible to rear these insects in unlimited numbers at costs that may be well below $5 per 1,000 insects (Henneberry and Kishaba, 1966; Raulston and Shaver, 1970; Mangum *et al.*, 1969; Burton, 1967, 1969). Substantial progress has also been made in methods of sterilization by radiation and chemical means (North and Holt, 1968; Proshold and Bartell, in press). I have given a great deal of study to the feasibility of area suppression of the corn earworm (bollworm) (*H. zea*), tobacco bud worm (*H. virescens*), and the cabbage looper by the use of sterile or partially sterile males (Knipling, 1970).

It may be entirely feasible to suppress populations of *H. zea* and

H. virescens (two of the nation's most destructive insects) on a regional basis early in the season, when populations are at a minimum, by releasing sterile or partially sterile males. If the technique could be developed, effective control might be achieved at costs that would be only a fraction of present costs with chemicals. Of even greater significance, this would be a selective method that would replace the use of broad-spectrum insecticides that are now used extensively for their control. However, there is no way to determine the feasibility of the genetic method until proper pilot tests are carried out. It might be necessary to first suppress natural populations by other means. The release of mass-produced parasites might be feasible for the suppression of high populations. The integration of biological agents and the genetic method may be more practical than either method employed alone (Knipling, 1971). Both techniques should be completely selective and should cause no adverse effects to nontarget organisms.

Other important lepidopterous pests such as the cabbage looper and the fall armyworm are also excellent candidates for area-suppression by genetic means. There is evidence that these two pests along the eastern seaboard can successfully overwinter only in Florida and perhaps parts of adjacent states. The natural populations may be relatively low during the winter months. Thus, it may be feasible to use the genetic method to suppress these insects in the greatly restricted area of distribution in the winter and protect agricultural crops as far north as New York during the rest of the season. Such an approach to insect population management may cost only a fraction of what current practices cost and may also reduce environmental pollution.

GENERAL COMMENTS

In the genetic approach to suppression or elimination, each insect problem has its special problems and requirements. Space will not permit a discussion of all of the insects named in the list or others in which the genetic technique might eventually prove useful. However, the insect problems discussed in the very brief and oversimplified manner should provide the reader with an overview of the status of research and development of the sterilization technique. Admittedly, there are many major problems that need to be investigated and resolved, but the principles are sound and the possibilities are highly encouraging.

MAJOR REQUIREMENTS TO FULLY DEVELOP GENETIC TECHNIQUES

More and more scientists in various research institutions throughout the world are devoting their efforts to the development of sterility and other genetic techniques for insect control. The Entomology Research Division has placed a high priority on this general approach to the control of a number of major pests. The International Atomic Energy Agency and the World Health Organization have likewise placed a high priority on research involving this approach to insect control. Substantial progress has been made in all of the areas of research that are basic to the development of the technique. As already stated, much remains to be done, however, in the years ahead. A few comments on the major areas of research needing attention seem appropriate.

MASS-REARING INSECTS

As noted earlier, the ability to rear insects is a basic requirement for application of the autocidal method, whether the technique involves sterilizations, translocations, conditional lethal factors, or other genetic methods. While the rearing of large numbers of insects at low cost is often difficult, outstanding progress has been made in the last decade. If such progress could continue in the next decade, we should be in a position to think in terms of rearing insects of many species by the hundreds of millions or even billions. This capability already exists for the screwworm and several species of tropical fruit flies, the pink bollworm moth, and the boll weevil. However, to attain this goal for other major pests and for parasites and predators, the concerted efforts of scientists in various disciplines will be needed. Assistance will be needed from insect behaviorists, agricultural engineers, insect nutritionists, and insect pathologists. Major problems that must be resolved before we can mass produce some of the most important species include mechanization procedures to reduce labor costs and proper but low-cost nutritional elements that will assure the production of vigorous, healthy insects that can accomplish their mission when released. Also, one of the most important problems is to prevent or control diseases in laboratory insect colonies. This has been a particularly serious problem in developing rearing methods for lepidopterous insects. The ability to rear insects in large numbers is basic to the development of genetic methods of insect population management, and efficient mass production of insect parasites and predators will be the key to successful use of programed releases of natural biological agents in insect management systems or for use in conjunction with genetic techniques.

STERILIZATION AND OTHER GENETIC DEFICIENCIES

The Agricultural Research Service, Metabolism and Radiation Research Laboratory, Fargo, North Dakota, has been the leading research institution in the world investigating various genetic methods that may be useful in insect population suppression. The use of atomic radiation, sterilizing chemicals, and various conditional lethal traits in insects are under study. However, more and more geneticists throughout the world are devoting their efforts to research on various genetic mechanisms for the management of insects. As research in this area progresses, scientists may find that other genetic techniques will be much more effective than the sterilization procedure for suppressing certain insect populations.

BIOLOGICAL AND ECOLOGICAL STUDIES

One of the most important requirements in applying the genetic techniques for insect control is detailed knowledge of the biology, dispersal behavior, and population dynamics of various insects. It must be emphasized that the genetic technique will not be applicable for suppressing insect populations when they exist at high levels. Other methods of suppression are potentially far more efficient in suppressing high populations than is the genetic technique. Most research on the genetic techniques has been conducted on sterilization and rearing procedures. The field aspects have received much less attention. Yet, ecological information about an insect in the field is equally vital to the development and application of the technique. Field information basic to practical development of technique includes (a) reasonably good estimates of absolute numbers of the target insect throughout the ecosystem where suppression is required; (b) an understanding of the potential rates of increase of populations at low density levels, as a basis for determining the rate of overflooding with genetically altered insects that will be necessary; and (c) information on the distance and rate of spread of insects as a basis for determining the size of an area that must be included in a suppression program.

PILOT TESTING

Without question, the greatest need to ensure more rapid development of the genetic technique is to undertake the research and development phases of this approach to insect control. It is not possible to determine the requirements or costs of mass-rearing insects until mass-rearing facilities are available and mass-production methods are developed. Pilot tests cannot be conducted to determine the effectiveness of genetically altered insects until adequate numbers of insects are available for tests in a suitable area. Since most of our major pests are highly mobile, it is necessary that appropriate tests be carried out against isolated populations or against populations in large nonisolated areas.

NECESSITY FOR UNDERSTANDING OF BASIC PRINCIPLES

While an increasing number of scientists are devoting their thoughts and efforts to the development of the genetic techniques, there are still too many members of the scientific community who do not yet fully understand the basic principles involved in the development and application of genetic techniques for insect population management. The benefits to be gained by the integration of different systems of insect population suppression is not generally realized. The efficiency of different techniques of insect control vary depending on the density of the insect to be controlled. For example, programed releases of parasites can be expected to have maximum efficiency when the target pest exists at the highest density. In contrast, the genetic system has its greatest efficiency when the target pest exists at the lowest density level. Thus, an integration of the two systems promises to become the best strategy to employ for the management of certain important pests.

Effective and efficient pest-management strategies in the future will not only require the development of various techniques for insect population suppression but also require an appreciation of how and when to employ techniques alone, in combination, or in sequence to achieve optimum results. Thus, in planning, developing, and applying insect-management systems, the need for a better understanding of the principles of insect population suppression by various techniques cannot be overemphasized.

REFERENCES

Adkisson, P. L., and W. L. Sterling. 1965. Two-phased control program for reducing diapause boll weevil populations in the high plains of Texas in 1965. Texas A&M Univ./Tex. Agric. Exp. Sta. Misc. Publ. 511.

Baumhover, A. H., A. J. Graham, B. A. Bitter, D. E. Hopkins, W. C. New, F. H. Dudley, and R. C. Bushland. 1955. Screwworm control through release of sterilized males. Science 122(3163):287–28.

Baumhover, A. H. 1966. Eradication of the screwworm fly—an agent of myiasis. J. Am. Med. Assoc. Vol. 196, p. 240–248, April 18, 1966.

Brazzel, J. R., T. B. Davich, and L. D. Harris. 1961. A new approach to boll weevil control. J. Econ. Entomol. 54:723–730.

Burton, R. L. 1967. Mass rearing the fall armyworm in the laboratory. ARS 33-117, U.S. Department of Agriculture.

Burton, R. L. 1969. Mass rearing the corn earworm in the laboratory. ARS 33-134, Agric. Res. Service, USDA.

Bushland, R. C., and D. E. Hopkins. 1951. Experiments with screwworm flies sterilized by x-rays. J. Econ. Entomol. 44:725–731.

Butt, B. A., J. F. Howell, H. R. Moffit, and A. E. Cliff. Area-wide sanitation and chemical suppression of codling moth populations in preparation for a sterile moth release program. In press.

Coppedge, J. R., D. A. Lindquist, R. L. Ridgeway, C. B. Cowan, and L. A. Bariola. 1969. Side dress applications of Union Carbide U.C.-21149 for control of overwintered boll weevils. J. Econ. Entomol. 62(3):558–565.

Dame, David A. and Claude H. Schmidt. 1970. The sterile-male technique against tsetse flies, *Glossina* spp. Bull. Entomol. Soc. Am. (16)1:24–30.

Gast, R. T. 1968. Mass rearing of insects: Its concepts, methods, and problems. *In* Radiation, radioisotopes, and rearing methods in the control of insect pests. Proceedings of Panel, Tel Aviv. Joint FAO/IAEA Division of Atomic Energy in Food and Agriculture.

Graham, H. M., M. T. Ouye, R. D. Garcia, H. H. De La Rosa. Sterilizing dosages of gamma radiation for pink bollworm adults. In press.

Henneberry, T. J., and A. N. Kishaba. 1966. Cabbage loopers, p. 461–478. *In* Insect colonization and mass production. C. N. Smith (ed.). Academic Press, New York.

Klassen, W., J. F. Norland, and A. B. Borkovec. 1968. Potential chemo-sterilants for boll weevils. J. Econ. Entomol. 61(2):401–407.

Klassen, W., E. F. Knipling, and J. U. McGuire, Jr. 1970a. The potential for insect population suppression by dominant conditional lethal traits. Annals Entomol. Soc. Am. 63(1):238–255.

Klassen, W., J. F. Creech, and R. A. Bell. 1970b. The potential for genetic suppression of insect populations by their adaptions to climate. Misc. Publ. No. 1178, Agric. Res. Serv., USDA.

Klassen, W., and N. W. Earle. 1970. Permanent sterility induced in boll weevils with busulfan without reducing production of pheromone. J. Econ. Entomol. 63(4):1195–1198.

Knipling, E. F. 1964a. The sterility principle of insect population control. Pesticides Abstracts and News Summary, Sect. A, 1964. 10:587–603.

Knipling, E. F. 1964b. The potential role of the sterility method for insect population control with special reference to combining this method with conventional methods. ARS 33-98, USDA, Nov. 1964.

Knipling, E. F. 1968. Technically feasible approaches to boll weevil eradication. Summary Proceedings 1968 Beltwide Cotton Production-Mechanization Conf., Hot Springs, Ark. Jan. 11–12. Nat. Cotton Council of Am., Memphis, Tenn.

Knipling, E. F. 1970. Suppression of pest Lepidoptera by releasing partially sterile males—A theoretical appraisal. BioScience, p. 465–470, April 15, 1970.

Knipling, E. F. Use of population models to appraise the role of parasites of *Heliothis* larvae in suppressing *Heliothis* populations. USDA Tech. Bull. 1434.

Mangum, Charles L., W. O. Ridgway, and J. R. Brazzel. 1969. Large scale laboratory production of the pink bollworm for sterilization programs. ARS 8-35.

Raulston, J. R., and T. N. Shaver. 1970. A low agar casein wheat germ diet for rearing tobacco budworms. J. Econ. Entomol. 63(6):1743–1744.

Nash, T. A. M., A. M. Jordan, and J. A. Boyle. 1968. The large scale rearing of *Glossina austini* Newst. in the laboratory. IV—The final technique. Ann. Trop. Med. Parasit. 62:336–341.

North, D. T., and G. G. Holt. 1968. Inherited sterility in progeny of irradiated male cabbage loopers. J. Econ. Entomol. 61(4):928–931.

Proshold, F. I., and J. A. Bartell. Inherited sterility in progeny of irradiated male tobacco budworms. J. Econ. Entomol. In press.

Tumlinson, J. H., D. D. Hardee, R. C. Gueldner, A. C. Thompson, P. A. Hedin, and J. P. Menyard. 1969. Sex pheromones produced by boll weevil: Isolation, identification, and synthesis. Science 166:1010–1012.

Biological Control: Parasites and Predators

Richard L. Doutt

Biological control is the ingenious use of parasites and predators to suppress pest populations. It is a strategy designed to control a pest population in a way that is remarkably effective and especially consistent with the maintenance of environmental quality.

This initial concept of biological control should be carefully perpetuated and used in its pure form by the planners of future pest control strategies. The term should not be extended, as is currently the fashion, to embrace autocidal techniques, the use of sterile males, and nonchemical methods in general. This has the damaging effect of obscuring the unique functional and ecological basis of biological control. As a result the enormously powerful natural forces available for exploitation may not be adequately recognized by pest control planners and decision makers.

Biological control is seldom given the emphasis which it deserves. It is often categorized as an alternative method when its impressive credentials qualify it for primary consideration as the best foundation upon which to build environmentally safe, economic, and durable pest control programs.

Biological control has been widely tested during the past eight decades and has been spectacularly successful in a tremendous array of diverse situations. It has succeeded in both temperate and tropical climates; on continents, sub-continents, and islands; in northern latitudes and in regions south of the equator; in forests and in agricultural plantings. It has suppressed target pests of great taxonomic diversity;

it has worked well in depauperate biotas and in those of great complexity; it has been a boon to primitive agriculture and has reached pinnacles of economic success in modern, progressive, intensified agriculture. It has been used where chemical control measures were not feasible, and it has been used harmoniously with pesticide applications. Most significantly, it has worked everywhere that man has persevered in its use.

Biological control has reportedly been successful against 110 arthropod pests; in addition, various weeds have been controlled by this technique. This is especially remarkable when one considers the existing ratio of expenditure on pesticides and on research into biological control in its widest sense. "At present, in the non-Communist world, expenditure on biological control research is probably at a level of about 0.3 percent of the funds spent on pesticide research and use. In the light of present knowledge and problems, this ratio of expenditure is totally irrational." (Wilson, 1970)

Biological control should be in the forefront of any pest control strategy designed to be effective without impairment of environmental quality, yet it is not as widely adopted as it should be. This is puzzling until one analyzes why it is classified as an alternative method and is not given first consideration. There are two probable reasons for this. The first is that there seems to be almost a primitive gratification in taking some positive action against an organism that is viewed as an enemy or simply as a nuisance. Certainly, there is a strong inclination to choose an expedient, immediate, and very visible tactic. For this reason, the use of a powerfully toxic pesticide provides enormous satisfaction because it involves physical action and the expenditure of capital that is immediately rewarded by a highly visible chemical coverage and the comfortable imagination of enormous arthropod body counts. By contrast, biological control involves a scientific and intellectual activity, instead of a physical and muscular action. The parasites and predators are frequently microscopic, and the control process proceeds in almost an invisible fashion. Biological control is not flamboyant, noisy, and commercially exploitable. It is silent, ecologically sophisticated, and economically sound.

Through their experience with synthetic organic pesticides, entomologists have demonstrated that agroecosystems are composed of very complicated communities to which one cannot with impunity apply a powerful toxicant. There have been disruptions of ecosystems on every farm, field, and forest in America. The pest species exists not only in a physical environment but in a biotic one as well, and it constitutes only one part of a very complicated web of interacting forces. The application of chemical toxicants has, therefore, demonstrated, sometimes in an astonishing manner, the role of invertebrate components of the community. These are the parasites and predators whose key regulatory action is often invisible to us. In many cases, their role would probably have escaped our attention had it not been for the interfering action of powerful insecticides.

The easy availability of commercial pesticides and the simple procedures for applying them to agricultural crops have made the chemical control tactic very attractive. When disruptions did occur, fundamental reappraisals of programs were not made. Instead, the profession was conditioned to look toward a new, hopefully better, chemical. It is therefore understandable why biological control has commonly been neglected. This almost total preoccupation with chemical control so completely dominated the entomological profession that serious attention to other tactics was generally discouraged.

There is a second reason why biological control is not being as widely and intensely practiced as its merits justify. Casual attempts to undertake biological control are not likely to be successful. Brief, superficial projects have led to disappointments and then to the erroneous conclusion that biological control has extremely limited application. Such unsatisfactory operations are the result of a failure to realize that a minimum organizational structure is essential to biological control.

HISTORY OF BIOLOGICAL CONTROL

Biological control was conceived early in the nineteenth century when certain Europeans such as Erasmus Darwin in England and Vincent Kollär in Austria emphasized the particularly important role of entomophagous organisms in the economy of nature. This belief was shared by American entomologists who, because of their professional involvement in the rapid expansion of a new agriculture on this continent, were able to add dimensions of considerable magnitude to this phenomenon of entomophagy. It was clear to knowledgeable agricultural entomologists, such as Asa Fitch in New York and Benjamin Walsh in Illinois, that many pests in the New World were of foreign origin and that there was a marked contrast in the destructiveness of these alien pests in America compared to their innocuous state in their homeland. They attributed this difference to a lack of effective natural enemies of the pest—parasites and predators—on this continent. This led to specific suggestions for the importation of these parasites and predators, and it set the stage for the first dramatic demonstration of biological control of a key pest on a major commodity. The circumstances of this spectacular control of the cottony cushion scale, *Icerya purchasi* Maskell, are well documented, and the story contains elements of considerable human interest (Doutt, 1958). The fantastic success of this project on citrus in 1889 firmly established this pest control tactic in California. As a result, biological control has been strongly supported in California under conditions of intensified, progressive, and totally modern agriculture. A similar history of events occurred in Hawaii. Biological control has a remarkable record of successes in these two states. This is not because these areas supposedly have salubrious climates or are peculiar "ecological islands" as some ecologists have suggested. It is because these two states have recognized that a permanent organizational structure maintained solely to conduct biological control is a profitable investment in research dollars. The returned benefits are great, and the biological control operation amply repays its support. Estimates on expenditures and economic results of notably successful biological control projects in California during the period of 1923 to 1959 show savings to that state's agricultural industry of $110,990,973. These were the first 36 years that the University of California sponsored biological control research (DeBach, 1964). It is doubted if any investments by the University of California's Division of Agricultural Sciences can show greater returns than those for the biological control of insect and weed pests.

A permanent organization for conducting biological control is absolutely essential. This is because the steps through which each biological control project develops are very fragile. The first step is the foreign exploration for the promising natural enemies in the native home or center of origin of the pest species. This is followed by the careful screening and evaluation of the primary parasites or obligate predators under secure, quarantine conditions. The subsequent culture, colonization, and field evaluation of the candidate species complete the sequential phases of a conventional biological control project. Disciplined attention to details at each step and an intense personal dedication to the success of the project are absolutely essential. Otherwise it is highly probable that the entire project will be aborted, for carelessness at any one of the sequential phases can nullify everything that has preceded it.

There are important concepts that spin off a permanent biological control unit. For example, the concept that certain mortality agents function in a density-dependent manner as contrasted to mortality that is independent of density arose from biological control work. This very terminology, which has been adopted by ecologists throughout the world, came from Professor Harry Scott Smith, who headed biological control in California for many years. The refined concepts of delayed density-dependent action, the functional and numerical responses of predators, and the policy of multiple versus single species introductions are matters of daily concern to entomologists who continually work in biological control. This is not an activity that can be turned off and then turned on again. To be efficient and to achieve continuing successes in biological control there must be sustained effort.

There is no need, however, to provide this permanency of organization through the establishment of vast new facilities staffed by new personnel. Remodeling and reassignments within presently existing organizations would accomplish the same thing. Most entomologists entered their profession from a basic interest in biology and natural phenomena, not from a desire to spend their lives testing whether proprietary product A is slightly better than competitor's product B in causing temporary kill of *Lygus* bug on cotton. The recognition of biological control as a basic ingredient in pest management and the opportunity to use their talents freely in the development of pest management programs would be welcomed by most professional entomologists.

ECOLOGICAL BASIS FOR BIOLOGICAL CONTROL

It is a fundamental ecological principle that, barring any major change in a given environment, the resident insect species maintain a fairly constant quantitative relationship to one another. No single species continually increases or decreases in relation to the others. This is sometimes defined as the balance of nature. For example, if we were to take an annual census of the insects in a relatively stable habitat for 25 years, we would find that year after year some species are always abundant, others less so, and finally some species are so rare that we would consider them to be collector's items. We would find this to be true because each species maintains an equilibrium position in relation to the other species in its environment. At the same time, each species fluctuates positively and negatively from this equilibrium position because of the changing seasons, because of its reproductive cycle, or because of other factors, but these are only short-term departures. The species population always tends to return to its normal level of abundance. This reflects the inexorable, relentless truth that, on the average, only one progeny per parent can survive, otherwise the species would increase to infinity or decrease to extinction. This, then, is the static aspect of populations; it is the seeming paradox of stability in the midst of constant change. It can be compared to sea level, which furnishes us a zero basis for altitude and is one of the standard conditions in the most precise of physical sciences, although the sea is never level but is always in motion by waves or tides. In biological control, we are concerned with both the dynamic aspects of populations as well as the static characteristics. The equilibrium position concept is a very useful one. Intellectually, it is just as easy to adopt as a universal phenomenon among insects as sea level is an acceptable notion to apply to the oceans of the world.

The equilibrium position of each species in its particular environment is determined by the regulatory factors in that habitat. These are the mortality agents, usually biotic, which operate in a density-dependent manner, that is, as the population grows above its equilibrium position the probability of survival of any individual member of that population becomes increasingly less. The empirical work in biological control has demonstrated repeatedly that these regulatory factors acting on insect populations may commonly be entomophagous parasites and predators. These natural enemies tend to regulate the abundance of the host species in a given habitat and are therefore strongly influential in determining the equilibrium position of the host species. The major pest problems of the world are frequently caused by a great displacement in the equilibrium position of the pest species. If the pest has become established on this continent without natural checks that may have maintained it at a much lower population density in its endemic area, then there has been a major upward displacement in its equilibrium position. To regulate the pest at a lower and more tolerable equilibrium position, the attempt is made in biological control to introduce and establish its natural enemies. This is the ecological basis for conventional biological control. It is the basic explanation for the self-regulating, permanent feature of biological control. It also points up the difference between true control of a pest and merely temporary kill. Far too many entomologists equate temporary kill with control. These are very different things and planners of pest control strategies should carefully distinguish between the two effects on a pest population. Chemical toxicants are temporary suppressants. Biological control is usually a regulatory and controlling process.

The use of insects to achieve biological control has been directed at a remarkable diversity of target organisms including snails, noxious plants, and other arthropods. It is no longer novel to think of phytophagous insects as plant parasites fully capable of regulating the density of their host species, for this is precisely what has occurred in the spectacularly successful cases of weed control through the importation and colonization of carefully selected insect species. This emphasizes the basic principle of biological control which is to use organisms that regulate the population density, regardless of the taxonomic category of the target species. Noxious plants and noxious arthropods are therefore susceptible to control through this fundamental strategy because the same ecological principles apply.

Some of the most exciting results in biological control have been in the control of weed pests by imported phytophagous insects. In California, for example, Klamath weed has been cleared from thousands of acres of rangeland by imported chrysomelid beetles, and the plant now exists only as an occasional roadside inhabitant or a plant that grows on marginal areas of clearings. This example, coupled with many others throughout the world, provides remarkable evidence of the role that insects can play in determining the quantitative composition of the flora. Such examples of complete and enduring control of noxious pests, both plant and animal, reveal the enormously powerful natural forces available for exploitation by the strategists in future pest control programs.

APPLICATIONS OF BIOLOGICAL CONTROL

There are three major applications of biological control. One is the importation and colonization of parasites and predators against alien pests. There is a strong biological basis for reasoning that this procedure can also be used to suppress native weeds such as poison oak. The University of California's Division of Biological Control at Berkeley is presently involved in this particular project against poison oak, which is a public health problem in recreational areas. Native insect pests may also be controlled by importation of natural enemies. For example, the Levuana moth in Fiji was completely controlled by the introduction of parasites obtained from an ecologically similar but taxonomically distinct moth in Malaya.

This technique of using imported parasites and predators could be greatly expanded. It has not been adequately tested even against some of our major pests. For example, everyone knows that the codling moth is an alien, introduced, and major pest. Throughout the world there are approximately 120 species of parasites and 22 additional predators that are recorded as attacking the codling moth. Discounting some records that may be questionable or erroneous, this is still a very impressive total. Despite this tremendous list only two species of the codling moth parasites have been imported into the United States, and neither of these came from the area presumed to be the native home of the codling moth (Clausen, 1956). Therefore, in any consideration of pest control strategies for the future a high priority should be given to the increased importation and testing of parasites and predators of all major pests, both arthropods and weeds.

A second method of application of biological control is the mass culture and periodic colonization of parasites or predators. This has been used in a number of situations with varying degrees of success. It has enormous potentiality for use in closed systems such as greenhouses. A particularly outstanding example of the value of periodic colonization is the work of Howard B. Lorbeer and his insectary staff for the Fillmore Citrus Protective District in California. In his annual report for 1970 Mr. Lorbeer has shown that releases of *Metaphycus helvolus* (Compere), a parasite of the black scale, *Saissetia oleae* (Bernard), and releases of *Aphytis melinus* (DeBach), a parasite of the citrus red scale, *Aonidiella aurantii* (Maskell), reduced these two major citrus pests to a minor status throughout the district. From a total of 8,423 acres of citrus only 20 acres were sprayed for black scale and only 90 acres for citrus red scale. This is only 1.3 percent of the total acreage. The cost of the parasites was only $8 per acre. Planners of future pest control programs would do well to take note of this excellent example of the periodic colonization technique.

A third application of biological control is the conservation and augmentation of parasites and predators. As control techniques become better integrated into unified pest management schemes there will be an expansion of efforts to conserve and augment the existing natural enemies of crop pests. This can be accomplished through environmental manipulations that can be fitted into regular farm operations. The strip cutting of alfalfa, the judicious use of cover crops or sprinklers to change microclimates, and the establishment of refuges for beneficial insects are only a few of the examples of recent research in this direction. Artificial food supplements for parasites and predators are also being tested in the field. As the efficacy of the existing natural enemies is increased by such techniques the pesticide load can be correspondingly reduced. An indication of such benefits is suggested by the reduced expenditures for treatment of grape leafhoppers in California vineyards. As growers became more aware of the role of *Anagrus epos* Girault, the effective parasite of leafhopper eggs and began to use this knowledge in conjunction with a treatment action level based on nymphal leafhopper populations, there was a drop in pesticide application. Actual expenditures for grape leafhopper control dropped from a high of $3,383,982 in 1961 to $576,000 in 1969, the last season for which figures are available (Hawthorne, 1970). This reduction of over $2.8 million in annual costs cannot be attributed to the egg parasite so much as to a change of attitude among the growers. The conservation and augmentation of natural enemies is an important component in these pest management programs.

SUMMARY

Biological control, which is the ingenious and resourceful use of parasites and predators to suppress pest populations, should be given primary consideration in building safe, economical, and enduring pest control programs consistent with the maintenance of environmental quality. Its impressive record on many diverse pests throughout the world indicates that it should not be considered as merely a secondary, alternative method.

Biological control requires a particular organizational dimension that must be recognized by planners of future pest control strategies. This organizational structure is imperative if the maximum benefits are to be realized, but this can be provided by internal adjustments, by retooling, and by a realignment of priorities in existing governmental agencies.

Biological control has a unique ecological basis that imparts a high degree of permanency. It is based on the use of regulatory factors that really control a population instead of merely reducing it temporarily. The parasites and predators are density-dependent in action and can be used to regulate such diverse pests as arthropods and noxious weeds.

There are three major thrusts of biological control which cover the importation, colonization, conservation, augmentation, and periodic colonization of parasites and predators. These have enormous economic advantages either by themselves or integrated into pest management programs. The spectacular accomplishments of biological control in the past are but a mere prelude to the future of this still largely unexploited control method.

REFERENCES

Clausen, C. P. 1956. Biological control of insect pests in the continental United States. USDA Agr. Tech. Bull. 1939.

DeBach, P. 1964. Biological control of insect pests and weeds. Reinhold Publ. Corp., New York.

Doutt, R. L. 1958. Vice, virtue and the vedalia. Bull. Entomol. Soc. Am. 4:161–182.

Hawthorne, R. 1970. Estimated damage and crop loss caused by insect/mite pests–1969. Calif. Dep. Agr., mimeo E-82-12.

Wilson, F. 1970. Biology to combat pestilence. New Scientist, p. 72–74, October 8, 1970.

Insect Control by Microbial Agents

A. M. Heimpel

Insects, like all other animals, fall prey to infectious diseases caused by microorganisms of all types including bacteria, viruses, rickettsia, fungi, protozoa, etc. According to Steinhaus (1956), man has been aware of this fact since the time of Aristotle. Most of these early observations were made on beneficial insects such as bees and the silkworm; however, insect diseases in nature are so spectacular that man must have noted them in the past, even if he did not record them.

During the nineteenth and twentieth centuries, serious studies of the diseases of insects began to appear in the literature. There were no insect pathologists *per se* in those days and these studies were carried out by scientists scattered the world over and from a variety of backgrounds—from gentlemen "of independent means" indulging themselves in a hobby, such as A. Bassi (1835), to a biochemist named Louis Pasteur.

A Russian microbiologist, Eli Metchnikoff, was one of the first to suggest the use of insect diseases to control noxious insects (1880). He, together with his protégé Krassilstschick, actually produced a wide-spectrum fungus disease, *Metarrhizium anisopliae* (Metch.) Sor. on something resembling beer mash and used the harvested spores from this culture to successfully control the sugar beet curculio, *Cleonus punctiventris* Germ.

The Japanese have contributed much to our knowledge about the diseases of the silkworm, and the European Cornborer Commission coordinated experiments on control of *Ostrinia nubilalis* (Hübner) between several countries including Canada, France, and Hungary. One of these scientists, A. Paillot (1933), wrote one of the first comprehensive books on insect pathology entitled *L'infection chez les insectes.* Many other scientists added, piecemeal, to our general knowledge on this subject. In this country several extremely competent and dedicated scientists studied insect diseases and the possibility of microbial control of insects. These scientists included Snow, Forbes, Fawcett, Berger, Watson, White, and Glaser.

However, the main thrust in this field was initiated by the late Edward A. Steinhaus, whose efforts literally created an organized field of endeavor. In 1947, Steinhaus collected all of the scattered references to insect diseases in one volume entitled *Insect Microbiology.* Two years later (1949) he wrote a textbook, *Principles of Insect Pathology*, to aid in teaching the first *bona fide* course in insect pathology given in the world. Today, laboratories studying microbial control and insect pathology throughout the world are populated by Steinhaus' students.

Dutky carried out his studies on the milky disease of the Japanese beetle between the years 1934 to 1954. In 1947, Steinhaus began his work in the University of California, Berkeley, and Dr. J. J. De Gryse, in Canada, created a laboratory for the study of insect pathology in Sault Sainte Marie, Canada. In 1957, the Insect Pathology Pioneering Research Laboratory at Beltsville was created and began its work. Other laboratories sprang up throughout the world, as work intensified in this new field.

Tremendous strides have been made in developing microbial agents by a very small group of scientists and industrialists, mainly in the United States.

The research emanating from these centers of study led to establishing the basic principles of microbial control, which this paper will describe.

REQUIREMENTS FOR A MICROBIAL CONTROL AGENT

Efficacy An organism to control insects should be highly effective in killing the insect and it should do so quickly, especially when applied to rapidly growing food and fiber crops. In forest treatments, control of the insect is essential before damage to the tree causes loss of vigor.

Specificity The ideal microbial control agent should not be harmful for any other life form. It should attack one species only, or a group of insects, such as an order. Indeed this is a requirement insisted upon by most regulatory agencies in governments throughout the world.

Ease of Production The organism should be easily produced, and the cost of production should remain within reasonable limits.

Storage The organism should have, in its life cycle, some resting stage or stage resistant to adverse conditions in the environment that would permit it to be stored for months (preferably, years) without loss of viability or virulence.

Formulation The control agent should be amenable to formulation in any of the forms used for chemical insecticides, and should be readily disseminated by the available spraying or dusting equipment.

COVERAGE AND TIMING

Except for nematodes and fungi, which usually attack the insect externally, microbial control agents must be eaten by the insect. In all cases, it is essential to success that complete coverage of the plant to be protected is maintained. The target insect should be destroyed as soon as possible after hatching for the following reasons. In the first place, the first instar larvae are usually more susceptible to infection by most pathogens than are the later instars. Second, the insect should be removed before it achieves a size capable of causing economic damage to the plant. The timing of applications can be communicated in a variety of ways—the proper time to spray with a microbial agent can be announced on radio and TV, much the same as a weather report. Universities and state extension personnel could supply the proper information to the appropriate sources of information. Indeed, the companies that supply the materials might help supply this information.

THE TARGET INSECT

It is of the utmost importance that every facet of the target insect's habits, life cycle, and development be understood as completely as possible. For example, an insect may develop in the bud of a plant, or it may burrow into a leaf and mine it. A combination of both habits may occur so that the larva (usually in its younger stages) is exposed only very briefly to externally applied sprays. Angus (personal communication) attempted to control two insects in the summer of 1970, using *Bacillus thuringiensis* Berliner commercial products. Experiments on the spruce budworm, *Choristoneura fumiferana* (Clem.), on a 4-day treatment basis, from April 30 to May 7, yielded a mortality of 62 percent, increasing to 92 percent after May 7 and then dropping off gradually. The habits of the insect protected it during part of this program. On the other hand, three sprays were applied to Christmas-tree-size Scotch pine, to control the jack-pine budworm, *Choristoneura pinus* Freem. The insects were mainly in their 5th instar, starting around the 23rd of June. Almost 100 percent of the insects were killed during this test. This is the result of applying a microbial agent at a time of maximum exposure of the insect (regardless of larval age), and it also demonstrates the effectiveness of the products when good coverage is obtained and when killing doses are maintained.

METHODS OF CONTROL

There are two possible ways of carrying out control of insects with microbial agents: (a) There are a group of organisms that remain viable in the biocoenosis in sufficient quantity to ensure the death of the target insect should it attempt to develop in the contaminated area. These are termed "permanent control agents." There are a few excellent examples of this type of control. The virus disease of the European spruce sawfly, *Gilpinia hercyniae* (Htg.) virtually eliminated an introduced, burgeoning population of that insect in New Brunswick, Canada, from 1939 to 1942. The virus was detected in the United States by Dowden (1940), and the same level of control was demonstrated in this country.

Another famous example is the sporeformer *Bacillus popilliae* Dutky, the causal organism of the milky disease of the Japanese beetle, *Popilliae japonica* Newm. Dutky's research resulted in a combined federal–state control program during the years 1937 to 1948 that involved the dissemination of 184,000 lb of *B. popilliae* spore mixture on 194,000 sites in 14 states on the eastern seaboard and the government grounds in the District of Columbia. Permanent control was effected in most of the areas treated. This bacterium is now produced commercially but has been forbidden for use on pastures by the FDA pending further safety tests.

It might be apropos to point out that the permanent-control organism is amenable to government-sponsored control programs. It is quite within the realm of a federal or state responsibility to commit funds for a once-in-a-lifetime operation that would ensure the control of the population of a specific pest below its economic threshold level.

The majority of disease organisms available for microbial control do not, for one reason or another, effect permanent control. They must be applied annually to the crop in much the same way as do the chemicals. Into this class fall the control products based on *Bacillus thuringiensis* and the majority of viruses, some of the fungi, and the *Microsporidia* proposed for insect control.

Except for *B. thuringiensis* and some of the *Microsporidia*, the viruses and fungi remain in the environment and usually cause extensive mortality of their insect hosts annually. Unfortunately, these control agents do not reach the insects (from the soil or from foci of chronically infected insects) in sufficient quantity to cause significant mortality before the insects cause heavy damage to the crop, be it forest trees, or cabbage and spinach. This form of microbial control, then, is merely a process of producing an insect disease agent, isolated from the field, and spraying it on the plant as soon as possible after hatching to insure mortality of the insect, thus protecting the crop from economic damage.

This latter form of microbial control, employing multiple, annual treatments requires extensive and continuous production of the microbial agent. This is a task that is outside the purpose of government agencies and requires the interest and action from industry.

This, then, seems to be the winning pattern. Government research brings forth a new idea to the limits of its resources, and industry takes over and develops the idea into a reality. Yet the system is not flawless. To industry, time is money, and, in this field, industrial development of microbial agents has almost invariably outstripped the available and often necessary background information from basic research. A classical example of this lies in the story of the development of *Bacillus thuringiensis* as a commercial insect control agent.

J. M. Sudarsky, President of the Pacific Yeast Products Co. (PYPCO) met with Dr. E. A. Steinhaus and Dr. R. A. Fisher in 1956 to discuss the feasibility of producing *Bacillus thuringiensis* as a microbial control agent. PYPCO, later to become Bioferm Corporation, was eventually absorbed into the International Minerals and Chemical Co. (IMC) of Chicago, Illinois.

A remarkably short time after this meeting, the production of pilot-plant lots of *Bacillus thuringiensis* became available for field testing. Unfortunately, the company concentrated on spore production and this eventually led to the acceptance of the spore-count as a means of standardization of these products by the Pesticide Regulation Division, ARS, USDA. This was the first mistake. In 1954, Angus showed there was a toxic principle produced at the time of sporulation; one year previous, Hannay (1953) had demonstrated the presence of a diamond-shaped crystal produced along with the spore in the sporangium, and by 1956 there was sufficient data to show that the action of *B. thuringiensis* against lepidopterous larvae was from a toxic agent incorporated in this crystal. At this time, none of the companies involved, Bioferm Corporation, Merck Sharpe & Dohme, Rohm & Haas Co., Nutrilite Products, Inc., and Grain Processing Co., paid any real attention to this revelation. Neither did the Pesticide Regulation Division. This was most unfortunate, because, from 1956 to 1960 (when labels for products based on *B. thuringiensis* were granted to the various companies) the emphasis was on spore production and, as a consequence, many lots of *Bacillus thuringiensis* failed miserably. This not only reflected unfavorably on the general product, but it discouraged consumers who tried this product in good faith.

Two men were involved in rectifying this situation: Dr. Martin Rogoff, who joined Bioferm Corporation in 1961, and Dr. Howard T. Dulmage, who took the post of director of research at Nutrilite Products, Inc. Both men, good scientists, realized that the crystal toxin, the *B. T.* δ-endotoxin, was the primary agent against Lepidoptera and worked for maximum production of this toxic protein for insects. Almost immediately, the products made available by Nutrilite Products, Inc. and IMC (Bioferm Corp.) began to improve. By 1965, both products were dependable and in fairly wide use, especially in California. Both these scientists, through research determined that special nutritional requirements were necessary to produce the maximum toxicity, and determined a good fermentation process. Dulmage, who had more ideas on this subject, left Nutrilite Products, Inc., and joined the Entomology Research Division, ARS,

USDA, where he continued his research. He soon developed a new process for producing *Bacillus thuringiensis* and applied this to a selected strain of *Bacillus thuringiensis* var. *alesti*, serotype 3 ab Kurstaki, that has proved to be exceedingly toxic for lepidopterous insects. This strain was further developed, by Amdal of Abbott Laboratories, IMC, and by Nutrilite Products, Inc. The final product is, to date, the ultimate in a reliable, effective microbial control agent.

This, then, is another principle essential to microbial control. An organism is a pathogen for an insect because it has some special attribute, a toxin, a special enzyme for invasion, or an ability to grow vigorously in the gut contents or in the hemocoel. No attempt should be made to develop a microbial pathogen without studying the microorganism as carefully as the host. The mode of action and its special capabilities should be known in intimate detail before production commences.

The use of these products based on *B. thuringiensis* the world over in the calendar year 1970 was about 700,000-acre treatments, making a total of 245,000 acres treated at a cost of approximately $3,150,000 (my estimate). The market is expected to grow. The crops treated were, in order of importance, lettuce, cole crops, tobacco, greens, collards, spinach, and ornamental trees and shrubs. Foreign markets are developing in southeast Asia, for protection of cabbage, and in the Mediterranean, to protect prickly pine. All of the ornamental trees in the city of Amsterdam, Holland, are protected annually by applications of *Bacillus thuringiensis*. In the United States the products were used in the following states: California, Arizona, Florida, New Jersey, New York, Texas, Georgia, Wisconsin, South Carolina, New Mexico, Michigan, Delaware, Maryland, Ohio, North Carolina, and Arkansas.

It is indeed a pity that adequate funds and time were not available to conduct the necessary research before these products were first committed to the market. It is gratifying that both industrial and governmental scientists, working in tandem, solved most of the problems before the *Bacillus thuringiensis* project could die. As a result, we now have a formidable foe of the noxious lepidopteran insect in our arsenal of weapons. The industrial concerns involved are to be commended on their foresight and perseverance.

The question is, Did we learn anything from this on the necessity of preparedness? The answer is, probably, no. We are still far behind in basic information necessary to support the knowledgeable use of other potentially useful microbial agents for insect control.

VIRUSES

Let us turn to another group of microorganisms that show great promise, the insect viruses. To date, only four types of insect viruses have been extensively tested in the field. These are the nuclear polyhedrosis viruses (NPV), the Cytoplasmic polyhedrosis viruses (CPV), the granulosis viruses (GV), and the noninclusion viruses (NIV). Table 1 lists these along with some particulars on their size and morphology. During the past two decades there have been many tests of insect viruses as control agents. The NPV viruses were used predominately, while lesser tests were carried out using the CPV, the GV, and one NIV (the citrus red mite studies by ARS personnel on the West Coast). Examples of successful field tests may be listed as follows: Steinhaus and Thompson (1949) and Thompson and Steinhaus (1950) demonstrated good control of the alfalfa caterpillar *Colias eurytheme* Boisduval using the specific NPV. Other virus field tests included the wattle bagworm, *Kotochalia junodi* (Heyloerts) (Ossowski, 1957); the pine processionary moth, *Thaumetopoea pityocampa* (Denis and Schifermuller), (Grison *et al.*, 1959); the imported cabbage worm *Pieris rapae* (Linnaeus), (Tanada, 1956a, 1956b); the European cabbage worm *Pieris brassicae* (Linnaeus), (Billioti *et al.*, 1956); the cabbage looper *Trichoplusia ni* (Hübner), (McEwen and Hervey, 1959, and Semel, 1961); the cotton leafworm, *Spodoptera litoralis* (Boisduval), (Abdul-Nasr, 1959); the cotton bollworm *Heliothis zea* (Boddie), (Ignoffo *et al.*, 1965); the Great Basin tent caterpillar *Malacosoma fragile* (Stretch), (Clark and Thompson, 1954); the Douglas-fir tussock moth *Hemerocampa pseudotsugata* (McDunnough) (C. G. Thompson, personal communication).

There were several attempts to use insect viruses to control insects that have not succeeded. In some of these cases, bioassay of the virus prior to testing would have shown an organism of low virulence. In all these studies, selection for a virulent strain would have tendered much improved results. However, other factors enter the picture to cause failures. In the case of the gypsy moth tests with the specific virus (and with *Bacillus thuringiensis*), the cause for border-line results illustrates some of the principles of microbial control so well that it is too much of a temptation not to review the gypsy moth problem.

In the first place, the American strain of nuclear polyhedrosis virus specific for the gypsy moth appears to be of low virulence; killing doses of virus can be achieved but the amount of virus necessary

TABLE 1 Types of Insect Viruses Used in Insect Control

Type	Polyhedral Size[a]	Viral Particle Size	Predominant Nucleic Acid
Nuclear polyhedrosis virus (NPV)	0.3–15μ	20–50mμ × 200–400mμ; rod-shaped	DNA
Cytoplasmic polyhedrosis virus (CPV)	0.5–23μ	30–40mμ; spherical	RNA
Granulosis virus (GV)	119–300mμ × 300–600mμ	36–80mμ × 200–412mμ	DNA
Noninclusion virus (NIV)		30–65mμ; spherical	DNA

[a]Species and host-dependent.

is prohibitive. Scientists in the Forestry Service, USDA, are striving to overcome this problem by selection of virulent mutants from the native population and they are also testing the relative virulence of strains isolated from the gypsy moth in Japan, Spain, and Yugoslavia.

Aside from this question of virulence, there are the very important questions of coverage and maintenance of lethal doses on the trees. In all of the experiments to control the gypsy moth in the northeastern United States, only one or two applications of virus was made. Since this insect hatches over a period of 30 days or more in the field, and remains in the field as a larva for another 30 days, these applications were inadequate to maintain a lethal dose on the plant. Furthermore, the very nature of some of the hardwood forests treated makes food coverage impossible from the air using the available airplane spray equipment. Again the tendency is to label the virus or the bacterium as inadequate. In reality, one should blame lack of funds and time to carry out studies sufficient to rectify these flaws in our techniques. Control of the gypsy moth is possible, using microbial control agents. In the near future I personally expect to see great development in this area.

METHODS OF MANUFACTURE

Insect viruses are fastidious organisms that can only be grown in living insect tissue, consequently their manufacture up till now requires the mass-rearing of insects for their culture in the larvae of the host. Research over the last 25 years, mostly in the United States, has developed adequate artificial media for many Lepidoptera, some phytophagous Hymenoptera, some Diptera, and a few Coleoptera.

To digress slightly for a minute, the field of insect tissue culture has recently burgeoned. However, it was never possible, previously, to infect insect tissue culture cell lines with the inclusion body viruses; only the primary tissue culture cells were susceptible to virus infection. Since the source of primary tissue culture is so small (namely the ovaries from the full-grown larvae or pupae), there is no possibility of manufacturing insect viruses in primary cultures. Recently, Goodwin *et al.* (1970), working with the line of ovarian tissue from the fall armyworm *Spodoptera frugiperda* (J. E. Smith) developed by Vaughn (personal communication) have proven that, with proper manipulation of the nutriments in the tissue culture medium, infection by the specific nuclear polyhedrosis virus for the fall armyworm is possible. This definitely opens the door to the manufacture of insect viruses in tissue culture. However, a great deal of basic information must still be obtained.

Let us continue with a brief description of virus manufacture. Insect eggs are sterilized and placed in individual containers (if the insect is cannibalistic) or in groups in larger containers; they hatch, and the larvae feed until an appropriate size. The media is then sprayed with a precise dose of virus, and the insects contract the disease. Just before they begin to die of virus disease, they are harvested (this prevents bacterial growth which occurs rapidly after death) and are blended in water, filtered through a coarse filter to remove mouth parts and fragments of integument and then the suspension of virus polyhedra is dried or collected in a variety of ways. Careful checks for contaminating, undesirable biotypes are always made and animal tests by intraperitoneal injection are carried out. The yield from a mature *Heliothis zea* larva or a *Trichoplusia ni* larva should be in the vicinity of 6×10^9 polyhedra; this figure was thus selected as a Larval Equivalent (or 6×10^9 polyhedra). The Larval Equivalent can also be stated as dry weight of polyhedra, since 1×10^9 polyhedra of the above mentioned virus species, when scrupulously cleaned, weighed almost exactly 1 mg.

In the United States, when a living pathogen is proposed for microbial control and application is made to the regulatory agencies [now the Environmental Protection Agency (EPA)], there are usually four basic requirements necessary for registration and labeling:

1. A method of production of the microbial agent with demonstration of adequate techniques for standardization and uniformity of the product is required.
2. Evidence must be given that undesirable biotypes are nonexistent in the product.
3. The efficacy of the product must be demonstrated through data from thorough field testing against the specific host or hosts on all plants in several parts of the country.
4. The safety of the product for other life forms, including plants, must be established by applying a protocol of tests that has the approval of the regulatory agency.

A considerable amount of NPV specific for the larvae of the cotton bollworm *H. zea* was produced in the early sixties by IMC and by Nutrilite Products, Inc. The virus was formulated and was submitted for field trials. Extensive tests for safety were also carried out (Heimpel, 1967). Applications for registration of the *H. zea* nuclear polyhedrosis virus products to the Pesticide Regulation Division were made by both companies about 1965. Food and Drug Administration granted a temporary exemption from tolerance on December 2, 1970, and PRD granted a temporary label for extensive testing of these virus products during the 1971 season. If the products prove efficacious on cotton, registration should follow. Development of other viruses previously mentioned should follow.

STANDARDIZATION OF VIRUSES

In 1964, the Organization Internationale de Lutte Biologique (OILB) held a symposium on the Identification and Assay of Viruses and *Bacillus thuringiensis* Berliner used for Insect Control.

It was resolved that the polyhedrosis viruses and granulosis viruses be reasonably accurately standardized by (a) counting the polyhedra (or granules) microscopically; (b) bioassay of the virus using healthy insects reared, if possible, on clean (preferably sterile) semiartificial media.

Polyhedra can be readily counted by use of a hemocytometer. Bioassays can be carried out in containers partially filled with artificial media after the virus inclusion bodies have been either mixed with the media, or dispensed, as a suspension, on the surface of the medium. The dose is usually expressed as number of inclusion bodies per area-unit of media surface, or per volume of media. Thus the number of polyhedra required to kill a certain percentage (LD_{50} preferred) of the test population in a given time (usually 5 to 6 days) compared to a standard virus preparation is cited as an activity rating.

REMARKS ON THE USE OF VIRUSES

All of the principles previously mentioned apply to the use of viruses. Coverage, timing, and maintenance of lethal doses on the plant are very important.

Viruses are usually so specific that other pest insect species on the protected plant are not killed and may damage the crop. Consequently, a broad spectrum microbial agent such as *B. thuringiensis* might profitably be used in combination with the specific virus. In some cases viruses might be supplemented with specific, quick-breakdown chemical insecticides in order to kill sucking insects or other pests on the plant.

Since the virus is a living organism, it can be killed by ultraviolet radiation and heat. An understanding of the special requirements of a living control system is essential. Virus sprays or dust should be applied in the late afternoon when a negative thermocline would aid in covering the plant and when there is little chance of heat and ultraviolet radiation kill for several hours, during which time the feeding insect can ingest a lethal dose.

BACTERIA

The majority of bacteria suitable for microbial control are sporeformers. Nonsporeforming bacteria are very difficult to store and lose their viability very rapidly in the field. Consequently, the family Bacillaceae contains most of the microbial control agents. This family is composed of two genera, *Bacillus* and *Clostridium.*

It is extremely debatable whether the organisms causing milky disease in scarabaeid larvae actually belong in the genus *Bacillus.* They have all of the attributes of the genus *Clostridium.* This refers to the bacteria now known as *Bacillus popilliae* Dutky, *Bacillus fribourgensis* Wille, *Bacillus lentimorbus* Dutky, *Bacillus popilliae* var. *australis* Beard, and *Bacillus enloomarahae* Beard. Two Clostridia have been named by Bucher, namely *Clostridium brevifaciens* Bucher and *Clostridium malacosomae* Bucher, both pathogens of the tent caterpillars, Lasiocampidae. Both organisms are highly fastidious and are very difficult to culture; sporulation is almost impossible to achieve outside of the insect gut.

Bacillus thuringiensis var. *thuringiensis* Berliner (serological type 1 Berliner) is now considered by most insect pathologists as the type-species of the crystalliferous bacteria. There is some controversy among the pure taxonomists and academicians as to the connection between this bacterium and the closely related *Bacillus cereus* Fr. & Fr.; however, the debate is no more heated than that between the same taxonomists and the veterinarians who wanted to distinguish *Bacillus anthracis* from *B. cereus.* Twelve serotypes of *Bacillus thuringiensis* have been isolated to date and strains are now so numerous and diversified that products can be virtually "tailored" to take care of any susceptible insect species specifically and most efficiently, if the need arises.

PROTOZOA

The Protozoa and in particular the Microsporidia kill enormous numbers of insects annually, or they reduce the fecundity of adults and attenuate the larval life of insects, keeping them in the field exposed to parasites and predators far longer than is normal. Most protozoa do not kill decisively, rather they kill or damage the insect in an insidious way and are often found associated with viruses and other disease organisms in nature. It is not expected that the protozoa will be used in the near future for insect control. Furthermore, extensive testing for safety will always be required, since certain Microsporidia can cause disease in fish and mammals including man, and any protozoan recommended for dispersal as an insect control agent would have to be vigorously tested for safety to vertebrates.

RICKETTSIA

Several species of Rickettsia have been reported as killing insects. However, it is not anticipated that they will be seriously considered for microbial control for two reasons: (1) They require a long time to kill the host. Normally the prodromal phase of the disease may last from 14 days to several weeks; this lack of decisive control defeats the purpose of their use. (2) Isolates of these insect pathogens grow readily in mammalian tissue culture and cause death when injected, intracerebrally, into mice. It would be very improper, as a consequence, to propose their distribution in the field, because of potential hazards to other biotypes.

FUNGI

A great number of attempts have been made to control insects by disseminating fungal spores. In nature, tons of insects are killed each year in this country by fungal pathogens, and there is no doubt that fungi are effective pathogens of insects. However, fungi are highly dependent on the local weather, and even more dependent on the microenvironment in the immediate vicinity of the target insect. For example, fungi take a fearful toll of insects that dwell, for part of their life cycle, in the soil. However, I have personally spent three miserable summers in the larch swamps of northern Canada, spraying the larch sawfly *Pristiphora erichsonii* (Htg.) without increasing the mortality due to fungus more than a maximum of 3 percent. Perhaps this is due to a gross misunderstanding of the conditions required to use these pathogens effectively. Reports from the Soviet Union and from Austria give some encouragement supporting their use; however, fungi used in this country have consistently failed until recently. McCoy *et al.* (1971) have recently described a fungus *Hirsutella thompsonii* Fisher used against the citrus rust mite and have used considerable ingenuity in disseminating this organism.

Instead of culturing spore preparations, they prepared and stored suspensions of hyphae, the vegetative stage of the fungus. These preparations, when sprayed on orange tree foliage, with nutrient incorporated in the spray, caused the fungus to sporulate on the leaf and the fruit, where it is picked up by the mite traversing these contaminated areas. This is an entirely new concept of distribution of a fungal agent, and has shown considerable promise in field tests to date.

STANDARDIZATION AND APPLICATION

Nonfastidious fungi, which grow readily on artificial media, can be standardized in much the same manner as sporeforming bacteria. Spore counts and bioassay are adequate. Some of the fastidious fungi, such as the Entomophthoraceae, are difficult to assay because of the inability to determine viability of the spores, except by bioassay. These fastidious fungi must be reproduced on or in the living host. However, many species of entomogenous fungi can be propagated on artificial media in very large amounts and are therefore as readily obtainable as the insect viruses.

Fungi can be disseminated as water sprays or spore-dust preparations. Considerable success has been reported in the use of the fungus *Beauveria bassiana* spores with very low doses of DDT in the Soviet Union. Recent reports by Ferron (1971), who visited Russia, sound very encouraging concerning control of the Colorado potato beetle.

NEMATODES

Almost 1,500 species of insects have been reported as hosts to nematodes (Nemathelminthes) or round worms. Nematodes usually incapacitate the host if they do not kill it outright.

Obligate and semiparasitic nematodes of insects belong to the classes Nematoda, Nematopiorpha, and Acanthocephala. A relatively large number of obligate nematode parasites of insects are included in the families Mermithidae, Tetradonematidae, and Allantonnematidae. Of these, the Nematidae are the most prevalent and have received the most attention.

Some nematodes (particularly the genus Neoaplectana) can develop in the host after its death; thus, they are considered semiparasitic. The resulting progeny, as ensheathed second-instar larvae, break out of the host cadaver and survive as free-living entities, as long as moisture is present, until they encounter a new host.

CONCLUDING REMARKS

Microbial control is no longer in the experimental stage; it is a reality. Products based on this principle are now available and many more await development. Also, the possibility of partial conversion to microbial control in our fields and forests should be examined.

However, microbial control does have certain disadvantages. The microbial agent may be more expensive to apply and to manufacture. Since coverage is essential, several applications may be necessary since plants grow away from previously sprayed areas and rain can wash off the microorganisms applied. The second disadvantage may be of less significance now that the public is demanding specific chemical pesticides, and the broad-spectrum, persistent chemicals are being removed from the market. However, the microbial agents, with the exception of *Bacillus thuringiensis* are so specific that the cost of development and production may be considered slightly higher compared to a broad-spectrum chemical. Table 2 shows a comparison taken from a publication by Dulmage (1971). Costs are comparable, but the microbial agent (for example, a virus) attacks only one insect species; the chemical could be a broad-spectrum agent. The difference in potential market is considerable. Nevertheless, is it not time that we considered the total environment in our zeal to protect

TABLE 2 Comparison of the Costs of Research and Development of a Chemical and a Microbiological Insect Control Agent[a]

Insect Control Agent	Chemical	Microbiological
Synthesis and selection	$ 500,000	$ 80,000
Laboratory testing	200,000	480,000
Field trials	300,000	250,000
Toxicological studies	350,000	250,000
License	25,000	25,000
Formulation	80,000	100,000
Product depreciation	450,000	355,000
Equipment and plant depreciation	350,000	270,000
Market studies and demonstration trials	250,000	200,000
Total costs	$2,505,000	$2,010,000

[a]From Dulmage (1971).

a crop or a stand of forest trees? Won't the extra cost be paid back, with interest, to future generations?

And now, let us look at the advantages of microbial control:

1. The microbial control agents, when properly developed, are equal to the best insecticides in existence.
2. These agents have shown little change in virulence toward the host or target insect to the limits of our present knowledge. Insect resistance is not expected to develop as readily as it does to insecticides, if it develops at all.
3. Beneficial insects such as parasites and predators will add to the pressure against the target insect. Specific sprays of microbial agents do not adversely affect the biocoenosis in any way.
4. There will be no hazard to the operators and applicators using microbial agents.

No one connected with microbial control investigations would suggest that this technique offers the univeral panacea for solution of all insect problems. As biologists, we all know that there is no single answer to pest problems. Nevertheless, microbial control has a role to play in this complex problem. The sooner we invest wholeheartedly in the thorough study of these potentially useful organisms, the sooner our rewards will become evident.

REFERENCES

Abdul-Nasr, S. 1959. Further tests on the use of polyhedrosis virus in the control of the cotton leafworm *Prodenia litura* Fabricius. J. Insect Pathol. 1:99–106.

Angus, T. A. 1954. A bacterial toxin paralysing silkworm larvae. Nature 173:545.

Bassi, A. 1835. Del mal del segno calcinaccio o mossardino malattia che affligge i bachi da seta. Orcesi Lodi. 1–67.

Biliotti, E., P. Grison, and D. Martouret. 1956. Utilisation d'une maladie à virus comme méthode de lutte biologique contre *Pieris brassicae* L. Entomophaga. 1:35–44.

Clark, E. C., and C. G. Thompson. 1954. The possible use of microorganisms in the control of the Great Basin Tent caterpillar. J. Econ. Entomol. 47:268–272.

Dowden, P. B. 1940. Larval disease prevalent in heavy infestations of the European Spruce Sawfly in Southern New Hampshire and Vermont. J. For. 38: 970–972.

Dulmage, H. 1971. Economics of Microbial Control. *In* Microbial control of insects and mites. H. D. Burges and N. W. Hussey (eds). Academic Press, London.

Ferron, P. 1971. Augmentation de la sensìbilité des larves de *Melolontha* melolontha L. (Coléoptère, Scarabaeidae) à la mycose à *Beauveria tenella* (Delacr.) (Discussion) Proc. IV Intern. Colloquium Insect Pathol., College Park, Md., U.S.A., Aug. 25–28.

Goodwin, R. H., J. L. Vaughn, J. R. Adams, and S. J. Louloudes. 1970. Replication of a nuclear polyhedrosis virus in an established cell line. J. Invert. Pathol. 16:284–288.

Grison, P., C. Vago, and R. Maury, R. 1959. La lutte contre la processionaire de pin *Thaumetopoea pityocampa* Schiff. dans le massif du Ventoux. Essai d'utilisation pratique d'un virus specifique. Rev. Fôret Franc. 5:353–370.

Hannay, C. L. 1953. Crystalline inclusions in aerobic sporeforming bacteria. Nature 172:1004.

Heimpel, A. M. 1967. Progress in developing insect viruses as microbial control agents. Proc. Joint U.S.–Japan Seminar on Microbial Control of Insect Pests. Fukuoka, Japan.

Ignoffo, C. M., A. J. Chapman, and D. F. Martin. 1965. The nuclear polyhedrosis virus of *Heliothis zea* (Boddie) and *Heliothis virescens* (Fabricius) III. Effectiveness of virus against field populations of *Heliothis* on cotton, corn and grain sorghum. J. Invert. Pathol. 7:227–235.

McCoy, C. W., A. G. Selhime, R. F. Kanavel, and A. J. Hill. 1971. Suppression of citrus rust mite populations with applications of fragmented mycelium of *Hirsutella thompsonii* Fisher. J. Invert. Pathol. In press.

McEwen, F. L., and G. E. R. Hervey. 1959. Microbial control of two cabbage insects. J. Insect Pathol. 1:86–94.

Metchnikoff, E. 1880. Zur Lehre über Insecten Krankheiten. Zool. Anz. 3:44–47.

Ossowski, L. L. J. 1957. The biological control of the wattle bagworm (Kotochalia junodi Heyl.) by a virus disease 1. Small-scale pilot experiments. Ann. Appl. Biol. 45:81–89.

Paillot, A. 1933. L'infection chez les insectes. G. Patissier, Tréuoux (Ain).
Semel, M. 1961. The efficiency of polyhedrosis virus and *Bacillus thuringiensis* for control of the cabbage looper on cauliflower. J. Econ. Entomol. 54:698–701.
Steinhaus, E. A. 1947. Insect microbiology. Comstock Publ. Co. Inc., Ithaca, N.Y.
Steinhaus, E. A. 1949. Principles of insect pathology. McGraw-Hill, New York.
Steinhaus, E. A. 1956. Microbial control–the emergence of an idea. Hilgardia 26:107–159.
Steinhaus, E. A., and C. G. Thompson. 1949. Preliminary tests using a polyhedrosis virus to control the alfalfa caterpillar. J. Econ. Entomol. 42:301–305.
Tanada, Y. 1956a. Microbial control of the imported cabbage worm. Hawaii Form Sci. 4:6–7.
Tanada, Y. 1956b. Microbial control of some lepidopterous pests of crucifers. J. Econ. Entomol. 49:320–329.
Thompson, C. G., and Steinhaus, E. A. 1950. Further tests using a polyhedrosis virus to control the alfalfa caterpillar. Hilgardia 19:411–445.

Nonchemical Control of Nematodes and Root-Infecting Fungi

S. D. Van Gundy

Historically, chemical soil fumigation began in the early 1860's when carbon disulfide was used to control *Phyloxera* in French vineyards. By 1870, more than 10,000 acres were treated annually. The next advance came in the late 1930's and early 1940's when several thousand tons of chloropicrin were applied to pineapple land in Hawaii. Ethylene dibromide and D-D (1,3-dichloropropane-1,3-dichloropropene mixture) soil fumigants were discovered in the 1940's and proved to be one of the outstanding advances in nematology. These latter two, plus methyl bromide and DBCP (1,3-dibromochloropropane) constitute the bulk of chemicals now being applied against soil-borne pests and plant pathogens. Even though they are halogenated hydrocarbons, relatively few cases of persistent or phytotoxic residue buildup in soil or crop have been reported. Of 260 million acres of cropland in the United States, only about 1.75 million acres are fumigated annually at a cost of $60 million.

Chemical soil fumigation appeals to growers because it is a direct and, in many ways, a very successful way to solve complex disease problems. Nevertheless, some instances of elemental bromine buildup in soil have been found, and increasing attention to the ecological aspects of chemicals in the total environment must figure prominently in the future. Even with intensive present-day fumigation, complete eradication of target organisms in the field has not been possible. Often, extensive disruption of normal soil microfloras causes an ecological "vacuum" in which a few surviving (or reintroduced) plant pathogens build up rapidly in the absence of normal competitors. Therefore, once a soil-borne pathogen has been introduced into a cropping ecosystem, it must be considered at least a potential threat in all future uses of that plot of soil.

In view of these facts, I predict that in the future, chemical fumigants will probably be used on greater acreages in this country but with much more biological finesse. The aim will be for "tolerable losses," "economic control," "pathogen inhibition," and "long-range control" using minimum amounts of fumigant, or not using them at all whenever that is feasible. This will contrast sharply with the higher rates of fumigant application and accompanying "ecological splatter" often encountered today.

Such an approach to integrated control will represent a systematized effort to bring together suitable cultural practices, plant resistance, and chemicals. We must be able to define our objectives in terms of permissible economic loss and to develop a control system that will manage the pathogen populations at or below the acceptable level. True management of soil pathogens requires thorough understanding of the dynamics and ecology of populations.

Today, plant pathologists and nematologists often think an integrated control program employs only chemicals and host resistance. In the past they have been our strongest weapons in combating soil pathogens; however, they have also obscured our need to blend into our control programs a variety of cultural practices designed to manage population levels of the pathogen. I suspect that these attitudes will prevail until the general public is prepared to accept a trade-off of higher prices and lower quality for a reduction or a discontinuance of chemicals in our control programs. Because even under the best of integrated control programs there are often some crop losses and some reduction in the quality of produce, the increasing awareness by the general public and scientists of the potential dangers of environmental pollution now demands a renewed interest and emphasis on nonchemical means of controlling our most important soil pathogens.

Many of the specific principles involved in the control of soil pathogens by land management and cultural practices are well known and are used in many countries where the cost of chemical control is prohibitive or, in this country, on crops where chemical control is uneconomical. There are, however, important limitations that appear to be basic to all environmental pollution problems, i.e., time, economics, and attitudes. Scientific technology already has many of the answers to pollution control, but the general public is not ready to accept and pay for control.

Nematodes are primarily soil-borne pests and, together with root-infecting fungi, should be considered an integral part of the soil biosphere that interacts continuously with fluctuating physical and chemical factors in soil microenvironments. Therefore, every manipulation of the soil affects the level of activity or balance of its microbial population and thereby the fundamental basis for a nonchemical control program.

A model of the ecosystem that illustrates some of the known complex interrelationships among soil pathogens, plant, climate, and soil environment is illustrated in Figure 1. This model provides a hypothesis to develop a definition of microhabitats and to quantitatively define biological and physical factors comprising the environment that should be considered in any soil-borne disease control program.

Disease development results from the association of a pathogen and its host in a suitable environment. Nonchemical control developed around this principle is preventive in nature and uses alterations in the environment, the condition of the host, and/or the behavior of the pathogen. Intelligent application of cultural control depends on knowledge of the life cycle of the pathogen and the primary factors that influence its growth, reproduction, survival, and pathogenesis in

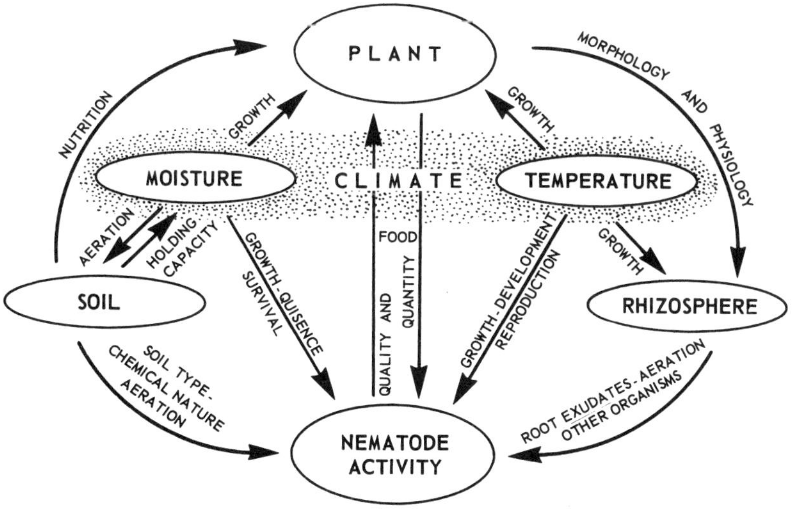

FIGURE 1

a particular ecological setting. Therefore, the development of any plant disease control program is dependent on knowledge and ability to manipulate pathogen, host, or environment singly or in combinations. This has been expressed in a simple equation:

$$\text{Disease severity} = [\text{inoculum density} \times \text{capacity}] \times [\text{proneness} \times \text{susceptibility}].$$

Capacity is the effect of the environment on energy for colonization of the pathogen while proneness refers to the effect of the environment on the host.

MANIPULATION OF THE PATHOGEN

Both nematodes and root-infecting fungi are metabolically relatively inactive in the absence of host roots and are reactivated by the appearance of a food source in the environment. The death rate of these soil-borne pathogens depends on many variables. However, most of them have long-lived stages in their life cycle, often enabling survival without host contact for several years. Prediction of disease potential is based on definite knowledge of population dynamics of the pathogens in soil and the techniques for establishing economic thresholds. Both plant pathology and nematology lag far behind entomology, wild life management, and fisheries in the application of systems analysis and design. Present techniques developed for quantitative analysis of soil pathogens are laborious, time consuming, expensive, and relatively insensitive. One of the most successful large-scale analyses of a soil pathogen has been developed to measure sugar-beet nematode cysts as the beets are dumped at the factory. Every field is sampled several times during harvest, and future plantings of sugar beets are allocated on the basis of these samples. Management of soil pathogens rests on future refinement of our soil sampling techniques and advisory services and on their application to mathematical systems and models. Since 1963 plant pathology has progressed a long way toward demonstrating the possibility of applying systems analyses to plant disease situations.

Activities of organisms now can be described in terms of time and space, and practically any model can be subjected to rigorous tests. Four different models of host–parasite interactions have been developed: (a) those involving nonmotile inoculum distributed about a fixed infection court on a below-ground hypocotyl; (b) nonmotile inoculum invaded by a moving infection court such as a growing tip; (c) motile inoculum about a fixed infection court; and (d) motile inoculum invaded by a moving infection court. On the basis of these models, predictions are that biological control of diseases involving moving infection courts might be difficult while biological control mechanisms affecting diseases might be more easily applied to interactions in which the infection courts are fixed. Unfortunately, these models have not been adequately tested and suffer from the lack of adequate and sensitive bioassays of the pathogen and disease rating. Much more needs to be done to establish sound economic thresholds.

In the last 10 years there has been rigorous application of the principal mathematical theories to nematode population dynamics. Two groups of nematode species can be distinguished: (a) sedentary species multiplying in separate generations (cyst nematodes) and (b) migratory or partly sedentary species multiplying continuously. These models have been used extensively to define variations in host status: good host, poor host, and nonhost. These same models could be used to study the influence of external conditions on populations. In conclusion, the density of the nematode populations depends on the quality and quantity of food available, as well as on the influence of environment, and population regulation must, therefore, take all of these factors into consideration in proposing any control program.

Acceptance of these theories and models depends on how well they predict the economic disease threshold in field situations. Furthermore, to successfully reduce the use of chemical applications during the cropping period and to intelligently plan programs designed to suppress soil pathogens, we may also need to define action thresholds. Action thresholds are levels of the pathogen at which corrective action must be taken to prevent the population from rising to the economic threshold where significant damage occurs. However, until we are able to identify economic thresholds there probably will be little attention given to action thresholds.

Two of the most important plant disease control principles involving the manipulation of the pathogen are exclusion and eradication. The old proverb "An ounce of prevention is worth a pound of cure" applies well to plant pathogens. The most satisfactory control program prevents the introduction and establishment of a pathogen in an uninvaded area. Currently several millions of acres of virgin land are being put into agricultural production under irrigation in the Pacific southwest and northwest. However, very little forethought has been given to the introduction of soil-borne pathogens in the irrigation water. Recently, in the Columbia River basin, it has been demonstrated that much downstream land is being thoroughly infested by many kinds of plant parasitic nematodes introduced by reuse of "high quality" irrigation drain water.

Another method of "exclusion," the use of disease-free planting stock, can be extremely important when planting in disease-free or chemically fumigated soil. In recent years many states have developed an inspection and certification program that governs the movement and marketing of disease-free planting stock in intrastate and interstate shipments. Many seeds, tubers, corms, and other propagative parts of plants can be treated with heat, gaseous fumigants, and chemical drenches to destroy root pathogens. Nursery stock can be grown in pathogen-free soil. Grower demand and willingness to pay premium prices for certified planting stock will stimulate establishment of inspection and certification programs in states now lacking those services.

The most drastic system for excluding plant pathogens is the enactment of quarantine restrictions. Unfortunately, quarantines have limited effectiveness because they are usually imposed after the pathogen has been introduced and has proved to be pathogenic. They do, however, decrease the number of introductions and thereby reduce the rate of spread. Plant quarantine inspectors of the USDA's Agricultural Research Service intercepted some 47,000 pests of significance during 1970. These inbound agricultural pests included some of the world's most destructive insects, nematodes, and plant diseases, capable of damaging millions of dollars' worth of U.S. crops, gardens, forests, and ornamentals. Examples of soil pathogens that are now being effectively controlled under rigid quarantine are the causal organisms of potato wart disease and the golden nematode.

The most successful and widely used techniques for eradication have been heat and chemical treatments, which usually are only effective in reducing or inactivating the pathogen. The nursery industry has pioneered in the use of heat to sterilize bench and potting soil. Steam and air mixtures have been developed that selectively kill plant soil pathogens at 50°–60° C in 30 min. These techniques to date have been uneconomical in the field except on very high value crops. In the future, waste energy from power plants may provide cheap energy so that treatment of soils on a field scale may be economical and practical. Another possibility includes the use of geothermal energy when available in such areas as the Imperial Valley of California.

One of the most talked-about means of pest management these days is biological control—reducing the activity of one pathogen by either biocidal or biostatic effects of another organism. The successes of biological control of insects have been striking and have received international acclaim. Attempts at biological control of soil-borne pathogens have not been as successful. These disappointing results are due largely to differences in environments. Soil is "biologically buffered" by its resident microbial population and selected competitors or antagonists introduced into this complex system must be highly competitive or they will be overwhelmed. Therefore, the most successful approaches to biological control in the soil environment have employed the principle of altering the soil environment through cultural practices which enhance natural biological control by the resident soil microflora–fauna. For example, there are over 50 species of fungi, two protozoans, numerous carnivorous nematodes, tardigrades, and other small invertebrates that occur in soils and that are known to destroy nematodes. Even though these organisms exist, our knowledge of the ecological factors affecting the biological equilibrium and biotic potential of these predators is still too fragmentary to support an effective biological control program. Much the same situation exists for soil-borne plant pathogenic fungi. Some successes in reducing population have been achieved in this area by selected cropping sequences, by the addition of large amounts of organic materials to soil, or by changes in the nitrogen sources in the soil, but these methods have not yet become economically important. Organic manuring sometimes results in reduced populations of root-knot nematodes and increased crop yields. The reasons for the reduction in nematodes are not clearly understood, but the reduction is believed to be caused, in part, by an increase in natural enemies and/or an increase in toxic compounds arising from the decomposition of the organic material. Similar examples exist for root-infecting fungi; *e.g.*, potato scab is effectively prevented by the toxic compounds produced during the decomposition of a soybean green-manure crop prior to planting potatoes and fusarium root rot of bean is avoided by reducing the available nitrogen through the decomposition of barley straw.

Future attempts to exploit the possibilities of biological control will depend on successes in understanding the complexities of soil ecology through further definition of microhabitats and our ability to manipulate and quantify the biological and physical factors operative within these environments.

MANIPULATION OF THE HOST

The manipulation of the host provides a direct and effective means of nonchemical control of soil-borne diseases. This is most successfully done by developing disease-resistant varieties. Since 1900, disease resistance has been a major factor in increasing and developing high levels of crop productivity. Most recently, the importance of crop breeding and the development of new disease-resistant lines of wheat has been recognized by the award of the Nobel Peace Prize to Dr. Norman Borlaug. Without resistant varieties, it would no longer be profitable to grow crops in some areas. Many examples of crop resistance to nematodes and root-infecting fungi exist for which the inheritance of disease resistance is clearly worked out. However, there are many soil-borne pathogens for which resistance has not yet been sought or where efforts to date have been unsuccessful. The nematode–fungus complexes have hardly been considered by the plant geneticist. This is one of the most important areas in which studies must be intensified if we expect to maintain production while decreasing the use of chemicals. In the past many breeding programs have not used the expertise of plant pathologists and nematologists and have succumbed to the plight of biotypes, races, and complexes. The keys to the future will be the team approach to breeding for disease resistance.

Crop rotation represents one of the oldest approaches to disease control and without question offers an effective, broadly applicable approach to nonchemical disease control with immense potential in an integrated control program for the future. Regardless of how successful it has been and can be, however, it may never survive the industrial revolution in agriculture unless education and grower concepts can be changed. Crop rotation requires time, never completely eradicates the pathogens, and often involves noncash crops. True pathogen management through host manipulation for some diseases is the only economic means of control and has worked very successfully for control of "take-all" disease in wheat and cyst nematodes on a number of crops. Combined with crop rotation, some successes have also employed trap-cropping and toxic plants. With some nematode pests, particularly the endoparasites, the nematode larvae enter the roots but fail to develop into adults. They are trapped within the root tissues and eventually die when the crop is harvested early and plowed under. Roots of several plants have also been shown to contain chemicals that, leached into the soil, are toxic to some plant parasitic nematodes. The most successful example of field control has been demonstrated by the use of crotolaria to trap root-knot nematodes and the French marigold interplanted with susceptible crop plants for suppressing the population of lesion nematode. Full application of crop rotation methods to disease control in the future awaits further knowledge regarding the biological changes induced in soil by different crop sequences and greater understanding of the behavior of each pathogen.

MANIPULATION OF THE ENVIRONMENT

Nonchemical control methods, to be competitive with chemical control methods, should work throughout the soil mass and over a wide range of microenvironmental niches and should be effective against specific soil pathogens. Of the naturally occurring factors in the environment that meet these criteria, temperature, soil moisture, soil aeration, and soil chemical solution have the greatest potential for manipulation in controlling diseases. Many cultural practices such as fallow, flooding, and plant nutrition by fertilizer practices have been used with varying degrees of success for controlling some diseases; however, they have limited use because of physical requirements, impairment of soil structure, and economic considerations. In most cases they have been discarded in favor of other cultural practices and chemical control. There is now some new information on the key soil factors involved and on the quantification of these factors controlling key processes in the ecology of soil pathogens, which may bring a fresh new approach to environmental control of soil pathogens.

Flooding of fields has been used to control nematodes and some root-infecting fungi; however, its use is limited because of difficulties in application and inconsistencies in control and cost effectiveness. Presumably, the principle involved was anaerobiosis. However, there is accumulating evidence that a shortage of oxygen in soil, especially in the presence of decomposable organic material, favors the production of toxic compounds such as fatty acids, alcohols, and other microbial metabolites. On the other hand there is evidence that low oxygen in soil actually favors the survival of some nematodes by inducing *cryptobiosis*, a state closely resembling suspended animation. In this state the nematodes are very resistant to adverse environmental factors and to chemical treatments. Although anaerobiosis due to flooding or excess irrigation may actually extend the nematode survival, it does reduce or inhibit movement, reproduction, and growth. The control of some nematodes is more effective in fallow soils where moisture is reduced and maximum aeration is induced in the soil. Aeration accelerates the depletion of stored food reserves, shortening survival time in the absence of the host plant. Dry-fallowing may also cause death by desiccation and is probably far more effective for control of nematodes than flooding.

Careful control of irrigation water is one of the most effective ways of keeping soil pathogens under control. In some cases reduced irrigation or rainfall prevents root-knot eggs from hatching, thereby reducing secondary and tertiary invasions of roots. Many root-infecting fungi that reproduce by motile spores are favored by free water in the medium. Growing plants on raised beds shortens the exposure period of susceptible plant parts to saturated soil. Extra or unnecessary irrigations often delay crop maturity and increase the likelihood of disease losses. Lack of aeration because of excessive amounts of soil moisture increases exudations of nutrients from roots and seeds which favors rapid development of root-infecting fungi and the attraction of nematodes and swimming zoospores. Often these weakened roots are easily attacked by soil-borne pathogenic organisms. As we better understand the effects of moisture and aeration on the biology of the organisms, we may be able to apply this knowledge to effective crop management.

The modification of soil pH and the use of fertilizers and other soil amendments may influence the incidence and severity of some diseases. Deficiencies of some mineral elements increase crop losses in the presence of root pathogens. A high pH may reduce the severity of club root in cabbage, while it may induce copper and zinc deficiencies in citrus infested with citrus nematodes. High applications of nitrogen fertilizer may reduce the carry-over in soil of *Rhizoctonia* root rot and *Ophiobolus*-caused "take-all" of wheat. It may increase fusarium rot of beans and reduce crop losses due to root-knot and lesion nematodes. The interactions between host, parasite, and nutrition are extremely complex, and we have hardly attempted to design fertilization programs to minimize nematode damage. First, we need a detailed understanding of the consequences of manipulating soil factors to control specific diseases. Although these environmental factors will seldom afford complete control, in combination with cultural practices they may provide an effective crop management system which will allow maximum economic return for investment in control of soil-borne pathogens.

For example, the host plant may be manipulated to take advantage of the local climatic factors to achieve disease escape. Good stands of cool-weather crops such as spinach and peas can be obtained by delaying planting in the fall until soils have cooled sufficiently to avoid attack by root-rotting fungi. Conversely, warm-weather crops such as cotton, beans, and melons can be protected if planting is delayed until the soils have warmed sufficiently. Other crops (such as crucifers) can be grown in fusarium-wilt infested soil during the summer in coastal areas where soil temperatures remain cool throughout the growing season. Similar situations may occur with nematodes where activity is directly dependent on temperature. Root-knot seldom affects winter potatoes grown in a cool soil temperature, which keeps the nematodes inactive. Summer potatoes grown in the same field may be seriously affected by root-knot nematodes. A classical example is the early planting of early potato varieties in Great Britain for the control of golden nematode disease. In this case the potato roots develop and grow at soil temperatures lower than those at which nematode eggs hatch and larvae develop.

INTEGRATED CONTROL SYSTEMS

The complexity of developing an integrated control system for soil pathogens is best illustrated with the potato crop. Potatoes are affected by more than 18 virus diseases, 46 fungus diseases, 6 bacterial diseases, 5 nematode diseases, 1 or more parasitic seed plants, and at least 39 nonparasitic conditions. Fortunately all of them do not occur in all parts of the potato-growing regions, nor do they have the same severity in all areas. Some are foliar diseases and others are tuber and seed diseases, either in or out of the soil. Therefore, the preparation of an integrated system must be tailored around all the diseases of a particular region and the relative importance of each. Such a program begins with clean seed that has been grown in a certification program. At planting time, chemical treatments, resistant varieties, and cultural practices may be employed to exclude and eradicate pathogens and to protect young plants from infection. Throughout the season, both chemical and cultural practices may be employed to reduce crop losses. Finally, at harvest time additional cultural practices are necessary to ensure proper produce quality. Along with the number of diseases involved, integrated control systems must also take into account all stages of crop production.

In another crop less profitable than potatoes, cultural practices are one of the bean growers' most economical weapons against fusarium root rot. Practices include seed treatment, shallow planting in a properly prepared seedbed, avoidance of excessive irrigation, rotation with a cereal crop, sanitation, and minimum nitrogen fertilization. Control of fusarium-wilt and root-knot nematode can be added by combining with root-knot nematode or fusarium-wilt resistant bean varieties and/or low overall dosages of soil fumigants for nematode control.

Recently, I had the opportunity to study a lesion nematode disease of wheat in Mexico. Obviously, chemical control was not an economical means of control, and we had to approach the problem from a cultural point of view. Chemicals did provide a technique for determining the severity of the disease. In this case, we found that if planting was delayed until soil temperatures dropped from 24° to 15° C and applied a nitrogen application at the first irrigation, there was insignificant crop loss. To this we hope to add resistant or at least tolerant wheat varieties and a rotation to soybeans or cotton. Proper management of crop planting such as this in conjunction with proper assessment of disease potential could probably be used much more effectively than it has been in the past if chemicals were not economical or considered the easy way out. As plant pathologists and nematologists, we need to approach the problem from the point of view that chemicals will only be used as a last resort to restore pathogen balance.

Our philosophy for the future has already been aptly stated by Professor Garrett. I would like to quote him:

> This world is that of the living soil, which is the mother of all terrestrial plant life and, therefore, of direct concern to every human being and of interest to every biologist. Inside the soil, many kinds of living organisms are interacting continuously with one another and with the root systems of green plants, and so disease of the root system cannot be studied in isolation as a simple interaction between root system and fungal pathogen; much more than this is involved, and so the root-disease investigator must be concerned with everything that goes on in the soil. As a plant pathologist, he is often perplexed and baffled by the intricacies of his problems, but as a biologist he is surprised and delighted by the insights he thus gains into the complex microcosmic world of the living soil. And so, I hope the zest of the biologist will compensate for the worries of the plant pathologist whose work is not finished when he has elucidated the bi-

ology of his disease; he must then attempt to translate this fundamental knowledge into successful practice, and this second problem is often more intractable than the first.

REFERENCES

Baker, K. F., and W. C. Snyder (eds). 1965. Ecology of soil-borne plant pathogens. Prelude to biological control. University of California Press, Berkeley.

Baker, R. 1968. Mechanisms of biological control of soil-borne pathogens. Ann. Rev. Phytopathol. 6:263–294.

Cook, R. J., and R. D. Watson (eds). 1969. Nature of the influence of crop residues on fungus-induced root diseases. Washington Agric. Exp. Stn. Bull. 716.

Garrett, S. D. 1970. Pathogenic root-defecting fungi. University Press, Cambridge, England

National Academy of Sciences. 1968a. Plant-disease development and control. Publ. 1596. Washington, D.C.

National Academy of Sciences, 1968b. Control of plant-parasitic nematodes. Publ. 1696. Washington, D.C.

Seinhorst, S. W. Dynamics of populations of plant parasitic nematodes. Ann. Rev. Phytopathol. 8:131–156.

Stolzy, L. H., and S. D. Van Gundy, 1968. The soil as an environment for microflora and microfauna. Phytopathol. 58:889–899.

Toussoun, T. A., R. V. Bega, and P. E. Nelson, 1970. Root diseases and soil-borne pathogens. University of California Press, Berkeley.

Nonchemical Weed Control

Boysie E. Day

Weeds may be controlled by direct action against offending plants or indirectly by manipulating the physical, chemical, and biological components of the environment. In agriculture, the objective is to create conditions unfavorable to weeds while maintaining conditions suitable for crops and other beneficial vegetation. Weeds are most efficiently controlled by the simultaneous application of a variety of practices, the total effect of which is usually greater than the effects of individual measures employed separately. This has long been recognized, and throughout history men have employed complex combinations of practices to control weeds.

Weed control methods may be classified as preventive, biological, managerial, physical, and chemical. All of these methods have been practiced by agriculturists for a long time. Herbicides, for example, in the form of ashes and brines have been used since ancient times. Advancing technology over the years has made available more procedures for possible use in combination with one another, yet, old techniques are rarely discarded but are retained as part of an increasing arsenal of weapons against weeds.

Although herbicides are of great importance in modern agriculture and land management, the principal methods of weed control are the basic cultural and managerial practices of "good farming." However powerful and effective our herbicides may be, they are virtually worthless in the absence of other sound management practices. Chemicals supplement traditional methods but do not serve as substitutes for them.

The most basic of all weed control methods is prevention, that is, those measures taken to forestall the introduction and spread of weeds. Such measures include production of weed-free seeds, cleaning of weed-infested seed, and proper quarantine and regulation of contaminated crop seed. It is surprising that in this modern day, the planting of contaminated crop seed continues to be a major source of spread of noxious weeds. In this way, weeds are spread from continent to continent, region to region, and farm to farm. Dissemination of weeds by this means is largely preventable; only common sense and diligence are needed to ensure success.

Other preventable means of dissemination include transport of seeds and plant parts from farm to farm on planting, tilling, harvesting, and processing machinery. Weeds are disseminated by livestock by adherence to their hair or feathers or passing through their digestive tracts. The seeds and vegetative parts of weeds are often transported in fill earth, potting soils, and manures. Weed seeds are often carried into new lands as contaminants of irrigation and drainage waters. Infested hay, grain, straw, and other agricultural products constitute a major means of spread of weeds; this is particularly true for animal feed transported from farm to farm, often over long distances.

The success of preventive weed control programs varies with the weed species, its means of dissemination, and the amount of effort applied. Preventive weed control programs are most successfully applied in those situations in which man himself is the vector. Preventive weed programs usually require community action through the enactment and enforcement of appropriate laws and regulations.

In the United States, both the federal government and the state governments now have clean-seed laws. These prohibit distribution of mislabeled or contaminated seed locally and in interstate commerce and regulate seed imported into the United States. Also, about half of the states have weed abatement laws and a staff of enforcement agents responsible for supervising control and eradication programs for specified noxious weeds. Most states have some form of quarantine laws but they are usually concerned only with the interception of insects and diseases and only rarely with preventing the introduction of weeds. Federal quarantine laws govern the importation of foreign plants or plant parts but in general are not aimed at preventing the introduction and spread of weeds. There is great need to strengthen federal quarantine laws in this respect.

The employment of natural enemies such as parasites and pathogens to reduce the population of a noxious plant is known as biological control of weeds. Grazing animals such as sheep, goats, and cattle have long been used in weed control; however, grazing management is not normally included in the definition of biological control. Biological control refers to actions brought about by introduced organisms such as insects, whose populations are reciprocally regulated by the abundance of their weed hosts. When suitable and effective biological agents can be obtained, this method is the cheapest of all means of weed control. On the other hand, biological control is best applied to individual species in mixed vegetation. It cannot be used to control a complex of many weeds. It has its greatest utility when directed against a single aggressive species in an alien environment where it has been introduced free from its normal enemies.

Needless to say, the introduced biological agent must be suitably specific to the host and for this reason most successful cases of biological control have been in plant families containing few species of economic importance. The closer the relationship of the weed to economic plants the fewer are the host-specific agents available. Most efforts in biological control of weeds have employed insects as the control agent. Insects are typically highly discriminatory in their choice of food and are often restricted to feeding on plants of a single genus or species. With notable exceptions, the host range of phytophagous insects follows taxonomic relationships.

Physical methods of weed control include hand and machine tillage, mowing, cutting, flooding, dredging, draining, dragging, crushing, smothering with nonliving materials, and the application of fire and heat. These are the traditional and predominant means of controlling weeds. These methods are components of the managerial techniques of farming and constitute the core technology to which chemical and other techniques play a supporting role. Of all methods of weed control, physical, chemical, biological, or otherwise, tillage remains the method supreme. The hoe and the plow are the very symbols of agriculture.

Tillage controls weeds by directly destroying them or by physically altering their relationship with the soil. The weeds may be cut or torn into parts or forcibly removed from the soil or have their tops pruned from their root systems, thus reducing their regenerative ability. Tillage may also bury weeds and incidentally also bury and preserve weed seeds as well as spread weeds by inadvertently transplanting roots and stem cuttings of noxious plants.

Motorized tillage with modern machinery is remarkably inexpensive. In skilled hands, the modern tractor-mounted cultivator can sweep out weeds between crop rows and at the same time bury weed seedlings directly in the crop row. But therein lies the fatal fault of machine tillage, the fact that it cannot completely remove weeds from within the crop rows. This is doubly unfortunate because weeds immediately adjacent to young crop plants are exceptionally competitive. This lack of ultimate selectivity in machine tillage usually requires supplemental hand work within the rows of intensively produced crops. Currently this need is more commonly met by hand treatments with pre-emergence herbicides.

Weed control by tillage has the advantage of economy and simplicity. However, in general, experiments show that under most soil conditions tillage is of little benefit other than for weed control. On soils of good tilth there is usually no advantage in cultivation if weeds can be controlled in some other manner.

In heavy soils tillage may offer some benefits beyond weed control but such situations are now known to be exceptional. The disadvantages of tillage are numerous including detrimental effects on soil physical properties, inability to cultivate on relatively wet soil, and ineffectiveness against weeds within the crop rows. For these reasons, the present trends are toward minimum tillage or noncultivation using chemicals for weed control.

Tillage machinery is exceptionally diverse and includes just about every imaginable way of digging, turning, and stirring the soil. The plow is the basic tillage implement and other machines include the cultivator, the harrow, the sweep, the rod weeder, and a wide range of rotary implements. The proper use of this array of implements constitutes an astonishingly complex technology often closely integrated with local weed problems and soil and cultural conditions.

Mowing and cutting are also important weed control methods dating back to prehistoric times. Manual cutting instruments are still widely used throughout the world. They include the scythe, the hoe, the machete, the ax, the saw, and a wide range of other implements. These hand implements are employed in areas inaccessible to power-driven equipment and where power equipment is not otherwise available. Motorized cutting machinery includes various kinds of power mowers and power saws. Machine mowing is a common method along roadsides and other areas where the objective is to establish and maintain sod. Mowing is effective against tall growing plants and ineffective against prostrate species. Thus, as a general rule, whatever vegetation is present initially, under frequent mowing a sod culture ultimately develops. The saw, often gang-mounted in power driven arrays, is the mowing machine for brushy vegetation.

Giant implements such as power shovels, draglines, dredges, bulldozers, huge crushing rollers, and various dragging implements are in increasingly widespread use for weed control in drainage canals, wet areas, and reclamation projects. Flow may be restored to weed-choked irrigation and drainage canals by dredging, draglining, or chaining. Under the latter procedure, a heavy anchor chain or other device is dragged along the bottom of the canal by crawler tractors driven along the opposite banks. The weeds broken loose by the chain are removed by hand or by power forks or by flushing into waste areas.

Weeds in canals and other aquatic areas may be controlled by draining, where conditions permit. The reverse is true with terrestrial weeds, which can often be killed by flooding. This is often an economical means of controlling noxious perennials that do not readily succumb to tillage and chemical practices. This may be accomplished by rotating the land to aquatic crops such as rice or the production of fish.

Perhaps the most neglected method of weed control is the use of fire and heat. In many situations, burning cannot be equaled for removal of existing uneconomic vegetation and for the removal of general noxious vegetation in a wide variety of situations. Grasslands can be protected against the invasion of brushy species, and certain coniferous forests can be kept free of undesirable understory vegetation by the timely and judicious use of fire. Dense and impenetrable stands of trees and brush can be restored to at least partial use as grazing lands by controlled burning.

Heat is used to control weeds by the steam sterilization of nursery soils and certain plant materials to kill noxious weed seed and prevent its dissemination. Another use of heat is the procedure known as flame cultivation. The flaming machinery is driven through cropland with the burners directed diagonally at the base of the crop row. The speed of travel and temperatures are adjusted such that tender weed seedlings are killed without injuring the larger crop plants. The flame does not kill by actual combustion but by raising the temperature of plant tissues slightly above the thermal death point.

Smothering by nonliving materials is increasingly used for weed control. Heavy mulches of straw, sawdust, bark, wood chips, gravel, concrete paving, and other organic or inorganic materials are employed in orchard and landscape situations for preventing the establishment and growth of weeds. These materials are usually effective against annuals but all except heavy paving are ineffective against creeping perennial weeds such as bindweed and the sod grasses. There is much current interest in sheet materials for mulching in high intensity crop production. Seed beds are covered with strips of black polyethylene 12 to 36 in. wide occupying about half of the soil area. Plants such as strawberries are transplanted through holes in the plastic. There are often additional environmental and cultural advantages to this practice other than weed control.

This cursory and incomplete survey of nonchemical methods of weed control outlines only a few of the parts of technology that must be put together in agricultural practice to make a system of weed management for a crop, a farm, a ranch, a railroad, or a forest.

The individual preventive, mechanical, biological, or chemical methods employed separately are but minor forces toward accomplishing the objective of most weed control programs. It is only as methods are combined in such ways as to reinforce one another that they begin to become major deterrents to invading vegetation. Hammers and saws and piles of bricks and lumber do not make dwellings; these are designed by men and executed by skilled hands. In weed control it is more the plan than the tools that determine success or failure. Every manageable factor in the habitat, then, should be reviewed as a potential weed control method and used accordingly.

The most important agent in cropland weed control is the crop itself. The vigor and competitive ability of the crop in terms of local conditions and the weed flora present are critical factors in determining whether or not the crop will capture and maintain the ascendancy. Every factor in the culture of the crop, the management of the forest or range, the mowing and harvesting practices, irrigation procedures, the timing of planting, the timing and methods of cultivation, grazing practices, the nature and timing of fertilizer applications, and so on, endlessly, are all weed control practices and are often the main weed control practices to which cultivation, herbicides, and other techniques are supplemental or supporting procedures. For example, in irrigated land the bean grower may begin the season by watering his soil deeply followed by creating a dry mulch on the surface by harrowing one or more times. The bean seeds are then planted deeply through the dry mulch into moist soil from which they germinate into a weed-free environment. By the time for the second irrigation the crop consists of robust plants large enough that the crop of weed-seedlings that germinate following the irrigation can be controlled by a cultivation procedure that simultaneously buries the small weeds in the crop row. Flame cultivation could also be employed within the crop row. The farmer in this instance uses irrigation in combination with tillage or tillage and flaming to establish his crop in a weed-free situation efficiently and inexpensively. He must understand his environment, the characteristics of his crop, and the weed flora present. If the field were heavily infested with bindweed, for example, this procedure would fail and his crop would be lost.

The principal basis for utilizing the crop as a weapon in weed control is crop rotation. Corn and the grain sorghums, for example, are highly competitive crops that, once established, tend to shade out weeds. Such crops can be used in rotation with flax, sugar beets, and low-growing vegetable crops to prevent excessive build-up of weed populations. Rotation procedures recognize that certain weeds are often associated with specific crops. Rotations among crops having drastically dissimilar life cycles or cultured conditions are among the most effective of all weed control methods.

Weed control in rangelands and forests must be based on less intensive and less costly methods. Preventive methods are particularly important on such land since the introduction of exotic species has often led to undesirable changes in the plant communities of vast areas of native vegetation. Much loss of land value has resulted from introductions of poisonous plants or aggressive species of low forage value. The present laws and regulatory procedures for the interception, detection, and eradication of alien plants are woefully inadequate and should be strengthened.

The most important agent in both the spread and the suppression of weeds on grazing lands is the grazing animal. Improperly managed grazing animals selectively destroy beneficial vegetation and weeds increase and become dominant. This force is always at work on rangelands but can often be largely negated by controlling the kind and number of grazing animals and the seasons of use. These basic practices are often supplemented by mechanical and chemical treatments, burning, reseeding, and the introduction of improved forage species. On the poorest of rangelands, grazing management may be the only economical procedure seemingly available to the rancher. When this fails to give adequate relief, it is often the case that one or more of the offending weedy species is alien and is a prime candidate for biological control. The record of success for biological control in such instances is quite creditable in terms of the research effort expended, and such efforts should be greatly expanded both nationally and internationally.

Forest land is rarely managed intensively, if at all, but where possible the intent is to provide favorable growing conditions for preferred species and reduce competition from unwanted species. Chemical methods are widely used for selective "release" of dominant conifer species from excessive competition by hardwoods. These techniques are so economical, effective, and free from hazard that I foresee no incentive or prospect for their replacement by nonchemical methods. Of nonchemical methods for control of forest weeds, fire has the greatest potential for further development and utilization. Many of our most useful forest species are fire-resistant, and understory growth can be controlled by periodic surface burning. This practice has a number of benefits, including preventing destructive crown fires due to the accumulation of understory fuel.

Aquatic weeds come in diverse forms—emergent, submersed, floating, planktonic, and emersed. The aquatic habitat is complex and our knowledge of it, as well as our ability to manipulate it, is less advanced than for terrestrial environments. This is equally true in our assessment of management objectives in terms of economic, social, and aesthetic values. Our scant knowledge about management of aquatic vegetation is acutely apparent when we reflect that rampant weed growth in our lakes and streams because of nutrient enrichment (eutrophication) is one of the major public issues of our times.

The usual physical, biological, and chemical methods offer no evident solution to the eutrophication problem other than the temporary clearance of navigation channels and drainage ways. The answer would appear to be the management of water quality on a national and international scale, surely a monumental weed control project.

The weed control measures applied to industrial and urban lands must be fitted to the intended use of the land. In most instances, ground cover is desirable both from a utilitarian and aesthetic viewpoint. Mowing is the indicated management procedure, and highway rights of way, parkways, airfields, and similar sites should be planned from the start to accommodate mowing machinery efficiently. Signboards, utility poles, and structures should be grouped and located to facilitate efficient mowing. Ground covers are not practical for transformer yards, lumber yards, and outside storage areas for pipe and other materials. Such areas are not suited to machine mowing or efficient hand work, and herbicides provide the only efficient means of weed control. Railroad rights of way present this problem on a vast scale. Chemicals seem ideally suited for controlling weeds beneath and around the rails, ties, and ballast. Barring complete redesign and reconstruction of the entire trackage system, one cannot imagine an effective mechanical means that would replace herbicides.

Let us now speculate on the prospects for further development in nonchemical methods of weed control. First of all, research on nonchemical methods has languished as available talent and resources have become increasingly involved in the spectacular success of chemical methods. This is particularly true of research workers who view weed science as their discipline. Yet this diversion of effort is more apparent than real. There may have been no comparable flagging of effort among crop ecologists, agronomists, range scientists, agricultural engineers, and aquatic biologists who concern themselves with improving a wide range of management procedures. These often turn out to be important in weed control, regardless of the original intent of the research. Improvements in crop variety, cultural and managerial technology, plant protection, and other technical and social changes provide new and often powerful options for action in growing better crops and fewer weeds.

V

Challenge: Effective Pest Management Strategies Consistent with Environmental Quality

Preventive Practices and Effective Pest Management

T. C. Byerly

In this paper, prevention is construed to include some prophylactic measures, for example, the use of endrin as a tree-seed coating in direct seeding programs to protect the seed from rodent depredation. The dipping of pine seedlings in DDT is a preplant prophylactic measure, found to be essential ad interim in Sweden and used in the southeastern United States by industry, and by state and federal agencies to prevent loss of pine reproduction from the pales weevil. This treatment is recommended as the one essential use of DDT for forestry by our Forest Service. Only a few pounds are required in a year, and environmental contamination is minimal.

One might include seed treatment as a preventive practice, e.g., compounds containing mercury, to prevent anthracnose in cotton or smut in cereals. But this seems to border too closely on treatment to be eligible for discussion as prevention. But no sharp line can be drawn. Treatment of cattle coming from cattle-tick infested areas with an arsenic dip is an essential adjunct of the quarantine procedures used by the Agricultural Research Service.

Current and historical preventive practices include isolation and quarantine. These practices continue to be important measures of pest control. They become even more important if some of the various proposals for controlled environments for food production are implemented. Greenhouse production and controlled environments for poultry and livestock production are already substantial industries. Only a few of them use fully effective isolation and quarantine procedures. Partial, transient isolation is achieved by crop rotation and, with greater but still transient effectiveness, by introduction of plants and animals, without their pests, into environments where such pests do not exist.

Other possibilities for prevention may include immunization or premunition. Such procedures do provide animals and man with protection against many pathogens, and there are a few tantalizing results with protozoan and helminth parasites. As for plants, we are beginning to learn about the nature of genetic resistance. We know that (1) feeding of European corn borer first generation larvae is deterred by DIMBOA, a chemical in corn plant, genetically controlled, (2) some corn earworm resistance is due to tight husk, a genetic character, and (3) that some wheat stem sawfly resistance is due to stem solidness.

We might foster natural balance among all levels of ecosystems. The primary producers, the secondary producers, the consumers and decomposers would cycle and recycle nutrients. No organism would emerge in pest numbers. Careful management of such a diverse system would include controlled harvest and replenishment or supplementation with mineral nutrients deficient or removed by harvest. Water management would be equally essential. Great adjustment in man's food habits might be required in order to adjust to the quality of product from such a system. Risk of failure from epiphytotic, weed, or insect population explosion would be present.

Alternatively, such a natural system with man only as an observer could be used to satisfy man's need for association with "nature." His food, fiber, and other needs for products now produced from terrestrial and aquatic environments could be satisfied by synthetic surrogates. Or he could adopt an intermediate stance—raise adequate food using the best agricultural technology available and enjoy the land not cultivated.

Now let's look at how man, the manager, develops and applies preventive measures.

In the center of origin, Anderson (1956) pointed out that weeds—and evolutionary change in plant species through hybridization—develop in disturbed places. While man is a major disturber, he was not the first. Rivers, glaciers, volcanoes, drought, wind, and other animals were here before him. Anderson (1956) also said, "All the major crops and most of the minor ones were domesticated in prehistoric times. Modern agriculture, classified solely by the plants it uses, is Neolithic agriculture."

Many of the crops and livestock and other species propagated by man flourish in areas far from their center of origin. There is a delightful little book by Edgar Anderson (1952) that recounts a few examples, e.g., sunflowers, which flourish in eastern europe but are a weed in Kansas near their center of origin, and rubber, which flourishes in Malaysia far from the Amazon forests where it originated.

Pests reach exotic crops too soon. Smith (1929) found that 81 of 183 crop insect pests he considered to be important in the United States were of undoubted foreign origin.

Genetic diversity is likely to be maximal. So, too, is the diversity of pests and of the predators, parasites, and diseases of those pests.

So, naturalists observe, a flea
Hath smaller fleas that on him prey;
And these have a smaller still to bite 'em,
And so proceed ad infinitum.
Thus every poet, in his kind
Is bit by him that comes behind.

Jonathan Swift, "On Poetry, A Rhapsody"

This is the balance of nature.

Many of our pests have followed their hosts to the United States. Others continue to threaten invasion. Our quarantine programs keep many of them out.

Clean seed stock and a clean environment provide a basic means of pest prevention. Consider the "beefsteak" tomato. This one is so highly blight-susceptible as to be generally unprofitable in field culture. Yet in isolation, e.g., in carefully fumigated soil in a carefully fumigated greenhouse, it can still be one of our best. Closed environments of many sorts offer opportunities for the future for isolation and quarantine in pest-free environments.

Crop rotation is another traditional method of prevention that is still being used, especially for soil-borne pathogens. Such rotations vary widely in length and effectiveness. A pine-wooded area may be converted to watermelons for two years then back to pine for the 20–50 years necessary to grow a crop of trees, followed by one year in a cover crop to get the land ready, two years of watermelons, and then back to pine trees. Cornstalk rot organisms are much reduced in fields rotated to other crops between successive corn crops, as compared to continual corn. Also, many seed crops are grown in isolated, irrigated valleys in the Northwest, far from the areas of commercial production. These valleys were initially free of pests that plagued commercial areas.

However, clean seed stock, in a clean environment, is still the method of choice for pest control wherever economically and technologically feasible. Seed laws generally require that seed offered for sale be free of noxious weed seeds.

In order to be effective, prevention measures must, in general, be integrated with biological, cultural, and chemical control. Quarantine, with its corollary inspection, provides effective means for preventing pest invasion of areas where they do not now exist.

The United States and other industrialized countries inspect imported plants, animals, and their products to a greater or lesser extent as a condition of importation. The United States, through the Agricultural Research Service* of the U.S. Department of Agriculture, in cooperation with the Customs Service, makes inspections at ports of entry and in some countries from which we import.

The task of inspection and quarantine of plant and animal pests is large. Some idea of its scope may be realized when one considers that the number of personal entries into this country each year approximates the population of the United States. This number of entries includes, of course, the daily commuters from Canada and Mexico. More than 40,000 plant pest and over 600,000 lots of prohibited plant materials interceptions have been made in a single year.

REQUIREMENTS APPLYING TO ALL REGULATED IMPORTS

Some general requirements apply to all imports of restricted plants, plant products, and other materials regulated by the plant quarantine laws and regulations. Six general entry requirements are as follows:

1. No restricted plants or plant materials may be imported except as provided by an official permit.

2. Requests for permission to import plant materials should be made by the importer in advance of shipment. The request should be made to the Plant Importations Branch, Agricultural Quarantine Inspection Division, Agricultural Research Service, U.S. Department of Agriculture, 209 River Street, Hoboken, N.J. 07030. Permit applications should include the following information: Kind (botanical

*Now the Animal and Plant Health Inspection Service.

name) of fruits, vegetables, plants, or plant products to be imported; the country and locality where the restricted agricultural, forest, or range products were produced; manner of importation (by baggage, cargo, or mail); port or ports of arrival, whether other shipments are contemplated; and the name and address of the importer. If applicant is uncertain what plants he will be importing, the permit application may say "admissible materials" in lieu of naming the kinds and varieties of plants to be imported. Permits are issued to persons in the United States who are to receive the plant material.

3. All agricultural, forest, and range imports are subject to inspection by plant quarantine inspectors.

4. Importers are liable for any costs incurred in meeting agricultural quarantine requirements. For example, charges for fumigating or otherwise treating infested materials are the responsibility of the importer. Similarly any costs, delays, and losses involved in unpacking the restricted goods for inspection and treatment or for rerouting to Agricultural Quarantine Inspection Division stations having inspection and treatment facilities must be borne by the importer.

5. If restricted materials entering the United States are found to be dangerous to plant culture, they may be seized, treated, ordered removed from the country, or otherwise disposed of as may be appropriate to eliminate the pest hazard.

6. Agricultural products must remain in Bureau of Customs custody until they have met agricultural quarantine entry requirements and are released by the plant quarantine inspector.

The Federal Noxious Weed Act is being proposed which will give our plant quarantine officials the authority to prevent the entry of weeds not yet in the United States that would pose some economic threat if they became widespread in this country.

PRE-ENTRY INSPECTION

In a number of special cases where it has been practical to work out necessary arrangements for safeguarding against pest entry, inspection and clearance of regulated plant materials or products are effected in countries where the plants are grown. In Belgium, France, Germany, Holland, Italy, Japan, and South Africa, flower bulbs are inspected by U.S. plant quarantine inspectors prior to the exportation to the United States. In Mexico, Haiti, Chile, Israel, and Japan, citrus and other fruits are treated under the supervision of the U.S. Department of Agriculture inspectors before shipping to the United States. And at Nassau and Bermuda, airplanes and passenger baggage are given preflight inspection prior to departure for this country. Similar service is provided for airplanes flying from Hawaii and Puerto Rico to the mainland.

When it is practicable, this system of preinspection and clearance of regulated items in the countries where they are produced has a double advantage: (a) By eliminating pests at their origin, it safeguards this country from their accidental entry; and (b) by stopping materials ineligible for entry prior to shipment, it saves shippers the expense of transporting items that would be refused admission because of pest risk.

PHYTOSANITARY EXPORT CERTIFICATION

Another service provided by the agricultural quarantine system is phytosanitary export inspection and certification.

To help American plant growers and exporters meet import requirements of foreign countries, the Agricultural Quarantine Inspection Division provides a service for inspecting plants and plant products offered for export and certifies shipments that meet admission requirements of importing countries.

Although exporters are not required by law to make use of this export inspection and certification service, it is widely used because certificates are usually necessary for importation abroad. Also, export inspection helps exporters avoid the unwarranted expense of shipping plants or plant products to foreign countries where they may be refused entry because of pest risk. In one year, the Agricultural Quarantine Inspection Division issued over 58,000 export certificates for shipments covering nearly 60 million containers and ranging in size from a single plant to shipload lots.

In order to certify that plant shipments meet the requirements of the countries where American plants are sold, the Agricultural Quarantine Inspection Division maintains up-to-date summaries of phytosanitary import requirements of about 145 countries. These summaries, which are approved by plant-pest authorities of the countries concerned, are available on request.

In addition to helping American shippers sell U.S. plants and plant products abroad and thus promoting a more favorable trade balance, phytosanitary export certification aids cooperative international pest control efforts to the mutual benefit of the United States and other agricultural countries.

Domestic ruminants and swine imported into the United States are limited to imports from those countries where foot-and-mouth disease and rinderpest do not exist. Those countries include Australia, Canada, Ireland, Japan, Norway, New Zealand, Mexico, Central America, and the British Virgin Islands and Santo Domingo.

Imports of livestock from countries where the cattle tick (*Boophilus annulatus*) occurs are prohibited by the Act of 1890, except for cattle originating in Mexico and the British Virgin Islands. Cattle from infested areas in Mexico may be imported into Texas only after they have met certain import requirements. We import about a half-million cattle a year from Mexico. Those coming into Texas from tick-infested Mexican areas must be dipped a week or two before entry under veterinary supervision, with an arsenious oxide solution. They are inspected at the border, must be found free from ticks, and must be dipped again prior to their entry into Texas. Cattle originating in the British Virgin Islands can be imported into the U.S. Virgin Islands for slaughter purposes only.

Cattle entering the United States at points west of Texas must come from Mexican areas that are not tick-infested. Cattle and sheep at these border points are treated with toxaphene in order to ensure freedom from scabies mites.

Wild animals may be imported from any country through stipulated quarantine stations and procedures. If they originate from foot-and-mouth disease- and rinderpest-infected countries, they must enter the United States at the port of New York and be processed through the Department-owned quarantine facility. Holding times, tests, inspection, and treatment at these points protect our livestock and wildlife from insect pests that sometimes arrive with such wild animal imports.

The Secretary of Agriculture is authorized to cooperate with the states and with Mexico and Canada in the control of incipient or emergency outbreaks of insect pests or plant diseases, including grasshoppers, Mormon crickets, and chinch bugs (U.S. Code, Section 148). He has specific authority with respect to golden nematode, "to eradicate, suppress, control, and prevent the spread of this pest" (U.S. Code, Section 150). He has similar authority with respect to Halogeton (U.S. Code, Section 165).

Quarantine measures are re-enforced with other preventive measures that are essential to effective quarantine. Such measures include disinsectization of cargo and passenger carriers; fumigation and other procedures applicable to plant materials intended for propagation; soil treatments in nurseries in quarantined areas as a condition of interstate movement of their products; treatment of campsites and trailer parks; treatment of loading and unloading areas, including airports; and eradication of leap-frogged infestations outside the quarantined areas.

INTERSTATE QUARANTINE

The Secretary of Agriculture, after public hearing, notification to public carriers, and publication of notice, is authorized to quarantine any portion of the United States. Such quarantines are authorized "to prevent the spread of a dangerous plant disease or insect infestation, new or not theretofore widely prevalent, or distributed within and throughout the U.S. States are also authorized to establish such quarantines and the Secretary of Agriculture is authorized to cooperate with them in connection with such quarantines" (U.S. Code, Section 161).

There are presently 13 federal interstate plant quarantines. They are as follows:

1. Black stem rust quarantine, No. 38.
2. Gypsy moth and brown tail moth quarantine, No. 45.
3. Japanese beetle quarantine, No. 48.
4. Pink bollworm quarantine, No. 52.
5. Mexican fruit fly quarantine, No. 64.
6. White fringed beetle quarantine, No. 72.
7. Khapra beetle quarantine, No. 76.
8. European chafer quarantine, No. 77.
9. Mediterranean fruit fly quarantine, No. 78.
10. Soybean cyst nematode quarantine, No. 79.
11. Witch weed quarantine, No. 80.
12. Imported fire ant quarantine, No. 81.
13. Cereal leaf beetle quarantine, No. 84.

The black stem rust quarantine program is one of our oldest. It is a useful component in the biological control of this very threatening wheat disease. While wheat breeders continue to produce genetic stocks of wheat resistant to virulent rust races as they emerge, their task is lightened by eradication of rust-susceptible barberry plants.

Not all barberries are rust-susceptible. A pamphlet containing a list of rust-resistant plants may be obtained from the Director of the Plant Protection Division, Agricultural Research Service, U.S. Department of Agriculture, Hyattsville, Maryland 20781. The susceptible ones have been almost eradicated from most of their range. Vigilance to prevent reintroduction and eradication of any that may have escaped previous notice continues. A program of search and destruction continues in southwestern Virginia, West Virginia, and Colorado.

Eradication of susceptible barberry from the spring wheat area saves 1–2 weeks' time in rust development there. Air-borne spores eventually come from the South. But when rust-susceptible barberry is present, spores go from that host to wheat a week or two before wind-borne spores arrive. That week or two may be critically important to wheat crop damage by black stem rust. More importantly, eradication of rust-susceptible barberry eliminates a major source in the development and emergence of new rust races, although it will not completely eradicate black stem rust. Eradication of rust-susceptible barberry species is surely of as much theoretical ecological concern as the possible elimination of species through chemicals.

Another preventive quarantine program, analogous to the black stem rust work, has been for white pine blister rust prevention. European black currant and wild gooseberries and currants are alternate hosts to the white pine blister rust disease. Eradication of these alternate hosts continues on a state-by-state basis where justified, but the federal interstate quarantine, no longer considered necessary, was revoked in October 1966.

The Japanese beetle program is a difficult one. There have been mistakes in the past which still haunt us. In order to stamp out a new infestation outside the then-quarantined area, dieldrin was used on many acres in Michigan. We wouldn't do that now.

Dieldrin is used for soil treatment in nurseries, as a protection against interstate movement of white fringed beetles outside the quarantined area. Some of our airports are still protected by dieldrin soil treatment applied several years ago.

Chlordane and milky white disease of Japanese beetles have been substituted. An increased trapping program at airports to increase our early warning of insect invaders has been implemented.

We hope that these measures will be fully effective. One deterrent is the supply of milky white disease spores. Those produced by culturing *Bacillus popilliae* on larvae of the Japanese beetle remain infective in the soil for several years. Our Northern Utilization Research Laboratory at Peoria has sought for several years to produce spores that will retain like infectivity in soil. Thus far they have no problem in propagating the bacteria, but the spore longevity is ephemeral.

Does quarantine work? We believe that it does. We have eradicated Khapra beetle, oriental fruit fly, Mediterranean fruit fly, and cattle tick from the United States. International quarantine has intercepted these and many other pests. But they have slipped through on several occasions. Interstate quarantine has been an essential component of eradication of such infestations before they have spread beyond the initial area of infestation. Quarantine of golden nematode has kept the infestation of this pest confined to a few thousand acres, most of which have now been treated. Eradication of this pest seems probable. Quarantine of the witch weed has confined the infestation to 39 counties in the Carolinas—the entire corn crop in all the states is vulnerable.

We may wish to construct and maintain barriers—*cordons sanitaires*—there is nothing new about this. We helped Mexico eradicate foot-and-mouth disease that way—by use of test and slaughter. We are using similar barriers south of the border with continued drops of radiation-sterilized insects to overwhelm fertile native flies of both the screwworm and the Mexican fruit fly. The use of zones imposes hardships. Segregation, isolation, and quarantine by area are powerful preventive tools to protect uninfested areas and for protection of areas in which eradication has been accomplished against reinfestation.

Not all our potential pests are here. Chance and the weather sometimes work in our favor. Hoja blanca disease of rice established itself in our fields in the late 1950's. We had studied this disease in Cuba and in other countries for several years. We had tested genetic material ready for introduction to make our rice crop resistant to Hoja blanca. Then we had a severe winter and apparently all of the insect vector, *Sogata orizicola*, froze to death. So, we were saved by the weather.

Without a host crop, pests must starve or adapt to an alternate host. This is the basis of crop rotation as a pest preventive method. On a grand scale, it could be used in pest management. How long would we have to stop growing cotton to eradicate the boll weevil? It has few alternate hosts. The bollworm could not be starved out in this way; it has many hosts.

There may be opportunities for preventive pest control through habitat management. Irrigation may provide habitat for mosquitoes and snails, vectors of animal and human diseases. Efficient water management, including drainage, may restrict that habitat. Management of grazing animals and restricting human exposure may limit infection.

SUMMARY

There are several identifiable systems of pest prevention:

1. "Man adapting," or laissez faire—accepting the yield limitation and inferior product quality caused by pests limited by natural biological control.
2. "Man managing"—controlling the use of the natural environment to prevent pests by law, regulation, enforcement, and compliance.
3. Abandoning to self-regulating ecological forces the natural environment, reserving to man a small part of it as a completely controlled environment.
4. Some combinations of the three above.

For man and for the environment, the route to be followed will depend on the time, the location, and the circumstances.

REFERENCES

Anderson, E. 1952. Plants, man, and life. Little Brown and Co., Boston.

Anderson, E. 1956. Man as a maker of new plants and new plant communities. *In* Man's role in changing the face of the earth. University of Chicago Press. Pp. 763–777.

Smith, H. S. 1929. On some phases of preventive entomology. Sci. Mon. 29: 177–184.

Role of Industry in Advancing New Pest Control Strategies

K. R. Fitzsimmons

The preparation by industry to meet the challenges of the future in advancing pest control strategies is based on a fact that all economic entomologists and most pesticide critics accept, namely, that chemical pesticides, including insecticides, will continue to be major weapons in our agricultural production arsenal.

Biological control of insects has made real contributions in the past and is likely to continue to do so. Yet the need to produce wholesome food economically has demanded increasingly large amounts of all kinds of pesticides. Since *Silent Spring* the use of pesticides in the United States has nearly doubled. Herbicide use has tripled. The end or leveling-off of this growth in demand is not yet in sight. Any general reduction in total chemical usage is uncertain as to time and extent. Meanwhile, economic factors and uncertain legislative outlook have caused several companies to withdraw from the market or drastically curtail their research and development programs. It is obvious that these trends place increased responsibility on those companies remaining in the production of agricultural chemicals for a continued supply of acceptable pesticides for chemical control.

On this basis, there are six ways in which industry is prepared to meet this continuing need:

1. Industry will continue to invent, synthesize, develop, and manufacture better chemical pesticides

Invention and synthesis of chemical pesticides are areas that depend on intensive and extensive research programs in chemistry. Although the U.S. Department of Agriculture and a very few state experiment stations have chemical research programs directed toward better pesticides, this area has been and is expected to continue to be the special province of the chemical industry. Of the approximately 500 pesticides now registered, all but a few natural products, some old inorganic compounds, and a mere handful of synthetic organic materials such as attractants and repellents have been invented and first synthesized in the laboratories of the chemical industry.

The chemical industry starts the new pesticides on their way through the research programs in organic synthesis. The most important steps in the progress of the chemical from laboratory to farm are development and manufacture.

Development is slow and expensive. During the three to eight years of development, several million dollars are spent and no income is produced. Not only is effectiveness tested in all the major agricultural areas of the country, but also on many of the major crops. Industry has the field organization to organize and oversee such tests, although state and federal entomologists cooperate in the program. Even more important is the safety evaluation in terms of residue determinations, acute and chronic toxicity testing, and evaluation of effects on wildlife. Again, only industry has the complex development organization required to plan and expedite the many tests needed in safety evaluations. This involves coordination of programs in chemical synthesis, plant and animal metabolism, residue analytical method development, process development, pilot plant production, pharmacology, toxicology, formulation research, economic analysis, practical agriculture, wildlife biology, and patent and licensing arrangements.

Finally, if the development process appears favorable, manufacture is necessary. Sometimes existing manufacturing plants can be adapted to the new product in the initial stages, but sooner or later a new plant must be designed, financed, and built. Again it is only industry that has the experience, the money, and the profit incentive to see that the plant is built. The only products that are manufactured and sold by government are the raw materials for the generation of atomic energy. I doubt if either government or the public would wish to see a Pesticide Commission that functions analogously to the Atomic Energy Commission.

Thus the first contribution of industry, to continue to invent, synthesize, develop, and manufacture better chemicals, rests on several unique assets: an extensive research program in organic synthesis, an efficient organization for development and safety testing, and the engineering experience to build and operate manufacturing plants.

2. Industry will bear the major, but not the entire, responsibility for testing the new materials for safety to people, wildlife, and the environment in general

It has been an axiom of American business, since long before the current wave of consumerism, that the marketing of a product entails responsibility for its safety and effectiveness. We in industry have no wish to avoid that responsibility. Further, we know from past experience that a faulty or ineffective product hurts our business just as safe and effective materials improve our business.

For these reasons we make extensive tests of safety and effectiveness before a product is marketed and would do so whether or not federal and state laws required it. Our own standards of acceptability have become more demanding as previously unrecognized hazards are brought to light.

Until recently the safety requirements demanded by state and federal law were not significantly different from those we imposed on ourselves. There are indications at present that the public, and through them, the legislators, may wish to see the extent of required testing increased to the point where, we believe, the cost of development will be above what the farmer and public will pay for in product price. If the requirements for presale testing are to be increased primarily to satisfy public demand, it is proper that the public should bear a part of the cost. This is apparently the sort of cooperative effort envisioned in Recommendation 10 of the Mrak Commission Report, which states:

> Recommendation 10. Increase Federal support of research on all methods of pest control, the effects of pesticides on human health and on the ecosystems, and on improved techniques for prediction of human health effects.

While federal laboratories, especially in the Department of Fish and Wildlife, have assisted industry in safety evaluations in the past, there has been no overall assumption of responsibility by federal agencies to cooperate actively in all phases of safety testing. We feel we must be assured of the assumption of this responsibility by federal agencies if the new, more stringent, requirements now being discussed become regulations or laws.

3. Industry will guide the use of these products through trained and responsible field technologists

Unfortunately there seems to be no census of the number of entomologists, plant pathologists, nematologists, and weed control specialists employed by the agricultural chemicals industry, but based on our own company, which has about 30 such employees in the Agricultural Division, in addition to a like number in research, it is likely that there are more experts in industry than in the pest control staffs of either federal government or state experiment stations combined.

Anyone familiar with field conditions in major crop areas will recognize that the average farmer or contract spray operator more frequently talks to a pesticide salesman than to his university extension entomologist or his local farm adviser. It is obvious, therefore, that the field men employed by the agricultural chemicals industry exert a major influence on how pesticides are used.

Nearly all of these men are graduates of agricultural colleges and many have had experience as extension specialists and farm advisers. They are familiar with local problems and resources. It is inconceivable that modern pest control practices could be carried on without them.

Now we are discussing far more sophisticated programs, using many more specific pesticides, and integration of these chemicals with each other and with a great variety of nonchemical controls. A key to making integrated control programs work is almost daily surveillance by qualified personnel of pest and parasite levels, weather, crop development, and equipment availability, not to mention consideration of various economic factors and legal permissibility. We do not believe it is either necessary or desirable to transfer a thousand trained men from industry to the public payroll in order to get responsible advice to farmers. There is no other reservoir of trained and experienced pest control specialists. Unfortunately, applied pest control training is seriously lacking at the university level, which means that there is presently a shortage of trained replacements for industry, to say nothing of the expanded future need for pest control specialists by both industry and government.

So our third point is simply that industry offers the only effective available manpower for the implementation of the new pest control strategies. We are ready to cooperate with all federal, state, and local agencies and experts in sound, adequate efforts at such implementation.

4. Industry will cooperate fully with public agencies that may develop nonchemical methods so that chemical and nonchemical methods may be used jointly or separately to the maximum effectiveness of both approaches

So far we have talked about chemical control technologies. This is where our strength lies and this is where we honestly believe the greatest opportunity exists for improving our pest control strategies. But there have been some promising advances made in nonchemical pest control and we are promised more in the future. The public, and more importantly, the budget administrators, have demanded and supported maximum efforts to control pests without chemicals. There seems to have been a feeling that industry is not only unsympathetic but is actually denigrating this kind of research. We are frequently asked how much money we are spending for research in nonchemical control. Actually, industry is very active in the field. *Bacillus thuringiensis* would probably not have been developed without the industrial technology of the several companies involved. Companies like Zoecon and Schering are actively developing the insect hormones for practical pest control. Other tools of nonchemical control developed, produced, and marketed by industry range from spotted aphid resistant alfalfa seed to light traps and window screens.

Most of us are not convinced of the economic viability of unsubsidized projects to rear and sell lady-bird beetles, sterile males, or pheromones. That doesn't mean we are antagonistic or obstructionist. Such lines of research are outside our field of competence and their economic feasibility seems uncertain. If General Motors doesn't feel obliged to breed horses, or Consolidated Edison elects not to sell Coleman Lanterns, that's their prerogative. Ours is not to compete with the Department of Agriculture and the state experiment stations in biological control research.

We feel that nonchemical control research is needed and we are willing to see our tax money spent to support it. We follow it closely, prepared to adapt our chemical technologies to take full advantage of any new concept that will help the farmer.

5. When nonchemical methods are developed that are adaptable to commercial channels of distribution, we are ready to take over or participate in the business of supplying products

A major drawback to the chemical industry of publicly supported research is the imposed lack of profit incentive. No company can afford to underwrite substantial research and development expense without protection from competitors while such costs are being recovered. A biological control technique such as release of an imported parasite that is implemented without investment or effort is rare. A technique that depends on annual release of sterile males or parasites or nonviable diseases is expensive and cannot become self-supporting without insertion of the profit motive. There is no way to charge the cost of public benefits to the recipients except through public support. We in industry are, therefore, not set up to carry out programs that are public in their application. We do assume some responsibility for public education but we do not assume responsibility for public welfare or public defense whether against human or insect enemies.

This does not mean that we wish to contravene such public programs. We educate people, we contribute to public welfare, and we cooperate in public pest control programs. Furthermore, we will accept government research contracts, production contracts, or action program contracts which are negotiated to the mutual satisfaction of both parties. Such joint industry–government projects would be acceptable, or even welcome to industry but most companies would not see such projects as a substitute for manufacture and sale of conventional pesticides.

6. Industry will consider the assumption of pest management or even crop production contracts, whereby the chemical companies directly or through subsidiary companies or licensees would control and be responsible for major defined aspects of crop production

If pest control practices are to be ecologically integrated with respect not only to pests, parasites, and crops, but also to the whole biosphere, atmosphere, aquasphere, and lithosphere, they probably should be given to specially trained "pest managers." We doubt the economic and political feasibility of such prescription pest control. But to whatever extent such a move is deemed necessary, the logical people to be made responsible are employees of corporate farms, of agricultural cooperatives, or of the pesticide industry. We cannot see anyone controlling pest management costs to be paid by individual farmers without sharing in the profits to be derived from efficient pest control.

Whoever sells such a service would presumably be specially trained, licensed, and bonded. Agricultural pest control would then be handled as is human or veterinary medicine, the function of pharmacist and doctor being combined in a single individual, who might or might not be an employee of the pharmaceutical company.

Carrying the analogy one step further, we cannot subscribe to a system of "socialized" agricultural medicine where all pest control responsibilities are given to government employees who might well become legalized regional czars.

The recent development of "farm centers" indicates a trend toward prescription agriculture. Many cooperatives and agricultural chemical companies have such centers in operation or under development or study. So far these centers are operated as "service stations," chiefly for fertilizers, with charges made only for products sold and not for advice. Such centers could extend their services to include sale of advice on pest control, including advice on integrated control or nonchemical control. But if they are to be made legally responsible for the safety and effectiveness of their advice, and especially if their advice does not involve the sale of products offered for sale by the center, they must charge for their professional advice or management service.

The agricultural chemicals industry would prefer to maintain its role as supplier with responsibility for pest control practices remaining with the farmer. However, if the public and the regulatory agencies demand a radical change of system, industry will consider the assumption of pest management or even crop production contracts, provided that their extra services are paid for.

WHAT INDUSTRY EXPECTS FROM THE PUBLIC AGENCIES

Having stated what industry is prepared to do in order to implement new pest control strategies, we now stipulate what we must expect from state and federal agencies. The stipulations are based on the conviction that industry must have sympathetic cooperation from the public agencies in order to continue to provide safe and effective pesticides for agriculture and human and animal health.

1. A sympathetic consideration of our economics, public relations, and political problems

The wave of consumerism and Armageddonism that is rolling across the nation seems to be engendering a notion that industry is ethically irresponsible. There is fortunately little evidence that public agencies feel this way. It was mentioned earlier that some companies were withdrawing from the agricultural chemicals business. An attitude that industry causes all the trouble and should be punished and disciplined is not likely to make the future of chemical control very attractive.

The voluntary compliance system of pesticide regulation practiced in the United Kingdom is perhaps a little too informal for this country, but the trust and goodwill that makes the system workable in England would certainly smooth government–industry relations in the United States. Not only do we seek this trust and goodwill from public agencies—but we also hope they will use their influence to minimize undeserved criticism from over-zealous critics.

2. Practicable federal regulations, and special efforts toward unification of state use, registration, and permit laws

Not many years ago federal registration and label approvals plus a few hundred dollars in state registration fees made it possible to sell pesticides throughout the country. Now new state regulations are being passed every few weeks. This complexity of regulation requires large and expensive staffs and many tens of thousands of dollars in state fees. We would, of course, prefer to deal with one agency instead of 51, but realize this is hardly in keeping with fundamental constitutional or superficial bureaucratic practice which gives the states the right to regulate their commerce. We believe the public and the environment could be adequately protected at far less public cost if reasonable federal legislation could be enacted with the states being discouraged from passing unnecessary or even contradictory legislation. We certainly don't wish to see small towns begin to regulate pesticides as they are now proposing to do for bottles and detergents.

3. Substantial federal assistance in financing and expediting the increasingly stringent toxicological studies anticipated

Industry readily assumes major responsibility for safety testing for their own satisfaction and protection. But where such extensive testing is proposed that it would cause potentially valuable compounds to be dropped, it will be to the public benefit to assist industry. As an example, the Delaney Amendment, if rigorously applied, would remove a number of useful and apparently innocuous pesticides and food additives from the market. This why the Mrak Commission (Recommendation 8) and several scientifically trained government administrators have called for its modification. If legislators insist on its retention, they must be prepared to see that the administrative agencies bear the burden of developing the protocols and the statistically significant data needed to decide on the carcinogenicity, mutagenicity, or terratogenicity of a product. The same applies to any other requirement for unreasonably long and expensive product testing.

We particularly need better communication with registration agencies to avoid the delays and repeated requests for more and more information which have been far too common in recent years.

4. Better state and federal advisory services to users

We have said that we in industry could take over the advisory services to growers either by expansion of the present field technologist–distributor system or by service contracts. However, if pesticide use is to be more carefully monitored than at present, full use must be made of every trained entomologist, horticulturist, agronomist, or plant pathologist in the country—state, federal, and industrial. They should not all be on industry payrolls. But not all of them should be federal or state employees either. This means a cooperative system, made up of the same people we have now, working in even closer cooperation than at present. The company salesmen, distributors, and technologists, the county farm advisers, the extension specialists from the state experiment stations, research pest control specialists in state and federal field stations, the contract spray operators and airplane applicators, the mosquito abatement district staffs, the grower associations and cooperatives, and the managers of large farming operations are all qualified or could become so with short course training. All these people are working now, guiding pest control practices.

If we are to have closer supervision of chemical usage, and particularly if we are to start using more complicated integrated control programs, we will have to have better state and federal advisory services. Better advice requires more knowledge and this implies more research. So we can't get our increased number of advisers by shifting people from research to extension. This is one of the reasons why industry, while subscribing to the principles of integrated control and prescription entomology, is not optimistic about early application except in a few limited areas on a few limited crops with simple pest problems.

In the enthusiasm for new concepts in pest control strategies we must recognize that for many years to come wise and carefully monitored use of chemicals will be the best of all strategies. So we must do chemical control research in the tax-supported laboratories. Basic research is useful. Biological control research is especially important because it will only get done by public laboratories. But the bread-and-butter business of killing weeds, insects, and fungi will be done with chemicals and the agricultural chemicals industry cannot provide the chemicals needed without the participation of the state and federal laboratories in the testing and development of new pesticides and in training the needed pest control specialists.

We have a difficult time ahead in educating the public, funding our programs, satisfying our critics, and coping with the politics of an election year. Among all these problems, this country must find time to devise some new pest control strategies. Insofar as these new strategies depend on chemicals, industry will be in the forefront. Insofar as the strategies can be implemented without chemicals, industry will be following closely.

Challenge to Public Agencies

N. C. Brady

This symposium has revealed a wide range of interests in the common goal of protecting man and his food and fiber supply from the damage and annoyance of insect, disease, and weed pests. Despite the diversity of views and approaches suggested here, there appears to be agreement on three basic themes:

1. Mankind continues to be in competition with a host of pest species that threaten his health and comfort and that interfere with his ability to produce food and fiber,
2. There is and will continue to be needs for effective controls and means of managing these pests through a combination of methods, and
3. Pest control must be implemented by methods that minimize adverse effects on other than the target organism.

These basic themes suggest that the problems of pest control are problems of our entire society and that their solution is not only a legitimate concern for public agencies but should be a mandated one. The public sector should have major responsibility for pest control and regulation and for maintenance of safeguards and it should mobilize for the effort and resources that these vexing problems will require for satisfactory solution.

Before we explore these challenges to public agencies, it seems useful to take stock of our present situation, a situation of changing directions, re-evaluation, and searching for new concepts.

THE PESTICIDE DILEMMA

We acknowledge that modern commercial agriculture is currently locked into a chemical war on pests. Agriculture, forestry, and many municipal pest control programs are heavily if not entirely dependent on the vast arsenal of pesticides that science and invention have produced. These pest controls have been developed to meet many important needs, they are used with a high degree of technology, and they produce benefits that are worth many times the direct costs to their users. Our marketing system has made them readily available. Conscientious efforts are made to see that they are properly used to minimize danger to the users or the consuming public.

Like so many of our technological advances, however, some of these tools of man have undesirable features. Some of them kill organisms well beyond their target populations; some of them are persistent and last much longer than their intended period of usefulness; still others are hazardous to use unless rather stringent precautions are exercised. Also, public awakening to the general impact of technology on our world has brought an unparalleled concern for the side effects of all chemicals added to the environment, especially pesticides. People are questioning our past use of "first cost" economics as a prime determinant of decisions with respect to pest control. A more comprehensive "total cost" concept is emerging—a concept whose implementation will likely place an ever-widening burden on public agencies.

As is often the case, the first public reaction to malfunctions and excesses in our society is to invoke regulatory and restrictive rules. It is well to remember, however, that restriction and regulation of pesticide use does not solve the basic problem of pest control. The taking away of certain compounds and restricting the uses of others is not accompanied by the elimination of insects, weeds, or disease pests. On the contrary, we have already witnessed such ramifications as the spread of gypsy moth in the Northeast since the use of DDT has been curtailed, and of malaria epidemics in India where mosquito control programs have been relaxed. Quite likely, there will be more similar cases in the future.

Thus, losses from pests are still very much with us, and as before, science is being called upon to provide new solutions to these problems. Should these problems become more acute or affect a greater segment of our society, there will be additional demands on government to "do something." This combination of presently known needs and anticipated public pressures in the realm of pest control suggests that some very real challenges lie just over the horizon. Most of these challenges will require expanded public agency involvement.

PUBLIC SECTOR INVOLVEMENT

The need for public agency involvement in pest control has been assumed or implied in almost every paper presented in this symposium. The intent of this summary paper is to identify the areas wherein public agency input is most critical and to suggest steps to encourage greater public sector involvement.

Preceding papers have suggested five obvious areas for involvement by the public sector:

1. Discovery and development of improved pest control techniques (research and development)
2. Dissemination of information about these techniques (education)
3. Direction and coordination of group or community actions required to implement sophisticated (and probably integrated) pest control methods (catalysis)
4. Development and implementation of regulatory procedures to ensure optimum interrelationships between pest control, health, and environmental quality (regulation)
5. Observation of the levels in food, soil, water, and fish and wildlife of the chemicals used in pest control (monitoring)

As has been demonstrated in the past, each of these areas requires major public agency inputs; and private sector involvement is likewise critical to their success. Thus, the discussion of public agency inputs that follows must be considered in the light of one overriding requisite—the need to integrate and coordinate public and private efforts.

NEW DIRECTIONS OF RESEARCH AND DEVELOPMENT

In the past, private industry has carried the primary research and development burden relating to pest control. Our heavy dependence on pesticides produced almost exclusively by the private sector has dictated this situation. There are reasons to believe, however, that future pest control efforts will require a relatively greater research and development input from public agencies than has been the case in the past (von Rumker *et al.*, 1970). Several factors support this view.

Changes in legal statutes and regulatory policies governing pesticide use make these chemicals less attractive to the private sector than in the past. Also, industry research and development costs for pesticides have been increased markedly by more stringent "burden of proof" requirements relating to health and environmental-quality considerations. Simultaneously, demands for chemicals with greater target specificity will likely narrow the range of usefulness for each new pesticide, thereby reducing its expected market potential. Rising research and development costs and shrinking potential income will likely discourage some private sector inputs. While there may be valid justification for the regulatory actions that in turn reduce the commercial attractiveness of pesticides to industry, we must also accept the economic fact-of-life consequences that these actions imply for the future of target-specific pesticides.

Even though pesticides may play a diminishing role in future pest-control strategies, these chemicals are likely to remain as the primary tools of control for some time to come. Because there is no readily available arsenal of target-specific pesticides and the prospects of commercially developing them are not promising, there is a decided need to supplement industrial research and development inputs with those by key public agencies. This need will be especially critical for chemicals to protect minor (in terms of acreage) but essential (in terms of human health and welfare) crops from pest damage. Thus, industry can justify greater research and development inputs for chemicals to control pests of multimillion-acre corn or cotton crops than for limited-acre crops such as carrots, celery, or strawberries. For these minor crops, evaluation of chemicals for efficacy and human safety will have to be assumed by public agencies to ensure that they will continue to be developed and made available. Unless such profit and loss considerations are given more attention than now seems intended, specific crop-chemical combinations will be of decreasing interest to the private sector. To further encourage the availability of pest controls having only limited demand, public agency research and development involvement may also require critical re-evaluation of current patent and licensing policies.

ALTERNATIVES TO PESTICIDES

As we seek alternatives to the use of pesticides, the research and development role of public agencies looms even greater. Many of the alternatives do not provide the sales potential needed to encourage private research and development investments. Others require long-term inputs with no assured return on investment. Some require group action or integrated control techniques that are most logically implemented by public bodies (National Academy of Sciences, 1969).

Biological control techniques involving the use of predators, parasites, or pathogens will likely be dependent largely on public agency inputs. In some cases, they will require the importation of biological materials from overseas through mechanisms best handled by public agencies. In others, they will require mass rearing and release of the agents through operations that may provide public good but no ready mechanism by which private industry could "sell" a product to the beneficiaries. While private industry can indeed be expected to contribute through the development of some specific pathogen agents, the prime responsibility for the development of these biological controls will likely rest with government research and development agencies. The most noted example of public agency involvement in "biological" control is the famous screwworm fly eradication program which was developed and implemented with public funds.

The discovery, isolation, synthesis, and testing of insect repellents and attractants may offer some potential for private industry since these products may be sold commercially. Even here, however, government research and development inputs will likely dominate. Basic research efforts of universities and federal laboratories will probably be the source of most of the chemicals. Their evaluation under field conditions will be expedited by the availability of sizable trial areas involving several farmers or perhaps hundreds of farms in an ecological region. These efforts can be implemented best by public agencies alone or in cooperation with industry.

As with many of the other biological means of pest control, the use of techniques to modify insect genetics appears to offer little proprietary protection which industry might use to recoup its investment. The difficulties, uncertainties, and high-risk characteristics of this technique of incorporating deleterious genes into field populations are not conducive to commercial involvement. And the probable operational requirements for insect control by autocidal methods further denies industry the ready means to "profit" from its involvement. Thus, unless some of these inherent problems can be overcome, the situation clearly points to the public sector as the prime mover for the development of insect controls using genetic principles.

The development of host resistance to pest attack through plant breeding programs traditionally has been a shared responsibility between the public and private sectors. Initially the leadership was with the U.S. Department of Agriculture and the state agricultural experiment stations. In time, however, commercial seedsmen employed research and development teams who have assumed primary responsibility for plant breeding of large-acreage crops such as cotton, corn, wheat, and sorghum. The public sector has turned more of its attention to basic genetics research and to the breeding of minor crop varieties for which industry has less incentive.

Plant breeding research aimed at developing resistance to pests offers one of our most promising long-term approaches to pest control. This technique has long been a primary means of improving yields and of developing resistance to plant diseases. Restrictions on the use of some insecticides are focusing new attention on the potential of host resistance to insects. Plant breeding research in this area has lagged far behind that for resistance to disease and for increasing the yield potentials, but the next decade will probably bring a marked increase in research on host resistance to insects. In 1970 about $1.1 million of U.S. Department of Agriculture and state experiment stations research funds were allocated to plant breeding research aimed at insect control (U.S. Department of Agriculture, 1970). About triple this amount was used to develop host resistance to disease while some $17 million was used to support research to increase productivity potential. It is obvious that insect resistance will require much greater funding in the future. It is also obvious that public funding for developing host resistance to both insects and disease is insignificant compared to inputs by industry to discover, evaluate, and produce chemical pesticides.

Because of the long-term nature of plant breeding research and the need to seek and maintain catalogs of genetic stock for testing and breeding purposes, public agencies must play a vital role. Such efforts will complement private sector inputs, which will also likely increase in the future, perpetuating the effective industry–government cooperation of the past.

The development and testing of integrated control techniques will require marked inputs by public agencies, hopefully in cooperation with industry. The testing of these techniques will likely require cooperation among groups of farmers, industry, and possibly regulatory agencies—cooperation which may be most easily arranged by public agencies.

There is a number of related researchable problems and opportunities that will require greatly expanded inputs by the public sector. Among the most important is the determination by economists, political scientists, and sociologists of the total costs of alternate means of pest control. In the past, the "first cost" to the user of a given pest control technique has been essentially our only criterion for determining whether the technique is to be used. Where "first cost" criteria are used, the figures on probable national losses resulting from discontinuance of a given technique are lacking or not firmly established. In any case, the secondary costs to society have not always been considered in deciding which pest controls were to be used. We need the econometrician and systems analyst to provide at least approximations of the total benefits and costs from alternate pest control regimes. It may well be that some of those with very high research and development requirements will, in the long run, provide the most public good in relation to total inputs. Such analyses are logical responsibilities of public agencies and are sorely needed.

While pest control strategies of the future vary greatly in their basic approaches and probable methods of implementation, they have one thing in common—a marked dependence on research and the knowledge that comes from research. This fact calls for a reassessment of past concepts as to the source of knowledge on which pest control technology can be based. It also calls for a reassessment of the budgetary inputs to support public research efforts. The first reassessment will demand the utilization of expertise from the entire scientific community in developing new and improved pest control strategies. The basic biologists and physical and social scientists must be attracted to the fascinating world of interaction among man and his environment and the pests that threaten him. Public agencies supporting these scientists must increasingly concern themselves with pest control problems and not shun them as being the proprietary concern of only the U.S. Department of Agriculture or the state agricultural experiment stations. The problems of pest control are much broader than agriculture and should be recognized as such. Agencies other than the USDA must concern themselves with pest control. Recent policy changes of the National Science Foundation to encourage and support interdisciplinary efforts to solve pressing human problems is a step in the right direction. It should be warmly supported by scientists, industry, executive and legislative bodies, and the public-at-large.

This logically leads to the second reassessment, that of public support for pest control research and development. Most pest control alternatives to broadly used pesticides will require greatly expanded research and development inputs from public agencies. Even though the total costs to society of these alternatives might well be less than that of persistent pesticides, their first cost in terms of discovery, testing, and practical implementation will be considerably higher. Furthermore, most of this first cost will likely be borne by the public rather than the private sector. Probably no other factor is more important in the long-run success of future pest control strategies than the grasping of this concept by the decision-makers in the executive and legislative bodies of government. Pest control strategies of the future are absolutely dependent on greatly expanded public support for research and development.

EDUCATION AND GROUP ACTION

Education will continue to play a vital role in pest control strategies of the future. Sophisticated chemical control methods will require user knowledge obtained through educational programs. Restrictions on chemicals and other techniques that call for licensed operators will necessitate greater educational inputs than heretofore. Biological control methods, especially those included in integrated pest control programs, will likewise require user knowledge and understanding for their success. While industry can be expected to disseminate information on products they produce, the prime responsibility for the total educational effort must rest on public agencies such as the Cooperative Extension Service.

Closely allied with educational efforts will be those of public or quasi-public organizations needed for the implementation of pest control strategies whose success depends upon group action. Such action might be encouraged by public agencies such as the Rural Environmental Assistance Program of the USDA which could provide cost sharing for efforts that clearly provide public benefits. Group action could be obtained through growers' cooperatives assisted by agencies such as the Farmers' Cooperative Service. Again, educational inputs from the Cooperative Extension Service would be essential to the success of such group activities. Such strong involvement of the public sector is essential in at least paving the way for cooperation among industry, regulatory bodies, and those who use pest control techniques.

REGULATORY AND CONTROL ACTIVITIES

Two major types of regulatory and control problems are of concern to pest control strategists. One is addressed to preventing the importation of pests and broadening interstate pest control efforts. The other is addressed to regulation and control of pesticides and other pest control chemicals or agents. Both are prime responsibilities of public agencies. Although the private sector is affected greatly by regulatory and control activities, its direct involvement in implementing regulatory programs is minimal.

Only brief mention will be made of efforts to prevent pest importation and those to prevent the spread of pests within the country. Both activities are vital to sensible national programs of pest control. At the same time, however, mass spraying with pesticides for insect eradication is one of those techniques most sharply criticized by environmentalists. Such conflicts will encourage public agencies to turn to biological control techniques, such as sterilization by irradiation, which has proved so successful in eradicating the screwworm fly in the Southeast. The need for back-up research in providing additional alternative techniques continues to be critical for such pest regulatory programs.

The most far-reaching regulatory responsibility of public agencies relates to pesticide use. To the requirement that pesticide use must not endanger human health has been added in recent years the concern over nontarget organisms. The protection of health and human comfort continues to be the primary objective of pest control strategies, but environmental concern will weigh much heavier in the coming decades than it has in the past.

As public concern over possible ill effects of pesticides has risen there has been a tendency to base decisions with respect to pesticide regulations more on emotion than on science. Evidence showing ill effects of one pesticide is accepted by some as evidence for the condemnation of all pesticides. Suggested modifications in current federal regulatory legislation would permit and perhaps encourage this trend. Public agencies must continue to rely heavily on scientific evidence as a basis for regulatory decisions.

Important guidelines for future regulations on chemicals in the environment can be obtained by measuring the presence of these chemicals in food, soil, water, and air. Progress has been made through "market basket" food sampling and through pesticide residue analyses in the United States. Limited analyses of soil and water samples also have provided some information on pesticides in these natural resources. However, a major recommendation of the recent study conference sponsored by the Massachusetts Institute of Technology (1970) called for a study of global monitoring as a means of continually evaluating the level of chemicals in the environment.

It is expected that pesticide regulatory activities at federal and state levels will be expanded greatly in the years ahead. This expansion will require a marked increase in back-up research aimed specifically at regulatory problems. It will also require critical evaluations to prevent an undue complexity in state-by-state pesticide regulations. The proliferation of uncoordinated restrictions on pesticides availability and use will likely lead to action at a federal or regional level to bring about greater uniformity.

As improvements in laws concerned with pesticide regulations are sought, attention should be given to the several recommendations in the Mrak Commission Report (U.S. Department of Health, Education and Welfare, 1969). For example, consider Recommendation 8 of the commission: "Seek modification of the Delaney clause to permit the Secretary of the Department of Health, Education and Welfare to determine when evidence of carcinogenesis justifies restrictive action concerning food containing analytically detectable traces of chemicals." Such consideration could provide a sounder basis for judging whether or not a chemical can be found in food than the sensitivity of analytical techniques for the chemical in question.

Coordination of actions, policies, and procedures among federal agencies and between federal and state agencies has long been a topic of serious concern. The combining into one federal agency (Environmental Protection Agency) of many of the federal regulatory responsibilities should expedite coordination of regulatory activities at the federal level; and may, in the long run, encourage better state–federal coordination. The Mrak Commission recognized these problems of agency cooperation and coordination and made strong recommendations for improvement.

CONCLUSIONS

In concluding this symposium on future strategies for pest control, we are obliged to make our projections and take our bearings from past experience and the present situation as we see it. Lessons of the past and the currently emerging trends should be useful in meeting the needs of this critical aspect of our future.

From past experience we know what can be accomplished when adequate support and incentives are provided. We know that effective results can be obtained; and we know that undesirable side effects are distinct possibilities. Often, the latter are the result of lack of knowledge and the very human habit of failing to consider the total picture beyond our immediate concern.

We are presently experiencing an awakening, a swing in the pendulum. The public is becoming aware of the shortcomings of some technology and the excesses of wasteful affluence and disregard. Concern for the use of pesticides is but one manifestation of a whole parcel of gnawing problems. Already, we are seeing that the public will take action by first attacking those elements that are close at hand, that have a handle to grasp, that are most easily controlled. As this public commitment grows, we will likely see more determined and resolute measures to restore the quality of our environment. Auto graveyards may disappear, our lakes and rivers may again sparkle, oil spills may become rare, and our inner cities may actually become pleasant and attractive places to live and work in.

But our concern here is for pest control. We know that the problem is far from solved and we know that it will not go away of its own accord. As long as we have commercial agriculture and seek to manipulate nature to our own ends, we will have pests competing with us for the harvest.

The challenge, then, is to fully develop those strengths and attributes that we have or can devise to bring to bear on the protection of our crops and animals, and on those vectors of disease that plague man. In so doing those jobs that we are best able to do, we must be mindful of their long-term effects, and of the possible ramifications of our actions to solve immediate problems.

These are not simple challenges, and they will not be met overnight. I am hopeful, even optimistic, that the suggestions offered during this symposium will be given due consideration as inputs and contributions to the total effort. Pesticides and pest control are in the public spotlight, let us now build a model and take the actions that others will see as worthy of emulation.

REFERENCES

Massachusetts Institute of Technology. 1970. Man's impact on the global environment. Report of the study of critical environmental problems.

National Academy of Sciences. 1969. Insect-pest management and control. Publ. 1695. Washington, D.C.

U.S. Department of Health, Education and Welfare. 1969. Report of the Secretary's Commission on pesticides and their relationships to environment health. Washington, D.C.

von Rumker, R., H. R. Guest, and W. M. Upholt, 1970. The search for safer, more selective, and less persistent pesticides. BioScience, 20 (18), Sept. 15, 1970.

U.S. Department of Agriculture. 1970. Central Research Information System (CRIS) special report on research activity and field of science.

Participants

MORTON BEROZA, Research Chemist, Agricultural Research Service, U.S. Department of Agriculture, Beltsville, Maryland

RAYMOND L. BISPLINGHOFF, Deputy Director, National Science Foundation, Washington, D.C.

JOHN L. BROOKS, Program Director, General Ecology Program, National Science Foundation, Washington, D.C.

NYLE C. BRADY, Director of Research and Director, Cornell University Agricultural Experiment Station, Ithaca, New York

ARTHUR H. BUNTING, Professor and Head, Department of Agricultural Botany, University of Reading, Berks, England

THEODORE C. BYERLY, Assistant Director of Science and Education, Office of the Secretary, U.S. Department of Agriculture, Washington, D.C.

GERALD CARLSON, Assistant Professor, Department of Economics, North Carolina State University, Raleigh

EMERY N. CASTLE, Professor, Department of Agricultural Economics, Oregon State University, Corvallis

DAVID E. DAVIS, Professor and Head, Department of Zoology, North Carolina State University, Raleigh

BOYSIE E. DAY, Associate Director, Agricultural Experiment Station (Universitywide), University of California, Berkeley

PETER R. DAY, Head, Department of Genetics, Connecticut Agricultural Experiment Station, New Haven

RICHARD L. DOUTT, Professor of Entomology, University of California, Berkeley

DONALD S. FARNER, Professor of Zoology, University of Washington, Seattle

K. R. FITZSIMMONS, General Manager, Agricultural Division, Shell Chemical Company, San Ramon, California

JOSEPH C. HEADLEY, Associate Professor, Department of Agricultural Economics, University of Missouri, Columbia

ARTHUR M. HEIMPEL, Principal Insect Pathologist, Insect Pathology Laboratory, Agricultural Research Service, U.S. Department of Agriculture, Beltsville, Maryland

JAMES G. HORSFALL, Director, Connecticut Agricultural Experiment Station, New Haven

L. L. JANSEN, Plant Physiologist, Plant Science Research Division, Agricultural Research Service, U.S. Department of Agriculture, Beltsville, Maryland

*JAMES B. KENDRICK, JR., Vice President, Agricultural Sciences, University of California, Berkeley

E. F. KNIPLING, Director, Entomology Research Division, Agricultural Research Service, U.S. Department of Agriculture, Beltsville, Maryland

JOHN J. McKELVEY, JR., Associate Director, Agricultural Sciences, Rockefeller Foundation, New York City

GEORGE L. McNEW, Managing Director, Boyce Thompson Institute for Plant Research, Yonkers, New York

ROBERT L. METCALF, Head, Department of Zoology, University of Illinois, Urbana

†CHARLES E. PALM, Professor of Entomology and Dean, New York State College of Agriculture, Cornell University, Ithaca, New York

WILLIAM E. ROBBINS, Director, Insect Physiology Laboratory, Agricultural Research Service, U.S. Department of Agriculture, Beltsville, Maryland

WARREN C. SHAW, Assistant Director, Plant Science Research Division, Agricultural Research Service, U.S. Department of Agriculture, Beltsville, Maryland

EDWARD H. SMITH, Director, Cooperative Extension, New York State Colleges of Agriculture and Human Ecology, Cornell University, Ithaca, New York

J. E. VAN der PLANK, Director of Research, Department of Agricultural Technical Services, Pretoria, South Africa

SEYMOUR D. VAN GUNDY, Professor of Nematology, University of California, Riverside

‡RICHARD H. WELLMAN, Vice President–General Manager, Process Chemicals Division, Union Carbide Company, New York City

HAROLD L. WILCKE, Vice President–Director of Research, Ralston Purina Company, St. Louis, Missouri

*Presided at session on "Alternative Means to Pest Control."

†Presided at session on "Strategies for Better Chemical Control."

‡Presided at session on "Challenge: Effective Pest Management Strategies Consistent with Environmental Quality."